Texts in Computing
Volume 4

The Haskell Road to Logic, Maths and Programming

Volume 1
Programming Languages and Operational Semantics
Maribel Fernández

Volume 2
An Introduction to Lambda Calculi for Computer Scientists
Chris Hankin

Volume 3
Logical Reasoning: A First Course
Rob Nederpelt and Fairouz Kamareddine

Volume 4
The Haskell Road to Logic, Maths and Programming
Kees Doets and Jan van Eijck

Volume 5
Bridges from Classical to Nonmonotonic Logic
David Makinson

Texts in Computing Series Editor
Ian Mackie, King's College London

The Haskell Road to Logic, Maths and Programming

Kees Doets
ILLC, Amsterdam

Jan van Eijck
CWI, Amsterdam

© Individual authors and King's College 2004. All rights reserved.

ISBN 0-9543006-9-6
King's College Publications
Scientific Director: Dov Gabbay
Managing Director: Jane Spurr
Department of Computer Science
Strand, London WC2R 2LS, UK
kcp@dcs.kcl.ac.uk

Cover design by Richard Fraser, www.avalonarts.co.uk
Printed by Lightning Source, Milton Keynes, UK

All rights reserved. No part of this publication may be reproduced, stored in a retrieval system or transmitted, in any form, or by any means, electronic, mechanical, photocopying, recording or otherwise, without prior permission, in writing, from the publisher.

Contents

Preface		**v**
1	**Getting Started**	**1**
	1.1 Starting up the Haskell Interpreter	2
	1.2 Implementing a Prime Number Test	3
	1.3 Haskell Type Declarations	8
	1.4 Identifiers in Haskell	11
	1.5 Playing the Haskell Game	12
	1.6 Haskell Types	17
	1.7 The Prime Factorization Algorithm	19
	1.8 The `map` and `filter` Functions	20
	1.9 Haskell Equations and Equational Reasoning	23
	1.10 Further Reading	25
2	**Talking about Mathematical Objects**	**27**
	2.1 Logical Connectives and their Meanings	28
	2.2 Logical Validity and Related Notions	38
	2.3 Making Symbolic Form Explicit	49
	2.4 Lambda Abstraction	57
	2.5 Definitions and Implementations	59
	2.6 Abstract Formulas and Concrete Structures	60
	2.7 Logical Handling of the Quantifiers	62
	2.8 Quantifiers as Procedures	67
	2.9 Further Reading	69
3	**The Use of Logic: Proof**	**71**
	3.1 Proof Style	72
	3.2 Proof Recipes	75
	3.3 Rules for the Connectives	78
	3.4 Rules for the Quantifiers	89

	3.5 Summary of the Proof Recipes	94
	3.6 Some Strategic Guidelines	97
	3.7 Reasoning and Computation with Primes	101
	3.8 Further Reading	110

4 Sets, Types and Lists — 111
- 4.1 Let's Talk About Sets — 112
- 4.2 Paradoxes, Types and Type Classes — 119
- 4.3 Special Sets — 123
- 4.4 Algebra of Sets — 125
- 4.5 Pairs and Products — 133
- 4.6 Lists and List Operations — 136
- 4.7 List Comprehension and Database Query — 141
- 4.8 Using Lists to Represent Sets — 146
- 4.9 A Data Type for Sets — 149
- 4.10 Further Reading — 155

5 Relations — 157
- 5.1 Relations as Sets of Ordered Pairs — 157
- 5.2 Properties of Relations — 162
- 5.3 Implementing Relations as Sets of Pairs — 171
- 5.4 Implementing Relations as Functions — 178
- 5.5 Equivalence Relations — 183
- 5.6 Equivalence Classes and Partitions — 187
- 5.7 Integer Partitions — 196
- 5.8 Further Reading — 199

6 Functions — 201
- 6.1 Basic Notions — 202
- 6.2 Surjections, Injections, Bijections — 213
- 6.3 Function Composition — 217
- 6.4 Inverse Function — 220
- 6.5 Partial Functions — 224
- 6.6 Functions as Partitions — 226
- 6.7 Products — 228
- 6.8 Congruences — 229
- 6.9 Further Reading — 231

7 Induction and Recursion — 233
- 7.1 Mathematical Induction — 233
- 7.2 Recursion over the Natural Numbers — 239
- 7.3 The Nature of Recursive Definitions — 245

7.4	Induction and Recursion over Trees	249
7.5	Induction and Recursion over Lists	258
7.6	Some Variations on the Tower of Hanoi	266
7.7	Other Data Structures	274
7.8	Further Reading	277

8 Working with Numbers — 279

8.1	Natural Numbers	279
8.2	GCD and the Fundamental Theorem of Arithmetic	283
8.3	Integers	287
8.4	Implementing Integer Arithmetic	291
8.5	Rational Numbers	293
8.6	Implementing Rational Arithmetic	299
8.7	Irrational Numbers	303
8.8	The Mechanic's Rule	306
8.9	Reasoning about Reals	308
8.10	Complex Numbers	313
8.11	Further Reading	322

9 Polynomials — 323

9.1	Difference Analysis of Polynomial Sequences	323
9.2	Gaussian Elimination	328
9.3	Polynomials and the Binomial Theorem	336
9.4	Polynomials for Combinatorial Reasoning	343
9.5	Further Reading	350

10 Corecursion — 351

10.1	Corecursive Definitions	352
10.2	Processes and Labeled Transition Systems	355
10.3	Proof by Approximation	361
10.4	Proof by Coinduction	368
10.5	Power Series and Generating Functions	373
10.6	Exponential Generating Functions	384
10.7	Further Reading	387

11 Finite and Infinite Sets — 389

11.1	More on Mathematical Induction	389
11.2	Equipollence	396
11.3	Infinite Sets	400
11.4	Cantor's World Implemented	406
11.5	Cardinal Numbers	408

The Greek Alphabet	411
References	412
Index	416

Preface

Purpose

Long ago, when Alexander the Great asked the mathematician Menaechmus for a crash course in geometry, he got the famous reply "There is no royal road to mathematics." Where there was no shortcut for Alexander, there is no shortcut for us. Still, the fact that we have access to computers and mature programming languages means that there are avenues for us that were denied to the kings and emperors of yore.

The purpose of this book is to teach logic and mathematical reasoning in practice, and to connect logical reasoning with computer programming. The programming language that will be our tool for this is Haskell, a member of the Lisp family. Haskell emerged in the last decade as a standard for lazy functional programming, a programming style where arguments are evaluated only when the value is actually needed. Functional programming is a form of descriptive programming, very different from the style of programming that you find in prescriptive languages like C or Java. Haskell is based on a logical theory of computable functions called the lambda calculus.

> Lambda calculus is a formal language capable of expressing arbitrary computable functions. In combination with types it forms a compact way to denote on the one hand functional programs and on the other hand mathematical proofs. [Bar84]

Haskell can be viewed as a particularly elegant implementation of the lambda calculus. It is a marvelous demonstration tool for logic and maths because its functional character allows implementations to remain very close to the concepts that get implemented, while the laziness permits smooth handling of infinite data structures.

Haskell syntax is easy to learn, and Haskell programs are constructed and tested in a modular fashion. This makes the language well suited for fast prototyping. Programmers find to their surprise that implementation of a well-understood algorithm in Haskell usually takes far less time than implementation of the same algorithm in other programming languages. Getting familiar with new algorithms through Haskell is also quite easy. Learning to program in Haskell is learning an extremely useful skill.

Throughout the text, abstract concepts are linked to concrete representations in Haskell. Haskell comes with an easy to use interpreter, Hugs. Haskell compilers, interpreters and documentation are freely available from the Internet [HT]. Everything one has to know about programming in Haskell to understand the programs in the book is explained as we go along, but we do not cover every aspect of the language. For a further introduction to Haskell we refer the reader to [HFP96].

Logic in Practice

The subject of this book is the *use* of logic in practice, more in particular the use of logic in reasoning about programming tasks. Logic is not taught here as a mathematical discipline per se, but as an aid in the understanding and construction of proofs, and as a tool for reasoning about formal objects like numbers, lists, trees, formulas, and so on. As we go along, we will introduce the concepts and tools that form the set-theoretic basis of mathematics, and demonstrate the role of these concepts and tools in implementations. These implementations can be thought of as *representations* of the mathematical concepts.

Although it may be argued that the logic that is needed for a proper understanding of reasoning in reasoned programming will get acquired more or less automatically in the process of learning (applied) mathematics and/or programming, students nowadays enter university without any experience whatsoever with mathematical proof, the central notion of mathematics.

The rules of Chapter 3 represent a detailed account of the structure of a proof. The purpose of this account is to get the student acquainted with proofs by putting emphasis on logical structure. The student is encouraged to write "detailed" proofs, with every logical move spelled out in full. The next goal is to move on to writing "concise" proofs, in the customary mathematical style, while keeping the logical structure in mind. Once the student has arrived at this stage, most of the logic that is explained in Chapter 3 can safely be forgotten, or better, can safely fade into the subconsciousness of the matured mathematical mind.

Pre- and Postconditions of Use

We do not assume that our readers have previous experience with either programming or construction of formal proofs. We do assume previous acquaintance with mathematical notation, at the level of secondary school mathematics. Wherever necessary, we will recall relevant facts. Everything one needs to know about mathematical reasoning or programming is explained as we go along. We do assume that our readers are able to retrieve software from the Internet and install it, and that they know how to use an editor for constructing program texts.

After having worked through the material in the book, i.e., after having digested the text and having carried out a substantial number of the exercises, the reader will be able to write interesting programs, reason about their correctness, and document them in a clear fashion. The reader will also have learned how to set up mathematical proofs in a structured way, and how to read and digest mathematical proofs written by others.

How to Use the Book

Chapters 1–7 of the book are devoted to a gradual introduction of the concepts, tools and methods of mathematical reasoning and reasoned programming.

Chapter 8 tells the story of how the various number systems (natural numbers, integers, rationals, reals, complex numbers) can be thought of as constructed in stages from the natural numbers. Everything gets linked to the implementations of the various Haskell types for numerical computation.

Chapter 9 starts with the question of how to automate the task of finding closed forms for polynomial sequences. It is demonstrated how this task can be automated with difference analysis plus Gaussian elimination. Next, polynomials are implemented as lists of their coefficients, with the appropriate numerical operations, and it is shown how this representation can be used for solving combinatorial problems.

Chapter 10 provides the first general textbook treatment (as far as we know) of the important topic of corecursion. The chapter presents the proof methods suitable for reasoning about corecursive data types like streams and processes, and then goes on to introduce power series as infinite lists of coefficients, and to demonstrate the uses of this representation for handling combinatorial problems. This generalizes the use of polynomials for combinatorics.

Chapter 11 offers a guided tour through Cantor's paradise of the infinite, while providing extra challenges in the form of a wide range of additional exercises.

The book can be used as a course textbook, but since it comes with solutions to all exercises (electronically available from the authors upon request) it is also well suited for private study. Courses based on the book could start with Chapters 1–7, and then make a choice from the remaining Chapters. Here are some examples:

Road to Numerical Computation Chapters 1–7, followed by 8 and 9.

Road to Streams and Corecursion Chapters 1–7, followed by 9 and 10.

Road to Cantor's Paradise Chapters 1–7, followed by 11.

Study of the remaining parts of the book can then be set as individual tasks for students ready for an extra challenge. The guidelines for setting up formal proofs in Chapter 3 should be recalled from time to time while studying the book, for proper digestion.

Exercises

Parts of the text and exercises marked by a * are somewhat harder than the rest of the book. All exercises are solved in the electronically available solutions volume. Before turning to these solutions, one should read the *Important Advice to the Reader* that this volume starts with.

Book Website and Contact

The programs in this book have all been tested with Hugs98, the version of Hugs that implements the Haskell 98 standard. The full source code of all programs is integrated in the book; in fact, each chapter can be viewed as a *literate program* [Knu92] in Haskell. The source code of all programs discussed in the text can be found on the website devoted to this book, at address `http://www.cwi.nl/~jve/HR`. Here you can also find a list of errata, and further relevant material.

Readers who want to share their comments with the authors are encouraged to get in touch with us at email address `jve@cwi.nl`.

Acknowledgments

Remarks from the people listed below have sparked off numerous improvements. Thanks to Johan van Benthem, Jan Bergstra, Jacob

PREFACE   ix

Brunekreef, Thierry Coquand (who found the lecture notes on the internet and sent us his comments), Tim van Erven, Wan Fokkink, Evan Goris, Robbert de Haan, Sandor Heman, Eva Hoogland, Rosalie Iemhoff, Dick de Jongh, Anne Kaldewaij, Breanndán Ó Nualláin, Alban Ponse, Vincent van Oostrom, Piet Rodenburg, Jan Rutten, Marco Swaen, Jan Terlouw, John Tromp, Yde Venema, Albert Visser and Stephanie Wehner for suggestions and criticisms. The beautiful implementation of the sieve of Eratosthenes in Section 3.7 was suggested to us by Fer-Jan de Vries. Jack Jansen, Hayco de Jong and Jurgen Vinju provided instant support with running Haskell on non-Linux machines.

The course on which this book is based was developed at ILLC (the Institute of Logic, Language and Computation of the University of Amsterdam) with financial support from the *Spinoza Logic in Action* initiative of Johan van Benthem, which is herewith gratefully acknowledged. We also wish to thank ILLC and CWI (Centrum voor Wiskunde en Informatica, or Centre for Mathematics and Computer Science, also in Amsterdam), the home institute of the second author, for providing us with a supportive working environment. CWI has kindly granted permission to reuse material from [Doe96].

It was Krzysztof Apt who, perceiving the need of a deadline, spurred us on to get in touch with a publisher and put ourselves under contract. Finally, many thanks go to Ian Mackie, Texts in Computing Series Editor, of King's College, London, for swift decision making, prompt and helpful feedback and enthousiastic support during the preparation of the manuscript for the *Text in Computing* series.

x

Chapter 1

Getting Started

Preview

Our purpose is to teach logic and mathematical reasoning in practice, and to connect formal reasoning to computer programming. It is convenient to choose a programming language for this that permits implementations to remain as close as possible to the formal definitions. Such a language is the functional programming language Haskell [HT]. Haskell was named after the logician Haskell B. Curry. Curry, together with Alonzo Church, laid the foundations of functional computation in the era Before the Computer, around 1940. As a functional programming language, Haskell is a member of the Lisp family. Others family members are Scheme, ML, Occam, Clean. Haskell98 is intended as a standard for lazy functional programming. Lazy functional programming is a programming style where arguments are evaluated only when the value is actually needed.

With Haskell, the step from formal definition to program is particularly easy. This presupposes, of course, that you are at ease with formal definitions. Our reason for combining training in reasoning with an introduction to functional programming is that your programming needs will provide motivation for improving your reasoning skills. Haskell programs will be used as illustrations for the theory throughout the book. We will always put computer programs and pseudo-code of algorithms in frames (rectangular boxes).

The chapters of this book are written in so-called 'literate programming' style [Knu92]. Literate programming is a programming style where the program and its documentation are generated from the same source. The text of every chapter in this book can be viewed as the documentation of the program code in that chapter. Literate programming makes it impossible for program and documentation to get out of sync. Program documentation is an integrated part of literate programming, in fact the bulk of a literate program is the program documentation. When writing programs in literate style there is less temptation to write program code first while leaving the documentation for later. Programming in literate style proceeds from the assumption that the main challenge when programming is to make your program digestible for humans. For a program to be useful, it should be easy for others to understand the code. It should also be easy for you to understand your own code when you reread your stuff the next day or the next week or the next month and try to figure out what you were up to when you wrote your program.

To save you the trouble of retyping, the code discussed in this book can be retrieved from the book website. The program code is the text in `typewriter` font that you find in rectangular boxes throughout the chapters. Boxes may also contain code that is not included in the chapter modules, usually because it defines functions that are already predefined by the Haskell system, or because it redefines a function that is already defined elsewhere in the chapter.

Typewriter font is also used for pieces of interaction with the Haskell interpreter, but these illustrations of how the interpreter behaves when particular files are loaded and commands are given are not boxed.

Every chapter of this book is a so-called Haskell module. The following two lines declare the Haskell module for the Haskell code of the present chapter. This module is called `GS`.

```
module GS

where
```

1.1 Starting up the Haskell Interpreter

We assume that you succeeded in retrieving the Haskell interpreter *hugs* from the Haskell homepage `www.haskell.org` and that you managed to install it on your computer. You can start the interpreter by typing `hugs` at the system prompt. When you start *hugs* you should see something like Figure 1.1. The string `Prelude>` on the last line is the Haskell prompt when no user-defined files are loaded.

1.2. IMPLEMENTING A PRIME NUMBER TEST

```
 __    __  __   __  __   ____   ___        ------------------------------------------
||    || ||  || ||  || ||__       Hugs 98: Based on the Haskell 98 standard
||___|| ||__|| ||__||  __||       Copyright (c) 1994-2003
||---||              ___||        World Wide Web: http://haskell.org/hugs
||    ||                          Report bugs to: hugs-bugs@haskell.org
||    ||  Version: November 2003  ------------------------------------------

Haskell 98 mode: Restart with command line option -98 to enable extensions

Type :? for help
Prelude>
```

Figure 1.1: Starting up the Haskell interpreter.

You can use *hugs* as a calculator as follows:

```
Prelude> 2^16
65536
Prelude>
```

The string `Prelude>` is the system prompt. `2^16` is what you type. After you hit the return key (the key that is often labeled with *Enter* or ↵), the system answers `65536` and the prompt `Prelude>` reappears.

Exercise 1.1 Try out a few calculations using * for multiplication, + for addition, - for subtraction, ^ for exponentiation, / for division. By playing with the system, find out what the precedence order is among these operators.

Parentheses can be used to override the built-in operator precedences:

```
Prelude> (2 + 3)^4
625
```

To quit the Hugs interpreter, type `:quit` or `:q` at the system prompt.

1.2 Implementing a Prime Number Test

Suppose we want to implement a definition of *prime number* in a procedure that recognizes prime numbers. A prime number is a natural number greater than 1 that has no proper divisors other than 1 and itself. The natural numbers are $0, 1, 2, 3, 4, \ldots$ The list of prime numbers starts with $2, 3, 5, 7, 11, 13, \ldots$ Except for 2, all of these are odd, of course.

Let $n > 1$ be a natural number. Then we use $\mathrm{LD}(n)$ for the least natural number greater than 1 that divides n. A number d divides n if there is a natural number a with $a \cdot d = n$. In other words, d divides n if there is a natural number a with $\frac{n}{d} = a$, i.e., division of n by d leaves no remainder. Note that $\mathrm{LD}(n)$ exists for every natural number $n > 1$, for the natural number $d = n$ is greater than 1 and divides n. Therefore, the set of divisors of n that are greater than 1 is non-empty. Thus, the set will have a least element.

The following proposition gives us all we need for implementing our prime number test:

Proposition 1.2

1. If $n > 1$ then $\mathrm{LD}(n)$ is a prime number.

2. If $n > 1$ and n is not a prime number, then $(\mathrm{LD}(n))^2 \leqslant n$.

In the course of this book you will learn how to prove propositions like this.

Here is the proof of the first item. This is a proof by contradiction (see Chapter 3). Suppose, for a contradiction that $c = \mathrm{LD}(n)$ is not a prime. Then there are natural numbers a and b with $c = a \cdot b$, and also $1 < a$ and $a < c$. But then a divides n, and contradiction with the fact that c is the smallest natural number greater than 1 that divides n. Thus, $\mathrm{LD}(n)$ must be a prime number.

For a proof of the second item, suppose that $n > 1$, n is not a prime and that $p = \mathrm{LD}(n)$. Then there is a natural number $a > 1$ with $n = p \cdot a$. Thus, a divides n. Since p is the smallest divisor of n with $p > 1$, we have that $p \leqslant a$, and therefore $p^2 \leqslant p \cdot a = n$, i.e., $(\mathrm{LD}(n))^2 \leqslant n$. ∎

The operator $\cdot$ in $a \cdot b$ is a so-called *infix* operator. The operator is written *between* its formal arguments. If an operator is written *before* its formal arguments we call this *prefix* notation. The product of a and b in prefix notation would look like this: $\cdot$ a b.

In writing functional programs, the standard is prefix notation. In an expression op a b, op is the *function*, and a and b are the *formal arguments*. The convention is that function application associates to the left, so the expression op a b is interpreted as (op a) b.

Using prefix notation, we define the operation divides that takes two integer expressions and produces a *truth value*. The truth values *true* and *false* are rendered in Haskell as True and False, respectively.

1.2. IMPLEMENTING A PRIME NUMBER TEST

The integer expressions that the procedure needs to work with are called the *formal arguments* of the procedure. The truth value that it produces is called the *value* of the procedure.

Obviously, m divides n if and only if the remainder of the process of dividing n by m equals 0. The definition of `divides` can therefore be phrased in terms of a predefined procedure `rem` for finding the remainder of a division process:

```
divides d n = rem n d == 0
```

The definition illustrates that Haskell uses = for 'is defined as' and == for identity. (The Haskell symbol for non-identity is /=.)

A line of Haskell code of the form `foo t = ...` (or `foo t1 t2 = ...`, or `foo t1 t2 t3 = ...`, and so on) is called a Haskell *equation*. In such an equation, `foo` is called the *function*, and `t` its *formal argument*.

Thus, in the Haskell equation `divides d n = rem n d == 0`, `divides` is the function, `d` is the first formal argument, and `n` is the second formal argument.

Exercise 1.3 Put the definition of `divides` in a file *prime.hs*. Start the Haskell interpreter *hugs* (Section 1.1). Now give the command `:load prime` or `:l prime`, followed by pressing *Enter*. Note that l is the letter *l*, not the digit 1. (Next to `:l`, a very useful command after you have edited a file of Haskell code is `:reload` or `:r`, for reloading the file.)

```
Prelude> :l prime
Main>
```

The string `Main>` is the Haskell prompt indicating that user-defined files are loaded. This is a sign that the definition was added to the system. The newly defined operation can now be executed, as follows:

```
Main> divides 5 7
False
Main>
```

What appears after the Haskell prompt `Main>` on the first line is what you type. When you press *Enter* the system answers with the second line, followed by the Haskell prompt. You can then continue with:

```
Main> divides 5 30
True
```

It is clear from the proposition above that all we have to do to implement a primality test is to give an implementation of the function LD. It is convenient to define LD in terms of a second function LDF, for the least divisor starting from a given threshold k, with $k \leqslant n$. Thus, $\text{LDF}(k)(n)$ is the least divisor of n that is $\geqslant k$. Clearly, $\text{LD}(n) = \text{LDF}(2)(n)$. Now we can implement LD as follows:

```
ld n = ldf 2 n
```

This leaves the implementation `ldf` of LDF (details of the coding will be explained below):

```
ldf k n | divides k n = k
        | k^2 >  n    = n
        | otherwise   = ldf (k+1) n
```

The definition employs the Haskell operation ^ for exponentiation, > for 'greater than', and + for addition.

The definition of `ldf` makes use of *equation guarding*. The first line of the `ldf` definition handles the case where the first argument divides the second argument. Every next line assumes that the previous lines do not apply. The second line handles the case where the first argument does not divide the second argument, and the square of the first argument is greater than the second argument. The third line assumes that the first and second cases do not apply and handles all other cases, i.e., the cases where k does not divide n and $k^2 < n$.

The definition employs the Haskell condition operator | . A Haskell equation of the form

```
foo t | condition = ...
```

is called a *guarded equation*. We might have written the definition of `ldf` as a list of guarded equations, as follows:

```
ldf k n | divides k n = k
ldf k n | k^2 >  n    = n
ldf k n               = ldf (k+1) n
```

1.2. IMPLEMENTING A PRIME NUMBER TEST

The expression condition, of type Bool (i.e., Boolean or truth value), is called the *guard* of the equation.

A list of guarded equations such as

```
foo t | condition_1 = body_1
foo t | condition_2 = body_2
foo t | condition_3 = body_3
foo t              = body_4
```

can be abbreviated as

```
foo t | condition_1 = body_1
      | condition_2 = body_2
      | condition_3 = body_3
      | otherwise   = body_4
```

Such a Haskell definition is read as follows:

- in case condition_1 holds, foo t is by definition equal to body_1,
- in case condition_1 does not hold but condition_2 holds, foo t is by definition equal to body_2,
- in case condition_1 and condition_2 do not hold but condition_3 holds, foo t is by definition equal to body_3,
- and in case none of condition_1, condition_2 and condition_3 hold, foo t is by definition equal to body_4.

When we are at the end of the list we know that none of the cases above in the list apply. This is indicated by means of the Haskell reserved keyword otherwise.

Note that the procedure ldf is called again from the body of its own definition. We will encounter such **recursive** procedure definitions again and again in the course of this book (see in particular Chapter 7).

Exercise 1.4 Suppose in the definition of ldf we replace the condition k^2 > n by k^2 >= n, where >= expresses 'greater than or equal'. Would that make any difference to the meaning of the program? Why (not)?

Now we are ready for a definition of prime0, our first implementation of the test for being a prime number.

```
prime0 n | n < 1     = error "not a positive integer"
         | n == 1    = False
         | otherwise = ld n == n
```

Haskell allows a call to the `error` operation in any definition. This is used to break off operation and issue an appropriate message when the primality test is used for numbers below 1. Note that `error` has a parameter of type `String` (indicated by the double quotes). The definition employs the Haskell operation < for 'less than'.

Intuitively, what the definition prime0 says is this:

1. the primality test should not be applied to numbers below 1,

2. if the test is applied to the number 1 it yields 'false',

3. if it is applied to an integer n greater than 1 it boils down to checking that $LD(n) = n$. In view of the proposition we proved above, this is indeed a correct primality test.

Exercise 1.5 Add these definitions to the file *prime.hs* and try them out.

Remark. The use of variables in functional programming has much in common with the use of variables in logic. The definition divides d n = rem n d == 0 is equivalent to divides x y = rem y x == 0. This is because the variables denote *arbitrary* elements of the type over which they range. They behave like universally quantified variables, and just as in logic the definition does not depend on the variable names. ∎

1.3 Haskell Type Declarations

Haskell has a concise way to indicate that `divides` consumes an integer, then another integer, and produces a truth value (called `Bool` in Haskell). Integers and truth values are examples of *types*. See Section 2.1 for more on the type `Bool`. Section 1.6 gives more information about types in general. Arbitrary precision integers in Haskell have type `Integer`. The following line gives a so-called *type declaration* for the `divides` function.

```
divides :: Integer -> Integer -> Bool
```

Integer -> Integer -> Bool is short for Integer -> (Integer -> Bool). A type of the form a -> b classifies procedures that take an argument of type a to produce a result of type b. Thus, `divides` takes an argument of type Integer and produces a result of type Integer -> Bool, i.e., a procedure that takes an argument of type Integer, and produces a result of type Bool.

1.3. HASKELL TYPE DECLARATIONS

The full code for `divides`, including the type declaration, looks like this:

```
divides :: Integer -> Integer -> Bool
divides d n = rem n d == 0
```

If `d` is an expression of type `Integer`, then `divides d` is an expression of type `Integer -> Bool`. The shorthand that we will use for

> d *is an expression of type* `Integer`

is: `d :: Integer`.

Exercise 1.6 Can you gather from the definition of `divides` what the type declaration for `rem` would look like?

Exercise 1.7 The *hugs* system has a command for checking the types of expressions. Can you explain the following (please try it out; make sure that the file with the definition of `divides` is loaded, together with the type declaration for `divides`):

```
Main> :t divides 5
divides 5 :: Integer -> Bool
Main> :t divides 5 7
divides 5 7 :: Bool
Main>
```

The expression `divides 5 :: Integer -> Bool` is called a *type judgment*. Type judgments in Haskell have the form `expression :: type`.

In Haskell it is not strictly necessary to give explicit type declarations. For instance, the definition of `divides` works quite well without the type declaration, since the system can infer the type from the definition. However, it is good programming practice to give explicit type declarations even when this is not strictly necessary. These type declarations are an aid to understanding, and they greatly improve the digestibility of functional programs for human readers. A further advantage of the explicit type declarations is that they facilitate detection of programming mistakes on the basis of type errors generated by the interpreter. You will find that many programming errors already come to light when your program gets loaded. The fact that your program is well typed does not entail that it is correct, of course, but many incorrect programs do have typing mistakes.

The full code for `ld`, including the type declaration, looks like this:

```
ld :: Integer -> Integer
ld n = ldf 2 n
```

The full code for `ldf`, including the type declaration, looks like this:

```
ldf :: Integer -> Integer -> Integer
ldf k n | divides k n = k
        | k^2 > n     = n
        | otherwise   = ldf (k+1) n
```

The first line of the code states that the operation `ldf` takes two integers and produces an integer.

The full code for `prime0`, including the type declaration, runs like this:

```
prime0 :: Integer -> Bool
prime0 n | n < 1      = error "not a positive integer"
         | n == 1     = False
         | otherwise  = ld n == n
```

The first line of the code declares that the operation `prime0` takes an integer and produces (or *returns*, as programmers like to say) a Boolean (truth value).

In programming generally, it is useful to keep close track of the nature of the objects that are being represented. This is because representations have to be stored in computer memory, and one has to know how much space to allocate for this storage. Still, there is no need to always specify the nature of each data-type explicitly. It turns out that much information about the nature of an object can be inferred from how the object is handled in a particular program, or in other words, from the *operations* that are performed on that object.

Take again the definition of `divides`. It is clear from the definition that an operation is defined with two formal arguments, both of which are of a type for which `rem` is defined, and with a result of type `Bool` (for `rem n d == 0` is a statement that can turn out true or false). If we check the type of the built-in procedure `rem` we get:

```
Prelude> :t rem
rem :: Integral a => a -> a -> a
```

1.4. IDENTIFIERS IN HASKELL

In this particular case, the type judgment gives a *type scheme* rather than a type. It means: if a is a type of class Integral, then rem is of type a -> a -> a. Here a is used as a variable ranging over types.

In Haskell, Integral is the class (see Section 4.2) consisting of the two types for integer numbers, Int and Integer. The difference between Int and Integer is that objects of type Int have fixed precision, objects of type Integer have arbitrary precision.

The type of divides can now be inferred from the definition. This is what we get when we load the definition of divides without the type declaration:

```
Main> :t divides
divides :: Integral a => a -> a -> Bool
```

1.4 Identifiers in Haskell

In Haskell, there are two kinds of identifiers:

- Variable identifiers are used to name functions. They have to start with a lower-case letter. E.g., map, max, fct2list, fctToList, fct_to_list.

- Constructor identifiers are used to name types. They have to start with an upper-case letter. Examples are True, False.

Functions are operations on data-structures, constructors are the building blocks of the data structures themselves (trees, lists, Booleans, and so on).

Names of functions always start with lower-case letters, and may contain both upper- and lower-case letters, but also digits, underscores and the prime symbol '. The following *reserved keywords* have special meanings and cannot be used to name functions.

case	class	data	default	deriving	do	else
if	import	in	infixl	infixr	infixr	instance
let	module	newtype	of	then	type	where
_						

The use of these keywords will be explained as we encounter them. _ at the beginning of a word is treated as a lower-case character. The underscore character _ all by itself is a reserved word for the wildcard pattern that matches anything (page 137).

There is one more reserved keyword that is particular to Hugs: *forall*, for the definition of functions that take polymorphic arguments. See the Hugs documentation for further particulars.

1.5 Playing the Haskell Game

This section consists of a number of further examples and exercises to get you acquainted with the programming language of this book. To save you the trouble of keying in the programs below, you should retrieve the module `GS.hs` for the present chapter from the book website and load it in *hugs*. This will give you a system prompt `GS>`, indicating that all the programs from this chapter are loaded.

In the next example, we use `Int` for the type of fixed precision integers, and `[Int]` for lists of fixed precision integers.

Example 1.8 Here is a function that gives the minimum of a list of integers:

```
mnmInt :: [Int] -> Int
mnmInt [] = error "empty list"
mnmInt [x] = x
mnmInt (x:xs) = min x (mnmInt xs)
```

This uses the predefined function `min` for the minimum of two integers. It also uses pattern matching for lists . The list pattern `[]` matches only the empty list, the list pattern `[x]` matches any singleton list, the list pattern `(x:xs)` matches any non-empty list. A further subtlety is that pattern matching in Haskell is sensitive to order. If the pattern `[x]` is found before `(x:xs)` then `(x:xs)` matches any non-empty list that is not a unit list. See Section 4.6 for more information on list pattern matching.

It is common Haskell practice to refer to non-empty lists as `x:xs`, `y:ys`, and so on, as a useful reminder of the facts that `x` is an element of a list of x's and that `xs` is a list.

Here is a home-made version of `min`:

```
min' :: Int -> Int -> Int
min' x y | x <= y    = x
         | otherwise = y
```

You will have guessed that `<=` is Haskell code for $\leqslant$.

Objects of type `Int` are fixed precision integers. Their range can be found with:

1.5. PLAYING THE HASKELL GAME

```
Prelude> primMinInt
-2147483648
Prelude> primMaxInt
2147483647
```

Since $2147483647 = 2^{31} - 1$, we can conclude that the *hugs* implementation uses four bytes (32 bits) to represent objects of this type. Integer is for arbitrary precision integers: the storage space that gets allocated for Integer objects depends on the size of the object.

Exercise 1.9 Define a function that gives the maximum of a list of integers. Use the predefined function max.

Conversion from Prefix to Infix in Haskell A function can be converted to an infix operator by putting its name in back quotes, like this:

```
Prelude> max 4 5
5
Prelude> 4 'max' 5
5
```

Conversely, an infix operator is converted to prefix by putting the operator in round brackets (p. 20).

Exercise 1.10 Define a function removeFst that removes the first occurrence of an integer m from a list of integers. If m does not occur in the list, the list remains unchanged.

Example 1.11 We define a function that sorts a list of integers in order of increasing size, by means of the following algorithm:

- an empty list is already sorted.
- if a list is non-empty, we put its minimum in front of the result of sorting the list that results from removing its minimum.

This is implemented as follows:

```
srtInts :: [Int] -> [Int]
srtInts [] = []
srtInts xs = m : (srtInts (removeFst m xs)) where m = mnmInt xs
```

Here `removeFst` is the function you defined in Exercise 1.10. Note that the second clause is invoked when the first one does not apply, i.e., when the argument of `srtInts` is not empty. This ensures that `mnmInt xs` never gives rise to an error.

Note the use of a `where` construction for the local definition of an auxiliary function.

Remark. Haskell has two ways to locally define auxiliary functions, `where` and `let` constructions. The `where` construction is illustrated in Example 1.11. This can also expressed with `let`, as follows:

```
srtInts' :: [Int] -> [Int]
srtInts' [] = []
srtInts' xs = let
                 m = mnmInt xs
              in  m : (srtInts' (removeFst m xs))
```

The `let` construction uses the reserved keywords `let` and `in`.

Example 1.12 Here is a function that calculates the average of a list of integers. The average of m and n is given by $\frac{m+n}{2}$, the average of a list of k integers $n_1, \ldots, n_k$ is given by $\frac{n_1 + \cdots + n_k}{k}$. In general, averages are fractions, so the result type of `average` should not be `Int` but the Haskell data-type for fractional numbers, which is `Rational`. There are predefined functions `sum` for the sum of a list of integers, and `length` for the length of a list. The Haskell operation for division `/` expects arguments of type `Rational` (or more precisely, of a type in the class `Fractional`, and `Rational` is in that class), so we need a conversion function for converting `Int`s into `Rational`s. This is done by `toRational`. The function `average` can now be written as:

```
average :: [Int] -> Rational
average [] = error "empty list"
average xs = toRational (sum xs) / toRational (length xs)
```

Again, it is instructive to write our own homemade versions of `sum` and `length`. Here they are:

1.5. PLAYING THE HASKELL GAME

```
sum' :: [Int] -> Int
sum' [] = 0
sum' (x:xs) = x + sum' xs
```

```
length' :: [a] -> Int
length' [] = 0
length' (x:xs) = 1 + length' xs
```

Note that the type declaration for `length'` contains a variable a. This variable ranges over all types, so [a] is the type of a list of objects of an arbitrary type a. We say that [a] is a *type scheme* rather than a type. This way, we can use the same function `length'` for computing the length of a list of integers, the length of a list of characters, the length of a list of strings (lists of characters), and so on.

The type [Char] is abbreviated as String. Examples of characters are 'a', 'b' (note the single quotes) examples of strings are "Russell" and "Cantor" (note the double quotes). In fact, "Russell" can be seen as an abbreviation of the list

['R','u','s','s','e','l','l'].

Exercise 1.13 Write a function count for counting the number of occurrences of a character in a string. In Haskell, a character is an object of type Char, and a string an object of type String, so the type declaration should run: count :: Char -> String -> Int.

Exercise 1.14 A function for transforming strings into strings is of type String -> String. Write a function blowup that converts a string

$$a_1 a_2 a_3 \cdots$$

to

$$a_1 a_2 a_2 a_3 a_3 a_3 \cdots .$$

blowup "bang!" should yield "baannngggg!!!!!". (Hint: use ++ for string concatenation.)

Exercise 1.15 Write a function `srtString :: [String] -> [String]` that sorts a list of strings in alphabetical order.

Example 1.16 Suppose we want to check whether a string `str1` is a prefix of a string `str2`. Then the answer to the question `prefix str1 str2` should be either yes (true) or no (false), i.e., the type declaration for `prefix` should run: `prefix :: String -> String -> Bool`.

Prefixes of a string `ys` are defined as follows:

1. `[]` is a prefix of `ys`,

2. if `xs` is a prefix of `ys`, then `x:xs` is a prefix of `x:ys`,

3. nothing else is a prefix of `ys`.

Here is the code for `prefix` that implements this definition:

```
prefix :: String -> String -> Bool
prefix [] ys = True
prefix (x:xs) [] = False
prefix (x:xs) (y:ys) = (x==y) && prefix xs ys
```

The definition of `prefix` uses the Haskell operator `&&` for conjunction.

Exercise 1.17 Write a function `substring :: String -> String -> Bool` that checks whether `str1` is a substring of `str2`.

The substrings of an arbitrary string `ys` are given by:

1. if `xs` is a prefix of `ys`, `xs` is a substring of `ys`,

2. if `ys` equals `y:ys'` and `xs` is a substring of `ys'`, `xs` is a substring of `ys`,

3. nothing else is a substring of `ys`.

1.6 Haskell Types

The basic Haskell types are:

- `Int` and `Integer`, to represent integers. Elements of `Integer` are unbounded. That's why we used this type in the implementation of the prime number test.

- `Float` and `Double` represent floating point numbers. The elements of `Double` have higher precision.

- `Bool` is the type of Booleans.

- `Char` is the type of characters.

Note that the name of a type always starts with a capital letter.

To denote arbitrary types, Haskell allows the use of *type variables*. For these, a, b, ..., are used.

New types can be formed in several ways:

- By list-formation: if a is a type, `[a]` is the type of lists over a. Examples: `[Int]` is the type of lists of integers; `[Char]` is the type of lists of characters, or strings.

- By pair- or tuple-formation: if a and b are types, then (a,b) is the type of pairs with an object of type a as their first component, and an object of type b as their second component. Similarly, triples, quadruples, ..., can be formed. If a, b and c are types, then (a,b,c) is the type of triples with an object of type a as their first component, an object of type b as their second component, and an object of type c as their third component. And so on (p. 136).

- By function definition: a -> b is the type of a function that takes arguments of type a and returns values of type b.

- By applying a *type constructor*. E.g., `Rational` is the type that results from applying the type constructor `Ratio` to type `Integer`.

- By defining your own data-type from scratch, with a `data` type declaration. More about this in due course.

Pairs will be further discussed in Section 4.5, lists and list operations in Section 4.6.

Operations are procedures for constructing objects of a certain types *b* from ingredients of a type *a*. Now such a procedure can itself be given a type: the type of a transformer from a type objects to b type objects. The type of such a procedure can be declared in Haskell as a -> b.

If a function takes two string arguments and returns a string then this can be viewed as a two-stage process: the function takes a first string and returns a transformer from strings to strings. It then follows that the type is String -> (String -> String), which can be written as String -> String -> String, because of the Haskell convention that -> associates to the right.

Exercise 1.18 Find expressions with the following types:

1. [String]

2. (Bool,String)

3. [(Bool,String)]

4. ([Bool],String)

5. Bool -> Bool

Test your answers by means of the Hugs command :t.

Exercise 1.19 Use the Hugs command :t to find the types of the following predefined functions:

1. head

2. last

3. init

4. fst

5. (++)

6. flip

7. flip (++)

Next, supply these functions with arguments of the expected types, and try to guess what these functions do.

1.7 The Prime Factorization Algorithm

Let n be an arbitrary natural number > 1. A *prime factorization* of n is a list of prime numbers $p_1, \ldots, p_j$ with the property that $p_1 \cdots p_j = n$. We will show that a prime factorization of every natural number $n > 1$ exists by producing one by means of the following method of splitting off prime factors:

$$\text{WHILE } n \neq 1 \text{ DO BEGIN} \quad p := \text{LD}(n); \; n := \frac{n}{p} \quad \text{END}$$

Here $:=$ denotes *assignment* or the act of giving a variable a new value. As we have seen, $\text{LD}(n)$ exists for every n with $n > 1$. Moreover, we have seen that $\text{LD}(n)$ is always prime. Finally, it is clear that the procedure terminates, for every round through the loop will decrease the size of n.

So the algorithm consists of splitting off primes until we have written n as $n = p_1 \cdots p_j$, with all factors prime. To get some intuition about how the procedure works, let us see what it does for an example case, say $n = 84$. The original assignment to n is called n_0; successive assignments to n and p are called $n_1, n_2, \ldots$ and $p_1, p_2, \ldots$.

$$
\begin{array}{lll}
 & & n_0 = 84 \\
n_0 \neq 1 & p_1 = 2 & n_1 = 84/2 = 42 \\
n_1 \neq 1 & p_2 = 2 & n_2 = 42/2 = 21 \\
n_2 \neq 1 & p_3 = 3 & n_3 = 21/3 = 7 \\
n_3 \neq 1 & p_4 = 7 & n_4 = 7/7 = 1 \\
n_4 = 1 & &
\end{array}
$$

This gives $84 = 2^2 \cdot 3 \cdot 7$, which is indeed a prime factorization of 84.

The following code gives an implementation in Haskell, collecting the prime factors that we find in a list. The code uses the predefined Haskell function div for integer division.

```
factors :: Integer -> [Integer]
factors n | n < 1     = error "argument not positive"
          | n == 1    = []
          | otherwise = p : factors (div n p) where p = ld n
```

If you load the code for this chapter, you can try this out as follows:

```
GS> factors 84
[2,2,3,7]
GS> factors 557940830126698960967415390
[2,3,5,7,11,13,17,19,23,29,31,37,41,43,47,53,59,61,67,71]
```

1.8 The `map` and `filter` Functions

Haskell allows some convenient abbreviations for lists: [4..20] denotes the list of integers from 4 through 20, ['a'..'z'] the list of all lower case letters, "abcdefghijklmnopqrstuvwxyz". The call [5..] generates an infinite list of integers starting from 5. And so on.

If you use the Hugs command :t to find the type of the function map, you get the following:

```
Prelude> :t map
map :: (a -> b) -> [a] -> [b]
```

The function map takes a function and a list and returns a list containing the results of applying the function to the individual list members.

If f is a function of type a -> b and xs is a list of type [a], then map f xs will return a list of type [b]. E.g., map (^2) [1..9] will produce the list of squares

[1, 4, 9, 16, 25, 36, 49, 64, 81]

You should verify this by trying it out in *Hugs*. The use of (^2) for the operation of squaring demonstrates a new feature of Haskell, the construction of sections.

Conversion from Infix to Prefix, Construction of Sections   If op is an infix operator, (op) is the prefix version of the operator. Thus, 2^10 can also be written as (^) 2 10. This is a special case of the use of sections in Haskell.

In general, if op is an infix operator, (op x) is the operation resulting from applying op to its right hand side argument, (x op) is the operation resulting from applying op to its left hand side argument, and (op) is the prefix version of the operator (this is like the abstraction of the operator from both arguments).

Thus (^2) is the squaring operation, (2^) is the operation that computes powers of 2, and (^) is exponentiation. Similarly, (>3) denotes the property of being greater than 3, (3>) the property of being smaller than 3, and (>) is the prefix version of the 'greater than' relation.

The call map (2^) [1..10] will yield

[2, 4, 8, 16, 32, 64, 128, 256, 512, 1024]

If *p* is a property (an operation of type a -> Bool) and xs is a list of type [a], then   map p xs will produce a list of type Bool (a list of truth values), like this:

1.8. THE MAP AND FILTER FUNCTIONS

```
Prelude> map (>3) [1..9]
[False, False, False, True, True, True, True, True, True]
Prelude>
```

The function map is predefined in Haskell, but it is instructive to give our own version:

```
map :: (a -> b) -> [a] -> [b]
map f [] = []
map f (x:xs) = (f x) : (map f xs)
```

Note that if you try to load this code, you will get an error message:

    Definition of variable "map" clashes with import.

The error message indicates that the function name map is already part of the name space for functions, and is not available anymore for naming a function of your own making.

Exercise 1.20 Use map to write a function lengths that takes a list of lists and returns a list of the corresponding list lengths.

Exercise 1.21 Use map to write a function sumLengths that takes a list of lists and returns the sum of their lengths.

Another useful function is filter, for filtering out the elements from a list that satisfy a given property. This is predefined, but here is a home-made version:

```
filter :: (a -> Bool) -> [a] -> [a]
filter p [] = []
filter p (x:xs) | p x       = x : filter p xs
                | otherwise = filter p xs
```

Here is an example of its use:

```
GS> filter (>3) [1..10]
[4,5,6,7,8,9,10]
```

Example 1.22 Here is a program `primes0` that filters the prime numbers from the infinite list `[2..]` of natural numbers:

```
primes0 :: [Integer]
primes0 = filter prime0 [2..]
```

This produces an infinite list of primes. (Why infinite? See Theorem 3.33.) The list can be interrupted with 'Control-C'.

Example 1.23 Given that we can produce a list of primes, it should be possible now to improve our implementation of the function LD. The function `ldf` used in the definition of `ld` looks for a prime divisor of n by checking $k|n$ for all k with $2 \leqslant k \leqslant \sqrt{n}$. In fact, it is enough to check $p|n$ for the *primes* p with $2 \leqslant p \leqslant \sqrt{n}$. Here are functions `ldp` and `ldpf` that perform this more efficient check:

```
ldp :: Integer -> Integer
ldp n = ldpf primes1 n

ldpf :: [Integer] -> Integer -> Integer
ldpf (p:ps) n | rem n p == 0 = p
              | p^2 > n      = n
              | otherwise    = ldpf ps n
```

`ldp` makes a call to `primes1`, the list of prime numbers. This is a first illustration of a 'lazy list'. The list is called 'lazy' because we compute only the part of the list that we need for further processing. To define `primes1` we need a test for primality, but that test is itself defined in terms of the function LD, which in turn refers to `primes1`. We seem to be running around in a circle. This circle can be made non-vicious by avoiding the primality test for 2. If it is given that 2 is prime, then we can use the primality of 2 in the LD check that 3 is prime, and so on, and we are up and running.

1.9. HASKELL EQUATIONS AND EQUATIONAL REASONING

```
primes1 :: [Integer]
primes1 = 2 : filter prime [3..]

prime :: Integer -> Bool
prime n | n < 1     = error "not a positive integer"
        | n == 1    = False
        | otherwise = ldp n == n
```

Replacing the definition of `primes1` by `filter prime [2..]` creates vicious circularity, with stack overflow as a result (try it out). By running the program `primes1` against `primes0` it is easy to check that `primes1` is much faster.

Exercise 1.24 What happens when you modify the defining equation of `ldp` as follows:

```
ldp :: Integer -> Integer
ldp = ldpf primes1
```

Can you explain?

1.9 Haskell Equations and Equational Reasoning

The Haskell equations `f x y = ...` used in the definition of a function `f` are genuine mathematical equations. They state that the left hand side and the right hand side of the equation have the same value. This is *very* different from the use of = in imperative languages like C or Java. In a C or Java program, the statement `x = x*y` does *not mean* that x and $x*y$ have the same value, but rather it is a command to throw away the old value of x and put the value of $x*y$ in its place. It is a so-called *destructive assignment statement*: the old value of a variable is destroyed and replaced by a new one.

Reasoning about Haskell definitions is a lot easier than reasoning about programs that use destructive assignment. In Haskell, standard reasoning about mathematical equations applies. E.g., after the Haskell declarations `x = 1` and `y = 2`, the Haskell declaration `x = x + y` will raise an error "x" multiply defined. Because = in Haskell has the meaning "is by definition equal to", while redefinition is forbidden, reasoning about Haskell functions is standard equational reasoning. Let's try this out on a simple example.

```
a = 3
b = 4
f :: Integer -> Integer -> Integer
f x y = x^2 + y^2
```

To evaluate `f a (f a b)` by equational reasoning, we can proceed as follows:

$$
\begin{aligned}
f\ a\ (f\ a\ b) &= f\ a\ (a^2 + b^2) \\
&= f\ 3\ (3^2 + 4^2) \\
&= f\ 3\ (9 + 16) \\
&= f\ 3\ 25 \\
&= 3^2 + 25^2 \\
&= 9 + 625 \\
&= 634
\end{aligned}
$$

The rewriting steps use standard mathematical laws and the Haskell definitions of a, b, f. In fact, when running the program we get the same outcome:

```
GS> f a (f a b)
634
GS>
```

Remark. We already encountered definitions where the function that is being defined occurs on the right hand side of an equation in the definition. Here is another example:

```
g :: Integer -> Integer
g 0     = 0
g (x+1) = 2 * (g x)
```

Not everything that is allowed by the Haskell syntax makes semantic sense, however. The following definitions, although syntactically correct, do not properly define functions:

```
h1 :: Integer -> Integer
h1 0 = 0
h1 x = 2 * (h1 x)

h2 :: Integer -> Integer
h2 0 = 0
h2 x = h2 (x+1)
```

The problem is that for values other than 0 the definitions do not give recipes for computing a value. This matter will be taken up in Chapter 7.

1.10 Further Reading

The standard Haskell operations are defined in the file *Prelude.hs*, which you should be able to locate somewhere on any system that runs *hugs*. Typically, the file resides in `/usr/lib/hugs/libraries/Hugs/`, on Unix/Linux machines. On Windows machines, a typical location is `C:\Program files\Hugs98\libraries\Hugs\`. Windows users, take care: in specifying Windows path names in Haskell, the backslash \ has to be quoted, by using \\. Thus, the Haskell way to refer to the example directory is `C:\\Program files\\Hugs98\\libraries\\Hugs\\`. Alternatively, Unix/Linux style path names can be used.

In case Exercise 1.19 has made you curious, the definitions of these example functions can all be found in *Prelude.hs*. If you want to quickly learn a lot about how to program in Haskell, you should get into the habit of consulting this file regularly. The definitions of all the standard operations are *open source code*, and are there for you to learn from. The Haskell *Prelude* may be a bit difficult to read at first, but you will soon get used to the syntax and acquire a taste for the style.

Various tutorials on Haskell and Hugs can be found on the Internet: see e.g. [HFP96] and [JR+]. The definitive reference for the language is [Jon03]. A textbook on Haskell focusing on multimedia applications is [Hud00]. Other excellent textbooks on functional programming with Haskell are [Tho99] and, at a more advanced level, [Bir98]. A book on discrete mathematics that also uses Haskell as a tool, and with a nice treatment of automated proof checking, is [HO00].

Chapter 2

Talking about Mathematical Objects

Preview

To talk about mathematical objects with ease it is useful to introduce some symbolic abbreviations. These symbolic conventions are meant to better reveal the **structure** of our mathematical statements. This chapter concentrates on a few (in fact: seven), simple words or phrases that are essential to the mathematical vocabulary: *not*, *if*, *and*, *or*, *if and only if*, *for all* and *for some*. We will introduce symbolic shorthands for these words, and we look in detail at how these building blocks are used to construct the logical patterns of sentences. After having isolated the logical key ingredients of the mathematical vernacular, we can systematically relate definitions in terms of these logical ingredients to implementations, thus building a bridge between logic and computer science.

The use of symbolic abbreviations in specifying algorithms makes it easier to take the step from definitions to the procedures that implement those definitions. In a similar way, the use of symbolic abbreviations in making mathematical statements makes it easier to construct proofs of those statements. Chances are that you are more at ease with programming than with proving things. However that may be, in the chapters to follow you will get the opportunity to improve your skills in both of these activities and to find out more about the way in which they are related.

```
module TAMO

where
```

2.1 Logical Connectives and their Meanings

Goal To understand how the meanings of statements using connectives can be described by explaining how the truth (or falsity) of the statement depends on the truth (or falsity) of the smallest parts of this statement. This understanding leads directly to an implementation of the logical connectives as truth functional procedures.

In ordinary life, there are many statements that do not have a definite truth value, for example 'Barnett Newman's *Who is Afraid of Red, Yellow and Blue III* is a beautiful work of art,' or 'Daniel Goldreyer's restoration of *Who is Afraid of Red, Yellow and Blue III* meets the highest standards.'

Fortunately the world of mathematics differs from the Amsterdam Stedelijk Museum of Modern Art in the following respect. In the world of mathematics, things are so much clearer that many mathematicians adhere to the following slogan:

> every statement that makes mathematical sense is either true or false.

The idea behind this is that (according to the adherents) the world of mathematics exists independently of the mind of the mathematician. Doing mathematics is the activity of exploring this world. In proving new theorems one discovers new facts about the world of mathematics, in solving exercises one rediscovers known facts for oneself. (Solving problems in a mathematics textbook is like visiting famous places with a tourist guide.)

This belief in an independent world of mathematical fact is called Platonism, after the Greek philosopher Plato, who even claimed that our everyday physical world is somehow an image of this ideal mathematical world. A mathematical Platonist holds that every statement that makes mathematical sense *has exactly one* of the two truth values. Of course, a Platonist would concede that we may not know which value a statement has, for mathematics has numerous open problems. Still, a Platonist would say that the true answer to an open problem in mathematics like 'Are there infinitely many Mersenne primes?' (Example 3.40 from Chapter 3) is either

2.1. LOGICAL CONNECTIVES AND THEIR MEANINGS

'yes' or 'no'. The Platonists would immediately concede that nobody may *know* the true answer, but that, they would say, is an altogether different matter.

Of course, matters are not quite this clear-cut, but the situation is certainly a lot better than in the Amsterdam Stedelijk Museum. In the first place, it may not be immediately obvious which statements make mathematical sense (see Example 4.5). In the second place, you don't have to be a Platonist to do mathematics. Not every working mathematician agrees with the statement that the world of mathematics exists independently of the mind of the mathematical discoverer. The Dutch mathematician Brouwer (1881–1966) and his followers have argued instead that mathematical reality has no independent existence, but is *created* by the working mathematician. According to Brouwer the foundation of mathematics is in the *intuition* of the mathematical intellect. A mathematical Intuitionist will therefore not accept certain proof rules of classical mathematics, such as proof by contradiction (see Section 3.3), as this relies squarely on Platonist assumptions.

Although we have no wish to pick a quarrel with the intuitionists, in this book we will accept proof by contradiction, and we will in general adhere to the practice of classical mathematics and thus to the Platonist creed.

Connectives In mathematical reasoning, it is usual to employ shorthands for *if* (or: *if...then*), *and, or, not*. These words are called *connectives*. The word *and* is used to form *conjunctions*, its shorthand $\wedge$ is called the *conjunction* symbol. The word *or* is used to form *disjunctions*, its shorthand $\vee$ is called the *disjunction* symbol. The word *not* is used to form *negations*, its shorthand $\neg$ is called the *negation* symbol. The combination *if...then* produces *implications*; its shorthand $\Rightarrow$ is the *implication* symbol. Finally, there is a phrase less common in everyday conversation, but crucial if one is talking mathematics. The combination *...if and only if...* produces *equivalences*, its shorthand $\Leftrightarrow$ is called the *equivalence* symbol. These logical connectives are summed up in the following table.

	symbol	*name*
and	$\wedge$	conjunction
or	$\vee$	disjunction
not	$\neg$	negation
if—then	$\Rightarrow$	implication
if, and only if	$\Leftrightarrow$	equivalence

Remark. Do not confuse *if...then* ($\Rightarrow$) on one hand with *thus, so, therefore* on the other. The difference is that the phrase *if...then* is used to construct conditional statements, while *thus* (*therefore, so*) is used to combine statements into pieces of mathematical reasoning (or: mathematical proofs). We will never write $\Rightarrow$ when we want to conclude from one mathematical statement to the next. The rules of inference, the notion of mathematical proof, and the proper use of the word *thus* are the subject of Chapter 3. ∎

Iff. In mathematical English it is usual to abbreviate *if, and only if* to *iff*. We will also use $\Leftrightarrow$ as a symbolic abbreviation. Sometimes the phrase *just in case* is used with the same meaning.

The following describes, for every connective separately, how the truth value of a compound using the connective is determined by the truth values of its components. For most connectives, this is rather obvious. The cases for $\Rightarrow$ and $\vee$ have some peculiar difficulties.

The letters P and Q are used for arbitrary statements. We use **t** for 'true', and **f** for 'false'. The set $\{\mathbf{t}, \mathbf{f}\}$ is the set of *truth values*.

Haskell uses `True` and `False` for the truth values. Together, these form the type `Bool`. This type is predefined in Haskell as follows:

```
data Bool     = False | True
```

Negation

An expression of the form $\neg P$ (*not P, it is not the case that P*, etc.) is called the *negation* of P. It is true (has truth value **t**) just in case P is false (has truth value **f**).

In an extremely simple table, this looks as follows:

P	$\neg P$
t	f
f	t

This table is called the *truth table* of the negation symbol.

The implementation of the standard Haskell function `not` reflects this truth table:

2.1. LOGICAL CONNECTIVES AND THEIR MEANINGS

```
not         :: Bool -> Bool
not True    = False
not False   = True
```

This definition is part of *Prelude.hs*, the file that contains the predefined Haskell functions.

Conjunction

The expression $P \wedge Q$ ((*both*) P *and* Q) is called the *conjunction* of P and Q. P and Q are called *conjuncts* of $P \wedge Q$. The conjunction $P \wedge Q$ is true iff P and Q are both true.

Truth table of the conjunction symbol:

P	Q	$P \wedge Q$
t	t	t
t	f	f
f	t	f
f	f	f

This is reflected in definition of the Haskell function for conjunction, && (also from *Prelude.hs*):

```
(&&)       :: Bool -> Bool -> Bool
False && x  = False
True  && x  = x
```

What this says is: if the first argument of a conjunction evaluates to false, then the conjunction evaluates to false; if the first argument evaluates to true, then the conjunction gets the same value as its second argument. The reason that the type declaration has (&&) instead of && is that && is an *infix* operator, and (&&) is its *prefix* counterpart (see page 20).

Disjunction

The expression $P \vee Q$ (P *or* Q) is called the *disjunction* of P and Q. P and Q are the *disjuncts* of $P \vee Q$.

The interpretation of disjunctions is not always straightforward. English has *two* disjunctions: (i) the *inclusive* version, that counts a disjunction as true also in case both disjuncts are true, and (ii) the *exclusive* version *either... or*, that doesn't.

Remember: The symbol $\vee$ will *always* be used for the **inclusive** version of *or*.

Even with this problem out of the way, difficulties may arise.

Example 2.1 No one will doubt the truth of the following:

for every integer x, $x < 1$ or $0 < x$.

However, acceptance of this brings along acceptance of every instance. E.g., for $x := 1$:[1]

$$1 < 1 \text{ or } 0 < 1.$$

Some people do not find this acceptable or true, or think this to make no sense at all since something better can be asserted, viz., that $0 < 1$. In mathematics with the inclusive version of $\vee$, you'll have to live with such a peculiarity.

The truth table of the disjunction symbol $\vee$ now looks as follows.

P	Q	$P \vee Q$
t	t	t
t	f	t
f	t	t
f	f	f

Here is the Haskell definition of the disjunction operation. Disjunction is rendered as || in Haskell.

```
(||)    :: Bool -> Bool -> Bool
False || x    = x
True  || x    = True
```

What this means is: if the first argument of a disjunction evaluates to false, then the disjunction gets the same value as its second argument. If the first argument of a disjunction evaluates to true, then the disjunction evaluates to true.

---

[1] := means: 'is by definition equal to'.

2.1. LOGICAL CONNECTIVES AND THEIR MEANINGS

Exercise 2.2 Make up the truth table for the *exclusive* version of *or*.

Implication

An expression of the form $P \Rightarrow Q$ (*if P, then Q*; *Q if P*) is called the *implication* of P and Q. P is the *antecedent* of the implication and Q the *consequent*.

The truth table of $\Rightarrow$ is perhaps the only surprising one. However, it can be motivated quite simply using an example like the following. No one will disagree that for every natural number n,

$$5 < n \Rightarrow 3 < n.$$

Therefore, the implication must hold in particular for n equal to 2, 4 and 6. But then, an implication should be *true* if

- both antecedent and consequent are false ($n = 2$),
- antecedent false, consequent true ($n = 4$),

and

- both antecedent and consequent true ($n = 6$).

Of course, an implication should be false in the only remaining case that the antecedent is true and the consequent false. This accounts for the following truth table.

P	Q	$P \Rightarrow Q$
t	t	t
t	f	f
f	t	t
f	f	t

If we want to implement implication in Haskell, we can do so in terms of `not` and `||`. It is convenient to introduce an infix operator `==>` for this. The number 1 in the `infix` declaration indicates the binding power (binding power 0 is lowest, 9 is highest). A declaration of an infix operator together with an indication of its binding power is called a *fixity declaration*.

```
infix 1 ==>

(==>) :: Bool -> Bool -> Bool
x ==> y = (not x) || y
```

It is also possible to give a direct definition:

```
(==>) :: Bool -> Bool -> Bool
True  ==> x = x
False ==> x = True
```

Trivially True Implications. Note that implications with antecedent false and those with consequent true are true. For instance, because of this, the following two sentences must be counted as true: *if my name is Napoleon, then the decimal expansion of π contains the sequence 7777777*, and: *if the decimal expansion of π contains the sequence 7777777, then strawberries are red.*

Implications with one of these two properties (no matter what the values of parameters that may occur) are dubbed *trivially* true. In what follows there are quite a number of facts that are trivial in this sense that may surprise the beginner. One is that the *empty set* $\emptyset$ is included in *every* set (cf. Theorem 4.9 p. 124).

Remark. The word *trivial* is often abused. Mathematicians have a habit of calling things trivial when they are reluctant to prove them. We will try to avoid this use of the word. The justification for calling a statement trivial resides in the psychological fact that a proof of that statement immediately comes to mind. Whether a proof of something comes to your mind will depend on your training and experience, so what is trivial in this sense is (to some extent) a personal matter. When we are reluctant to prove a statement, we will sometimes ask you to prove it as an exercise. ∎

Implication and Causality. The mathematical use of implication does not always correspond to what you are used to. In daily life you will usually require a certain causal dependence between antecedent and consequent of an implication. (This is the reason the previous examples look funny.) In mathematics, such a causality usually will be present, but this is quite unnecessary for the interpretation of an implication: the truth table tells the complete story. (In this section in particular, causality usually will be absent.) However, in a few cases, natural language use surprisingly corresponds with truth table-meaning. E.g., *I'll be dead if Bill will not show*

2.1. LOGICAL CONNECTIVES AND THEIR MEANINGS

up must be interpreted (if uttered by someone healthy) as strong belief that Bill will indeed turn up.[2]

Converse and Contraposition. The *converse* of an implication $P \Rightarrow Q$ is $Q \Rightarrow P$; its *contraposition* is $\neg Q \Rightarrow \neg P$. The converse of a true implication does not need to be true, but its contraposition is true iff the implication is. Cf. Theorem 2.10, p. 45.

Necessary and Sufficient Conditions. The statement P is called a *sufficient condition* for Q and Q a *necessary condition* for P if the implication $P \Rightarrow Q$ holds.

An implication $P \Rightarrow Q$ can be expressed in a mathematical text in a number of ways:

1. if P, then Q,
2. Q if P,
3. P only if Q,
4. Q whenever P,
5. P is sufficient for Q,
6. Q is necessary for P.

Equivalence

The expression $P \Leftrightarrow Q$ (P iff Q) is called the *equivalence* of P and Q. P and Q are the *members* of the equivalence. The truth table of the equivalence symbol is unproblematic once you realize that an equivalence $P \Leftrightarrow Q$ amounts to the conjunction of two implications $P \Rightarrow Q$ and $Q \Rightarrow P$ taken together. (It is sometimes convenient to write $Q \Rightarrow P$ as $P \Leftarrow Q$.) The outcome is that an equivalence must be true iff its members have the same truth value.

Table:

P	Q	$P \Leftrightarrow Q$
t	t	t
t	f	f
f	t	f
f	f	t

---

[2]'If Bill will not show up, then I am a Dutchman', has the same meaning, when uttered by a native speaker of English. What it means when uttered by one of the authors of this book, we are not sure.

From the discussion under implication it is clear that P is called a condition that is both *necessary* and *sufficient* for Q if $P \Leftrightarrow Q$ is true.

There is no need to add a definition of a function for equivalence to Haskell. The type `Bool` is in class `Eq`, which means that an equality relation is predefined on it. But equivalence of propositions is nothing other than equality of their truth values. Still, it is useful to have a synonym:

```
infix 1 <=>

(<=>) :: Bool -> Bool -> Bool
x <=> y = x == y
```

Example 2.3 When you are asked to prove something of the form P iff Q it is often convenient to separate this into its two parts $P \Rightarrow Q$ and $P \Leftarrow Q$. The 'only if' part of the proof is the proof of $P \Rightarrow Q$ (for $P \Rightarrow Q$ means the same as P only if Q), and the 'if' part of the proof is the proof of $P \Leftarrow Q$ (for $P \Leftarrow Q$ means the same as $Q \Rightarrow P$, which in turn means the same as P, if Q).

Exercise 2.4 Check that the truth table for exclusive *or* from Exercise 2.2 is equivalent to the table for $\neg(P \Leftrightarrow Q)$. Conclude that the Haskell implementation of the function <+> for exclusive or in the frame below is correct.

```
infixr 2 <+>

(<+>) :: Bool -> Bool -> Bool
x <+> y = x /= y
```

The logical connectives $\wedge$ and $\vee$ are written in infix notation. Their Haskell counterparts, `&&` and `||` are also infix. Thus, if `p` and `q` are expressions of type `Bool`, then `p && q` is a correct Haskell expression of type `Bool`. If one wishes to write this in prefix notation, this is also possible, by putting parentheses around the operator: `(&&) p q`.

2.1. LOGICAL CONNECTIVES AND THEIR MEANINGS

Although you will probably never find more than 3–5 connectives occurring in one mathematical statement, if you insist you can use as many connectives as you like. Of course, by means of parentheses you should indicate the way your expression was formed.

For instance, look at the formula

$$\neg P \wedge ((P \Rightarrow Q) \Leftrightarrow \neg(Q \wedge \neg P)).$$

Using the truth tables, you can determine its truth value if truth values for the components P and Q have been given. For instance, if P has value **t** and Q has value **f**, then $\neg P$ has **f**, $P \Rightarrow Q$ becomes **f**, $Q \wedge \neg P$: **f**; $\neg(Q \wedge \neg P)$: **t**; $(P \Rightarrow Q) \Leftrightarrow \neg(Q \wedge \neg P)$: **f**, and the displayed expression thus has value **f**. This calculation can be given immediately under the formula, beginning with the values given for P and Q. The final outcome is located under the conjunction symbol $\wedge$, which is the main connective of the expression.

```
   ¬   P   ∧   ((P   ⇒   Q)   ⇔   ¬   (Q   ∧   ¬   P))
       t       t        f             f             t
   f                f                         f
                                          f
                                  t
                          f
           f
```

In compressed form, this looks as follows:

```
   ¬   P   ∧   ((P   ⇒   Q)   ⇔   ¬   (Q   ∧   ¬   P))
   f   t   f    t    f    f    f   t    f    f   f   t
```

Alternatively, one might use a computer to perform the calculation.

```
p = True
q = False

formula1 = (not p) && (p ==> q) <=> not (q && (not p))
```

After loading the file with the code of this chapter, you should be able to do:

```
TAMO> formula1
False
```

Note that `p` and `q` are defined as constants, with values `True` and `False`, respectively, so that the occurrences of `p` and `q` in the expression `formula1` are evaluated as these truth values. The rest of the evaluation is then just a matter of applying the definitions of `not`, `&&`, `<=>` and `==>`.

2.2 Logical Validity and Related Notions

Goal To grasp the concepts of logical validity and logical equivalence, to learn how to use truth tables in deciding questions of validity and equivalence, and in the handling of negations, and to learn how the truth table method for testing validity and equivalence can be implemented.

Logical Validities. There are propositional formulas that receive the value **t** no matter what the values of the occurring letters. Such formulas are called (logically) *valid*.

Examples of logical validities are: $P \Rightarrow P$, $P \vee \neg P$, and $P \Rightarrow (Q \Rightarrow P)$.

Truth Table of an Expression. If an expression contains n letters $P, Q, \ldots$, then there are 2^n possible distributions of the truth values between these letters. The 2^n-row table that contains the calculations of these values is the *truth table* of the expression.

If all calculated values are equal to **t**, then your expression, by definition, is a validity.

Example 2.5 (Establishing Logical Validity by Means of a Truth Table)

The following truth table shows that $P \Rightarrow (Q \Rightarrow P)$ is a logical validity.

P	$\Rightarrow$	$(Q$	$\Rightarrow$	$P)$
t	t	t	t	t
t	t	f	t	t
f	t	t	f	f
f	t	f	t	f

2.2. LOGICAL VALIDITY AND RELATED NOTIONS

To see how we can implement the validity check in Haskell, look at the implementation of the evaluation `formula1` again, and add the following definition of `formula2`:

```
formula2 p q = ((not p) && (p ==> q) <=> not (q && (not p)))
```

To see the difference between the two definitions, let us check their types:

```
TAMO> :t formula1
formula1 :: Bool
TAMO> :t formula2
formula2 :: Bool -> Bool -> Bool
TAMO>
```

The difference is that the first definition is a complete proposition (type `Bool`) in itself, while the second still needs two arguments of type `Bool` before it will return a truth value.

In the definition of `formula1`, the occurrences of `p` and `q` are interpreted as *constants*, of which the values are given by previous definitions. In the definition of `formula2`. the occurrences of `p` and `q` are interpreted as *variables* that represent the arguments when the function gets called.

A propositional formula in which the proposition letters are interpreted as variables can in fact be considered as a *propositional function* or *Boolean function* or *truth function*. If just one variable, say p occurs in it, then it is a function of type `Bool -> Bool` (takes a Boolean, returns a Boolean). If two variables occur in it, say p and q, then it is a function of type `Bool -> Bool -> Bool` (takes Boolean, then takes another Boolean, and returns a Boolean). If three variables occur in it, then it is of type `Bool -> Bool -> Bool -> Bool`, and so on.

In the validity check for a propositional formula, we treat the proposition letters as arguments of a propositional function, and we check whether evaluation of the function yields true for every possible combination of the arguments (that is the essence of the truth table method for checking validity). Here is the case for propositions with one proposition letter (type `Bool -> Bool`).

```
valid1 :: (Bool -> Bool) -> Bool
valid1 bf = (bf True) && (bf False)
```

The validity check for Boolean functions of type Bool -> Bool is suited to test functions of just one variable. An example is the formula $P \vee \neg P$ that expresses the principle of excluded middle (or, if you prefer a Latin name, *tertium non datur*, for: there is no third possibility). Here is its implementation in Haskell:

```
excluded_middle :: Bool -> Bool
excluded_middle p = p || not p
```

To check that this is valid by the truth table method, one should consider the two cases $P := \mathbf{t}$ and $P := \mathbf{f}$, and ascertain that the principle yields $\mathbf{t}$ in both of these cases. This is precisely what the validity check valid1 does: it yields True precisely when applying the boolean function bf to True yields True and applying bf to False yields True. Indeed, we get:

```
TAMO> valid1 excluded_middle
True
```

Here is the validity check for propositional functions with two proposition letters, Such propositional functions have type Bool -> Bool -> Bool), and need a truth table with four rows to check their validity, as there are four cases to check.

```
valid2 :: (Bool -> Bool -> Bool)  -> Bool
valid2 bf =    (bf True  True)
            && (bf True  False)
            && (bf False True)
            && (bf False False)
```

Again, it is easy to see that this is an implementation of the truth table method for validity checking. Try this out on $P \Rightarrow (Q \Rightarrow P)$ and on $(P \Rightarrow Q) \Rightarrow P$, and discover that the bracketing matters:

```
form1 p q = p ==> (q ==> p)
form2 p q = (p ==> q) ==> p
```

2.2. LOGICAL VALIDITY AND RELATED NOTIONS 41

```
TAMO> valid2 form1
True
TAMO> valid2 form2
False
```

The propositional function `formula2` that was defined above is also of the right argument type for `valid2`:

```
TAMO> valid2 formula2
False
```

It should be clear how the notion of validity is to be implemented for propositional functions with more than two propositional variables. Writing out the full tables becomes a bit irksome, so we are fortunate that Haskell offers an alternative. We demonstrate it in `valid3` and `valid4`.

```
valid3 :: (Bool -> Bool -> Bool -> Bool) -> Bool
valid3 bf = and [ bf p q r | p <- [True,False],
                             q <- [True,False],
                             r <- [True,False]]

valid4 :: (Bool -> Bool -> Bool -> Bool -> Bool) -> Bool
valid4 bf = and [ bf p q r s | p <- [True,False],
                               q <- [True,False],
                               r <- [True,False],
                               s <- [True,False]]
```

The condition `p <- [True,False]`, for "p is an element of the list consisting of the two truth values", is an example of *list comprehension* (page 116).

The definitions make use of Haskell list notation, and of the predefined function and for generalized conjunction. An example of a list of Booleans in Haskell is `[True,True,False]`. Such a list is said to be of type `[Bool]`. If `list` is a list of Booleans (an object of type `[Bool]`), then and `list` gives True in case all members of `list` are true, False otherwise. For example, and `[True,True,False]` gives False, but and `[True,True,True]` gives True. Further details about working with lists can be found in Sections 4.6 and 7.5.

Leaving out Parentheses. We agree that $\wedge$ and $\vee$ bind more strongly than $\Rightarrow$ and $\Leftrightarrow$. Thus, for instance, $P \wedge Q \Rightarrow R$ stands for $(P \wedge Q) \Rightarrow R$ (and *not* for $P \wedge (Q \Rightarrow R)$).

42 CHAPTER 2. TALKING ABOUT MATHEMATICAL OBJECTS

Operator Precedence in Haskell In Haskell, the convention is not quite the same, for || has operator precedence 2, && has operator precedence 3, and == has operator precedence 4, which means that == binds more strongly than &&, which in turn binds more strongly than ||. The operators that we added, ==> and <=>, follow the logic convention: they bind less strongly than && and ||.

Logically Equivalent. Two formulas are called (logically) *equivalent* if, no matter the truth values of the letters $P, Q, \ldots$ occurring in these formulas, the truth values obtained for them are the same. This can be checked by constructing a truth table (see Example 2.6).

Example 2.6 (The First Law of De Morgan)

$\neg$	$(P$	$\wedge$	$Q)$	$(\neg$	P	$\vee$	$\neg$	$Q)$
f	t	t	t	f	t	f	f	t
t	t	f	f	f	t	t	t	f
t	f	f	t	t	f	t	f	t
t	f	f	f	t	f	t	t	f

The outcome of the calculation shows that the formulas are equivalent: note that the column under the $\neg$ of $\neg(P \wedge Q)$ coincides with that under the $\vee$ of $\neg P \vee \neg Q$.

Notation: $\Phi \equiv \Psi$ indicates that Φ and Ψ are equivalent.[3] Using this notation, we can say that the truth table of Example 2.6 shows that $\neg(P \wedge Q) \equiv (\neg P \vee \neg Q)$.

Example 2.7 (De Morgan Again)
The following truth table shows that $\neg(P \wedge Q) \Leftrightarrow (\neg P \vee \neg Q)$ is a logical validity, which establishes that $\neg(P \wedge Q) \equiv (\neg P \vee \neg Q)$.

$\neg$	$(P$	$\wedge$	$Q)$	$\Leftrightarrow$	$(\neg$	P	$\vee$	$\neg$	$Q)$
f	t	t	t	t	f	t	f	f	t
t	t	f	f	t	f	t	t	t	f
t	f	f	t	t	t	f	t	f	t
t	f	f	f	t	t	f	t	t	f

---
[3] The Greek alphabet is on p. 411.

2.2. LOGICAL VALIDITY AND RELATED NOTIONS

Example 2.8 A pixel on a computer screen is a dot on the screen that can be either *on* (i.e., visible) or *off* (i.e., invisible). We can use 1 for *on* and 0 for *off*. Turning pixels in a given area on the screen off or on creates a screen pattern for that area. The screen pattern of an area is given by a list of bits (0s or 1s). Such a list of bits can be viewed as a list of truth values (by equating 1 with **t** and 0 with **f**), and given two bit lists of the same length we can perform bitwise logical operations on them: the bitwise exclusive or of two bit lists of the same length n, say $L = [P_1, \ldots, P_n]$ and $K = [Q_1, \ldots, Q_n]$, is the list $[P_1 \oplus Q_1, \ldots, P_n \oplus Q_n]$, where $\oplus$ denotes exclusive or.

In the implementation of cursor movement algorithms, the cursor is made visible on the screen by taking a bitwise exclusive or between the screen pattern S at the cursor position and the cursor pattern C. When the cursor moves elsewhere, the original screen pattern is restored by taking a bitwise exclusive or with the cursor pattern C again. Exercise 2.9 shows that this indeed restores the original pattern S.

Exercise 2.9 Let $\oplus$ stand for exclusive or. Show, using the truth table from Exercise 2.2, that $(P \oplus Q) \oplus Q$ is equivalent to P.

In Haskell, logical equivalence can be tested as follows. First we give a procedure for propositional functions with 1 parameter:

```
logEquiv1 :: (Bool -> Bool) -> (Bool -> Bool) -> Bool
logEquiv1 bf1 bf2 = 
    (bf1 True  <=> bf2 True) && (bf1 False <=> bf2 False)
```

What this does, for formulas Φ, Ψ with a single propositional variable, is testing the formula $\Phi \Leftrightarrow \Psi$ by the truth table method.

We can extend this to propositional functions with 2, 3 or more parameters, using generalized conjunction. Here are the implementations of `logEquiv2` and `logEquiv3`; it should be obvious how to extend this for truth functions with still more arguments.

```
logEquiv2 :: (Bool -> Bool -> Bool) ->
                      (Bool -> Bool -> Bool) -> Bool
logEquiv2 bf1 bf2 =
  and [(bf1 p q) <=> (bf2 p q) | p <- [True,False],
                                  q <- [True,False]]

logEquiv3 :: (Bool -> Bool -> Bool -> Bool) ->
                  (Bool -> Bool -> Bool -> Bool) -> Bool
logEquiv3 bf1 bf2 =
  and [(bf1 p q r) <=> (bf2 p q r) | p <- [True,False],
                                      q <- [True,False],
                                      r <- [True,False]]
```

Let us redo Exercise 2.9 by computer.

```
formula3 p q = p
formula4 p q = (p <+> q) <+> q
```

Note that the `q` in the definition of `formula3` is needed to ensure that it is a function with two arguments.

```
TAMO> logEquiv2 formula3 formula4
True
```

We can also test this by means of a validity check on $P \Leftrightarrow ((P \oplus Q) \oplus Q)$, as follows:

```
formula5 p q = p <=> ((p <+> q) <+> q)
```

```
TAMO> valid2 formula5
True
```

Warning. Do not confuse $\equiv$ and $\Leftrightarrow$. If Φ and Ψ are formulas, then $\Phi \equiv \Psi$ expresses the statement that Φ and Ψ are equivalent. On the other hand, $\Phi \Leftrightarrow \Psi$ is just another formula. The relation between the two is that the formula $\Phi \Leftrightarrow \Psi$ is logically valid iff it holds that $\Phi \equiv \Psi$. (See Exercise 2.19.) Compare the difference, in Haskell, between `logEquiv2 formula3 formula4` (a true statement about the relation between two formulas), and `formula5` (just another formula).

2.2. LOGICAL VALIDITY AND RELATED NOTIONS

The following theorem collects a number of useful equivalences. (Of course, P, Q and R can be arbitrary formulas themselves.)

Theorem 2.10
1. $P \equiv \neg\neg P$ *(law of double negation)*,

2. $P \land P \equiv P$; $P \lor P \equiv P$ *(laws of idempotence)*,

3. $(P \Rightarrow Q) \equiv \neg P \lor Q$;
$\neg(P \Rightarrow Q) \equiv P \land \neg Q$,

4. $(\neg P \Rightarrow \neg Q) \equiv (Q \Rightarrow P)$;
$(P \Rightarrow \neg Q) \equiv (Q \Rightarrow \neg P)$;
$(\neg P \Rightarrow Q) \equiv (\neg Q \Rightarrow P)$ *(laws of contraposition)*,

5. $(P \Leftrightarrow Q) \equiv ((P \Rightarrow Q) \land (Q \Rightarrow P))$
$\equiv ((P \land Q) \lor (\neg P \land \neg Q))$,

6. $P \land Q \equiv Q \land P$; $P \lor Q \equiv Q \lor P$ *(laws of commutativity)*,

7. $\neg(P \land Q) \equiv \neg P \lor \neg Q$;
$\neg(P \lor Q) \equiv \neg P \land \neg Q$ *(DeMorgan laws)*.

8. $P \land (Q \land R) \equiv (P \land Q) \land R$;
$P \lor (Q \lor R) \equiv (P \lor Q) \lor R$ *(laws of associativity)*,

9. $P \land (Q \lor R) \equiv (P \land Q) \lor (P \land R)$;
$P \lor (Q \land R) \equiv (P \lor Q) \land (P \lor R)$ *(distribution laws)*,

Equivalence 8 justifies leaving out parentheses in conjunctions and disjunctions of three or more conjuncts resp., disjuncts. Non-trivial equivalences that often are used in practice are 2, 3 and 9. Note how you can use these to re-write negations: a negation of an implication can be rewritten as a conjunction, a negation of a conjunction (disjunction) is a disjunction (conjunction).

Exercise 2.11 The First Law of De Morgan was proved in Example 2.6. This method was implemented above. Use the method by hand to prove the other parts of Theorem 2.10.

```
test1  = logEquiv1 id (\ p -> not (not p))
test2a = logEquiv1 id (\ p -> p && p)
test2b = logEquiv1 id (\ p -> p || p)
test3a = logEquiv2 (\ p q -> p ==> q) (\ p q -> not p || q)
test3b = logEquiv2 (\ p q -> not (p ==> q)) (\ p q -> p && not q)
test4a = logEquiv2 (\ p q -> not p ==> not q) (\ p q -> q ==> p)
test4b = logEquiv2 (\ p q -> p ==> not q) (\ p q -> q ==> not p)
test4c = logEquiv2 (\ p q -> not p ==> q) (\ p q -> not q ==> p)
test5a = logEquiv2 (\ p q -> p <=> q)
                  (\ p q -> (p ==> q) && (q ==> p))
test5b = logEquiv2 (\ p q -> p <=> q)
                  (\ p q -> (p && q) || (not p && not q))
test6a = logEquiv2 (\ p q -> p && q) (\ p q -> q && p)
test6b = logEquiv2 (\ p q -> p || q) (\ p q -> q || p)
test7a = logEquiv2 (\ p q -> not (p && q))
                  (\ p q -> not p || not q)
test7b = logEquiv2 (\ p q -> not (p || q))
                  (\ p q -> not p && not q)
test8a = logEquiv3 (\ p q r -> p && (q && r))
                  (\ p q r -> (p && q) && r)
test8b = logEquiv3 (\ p q r -> p || (q || r))
                  (\ p q r -> (p || q) || r)
test9a = logEquiv3 (\ p q r -> p && (q || r))
                  (\ p q r -> (p && q) || (p && r))
test9b = logEquiv3 (\ p q r -> p || (q && r))
                  (\ p q r -> (p || q) && (p || r))
```

Figure 2.1: Defining the Tests for Theorem 2.10.

2.2. LOGICAL VALIDITY AND RELATED NOTIONS

We will now demonstrate how one can use the implementation of the logical equivalence tests as a check for Theorem 2.10. Here is a question for you to ponder: does checking the formulas by means of the implemented functions for logical equivalence count as a *proof* of the principles involved? Whatever the answer to this one may be, Figure 2.1 defines the tests for the statements made in Theorem 2.10, by means of *lambda abstraction*. The expression \ p -> not (not p) is the Haskell way of referring to the lambda term $\lambda p.\neg\neg p$, the term that denotes the operation of performing a double negation. See Section 2.4.

If you run these tests, you get result True for all of them. E.g.:

```
TAMO> test5a
True
```

The next theorem lists some useful principles for reasoning with $\top$ (the proposition that is always true; the Haskell counterpart is True) and $\bot$ (the proposition that is always false; the Haskell counterpart of this is False).

Theorem 2.12
1. $\neg\top \equiv \bot$; $\neg\bot \equiv \top$,
2. $P \Rightarrow \bot \equiv \neg P$,
3. $P \vee \top \equiv \top$; $P \wedge \bot \equiv \bot$ *(dominance laws)*,
4. $P \vee \bot \equiv P$; $P \wedge \top \equiv P$ *(identity laws)*,
5. $P \vee \neg P \equiv \top$ *(law of excluded middle)*,
6. $P \wedge \neg P \equiv \bot$ *(contradiction)*.

Exercise 2.13 Implement checks for the principles from Theorem 2.12.

Without proof, we state the following **Substitution Principle**: If Φ and Ψ are equivalent, and Φ' and Ψ' are the results of substituting Ξ for every occurrence of P in Φ and in Ψ, respectively, then Φ' and Ψ' are equivalent. Example 2.14 makes clear what this means.

Example 2.14 From $\neg(P \Rightarrow Q) \equiv P \wedge \neg Q$ plus the substitution principle it follows that
$$\neg(\neg P \Rightarrow Q) \equiv \neg P \wedge \neg Q$$
(by substituting $\neg P$ for P), but also that
$$\neg(a = 2^b - 1 \Rightarrow a \text{ is prime}) \equiv a = 2^b - 1 \wedge a \text{ is not prime}$$
(by substituting $a = 2^b - 1$ for P and a *is prime* for Q).

Exercise 2.15 A propositional contradiction is a formula that yields false for every combination of truth values for its proposition letters. Write Haskell definitions of contradiction tests for propositional functions with one, two and three variables.

Exercise 2.16 Produce useful denials for every sentence of Exercise 2.31. (A denial of Φ is an equivalent of $\neg\Phi$.)

Exercise 2.17 Produce a denial for the statement that $x < y < z$ (where $x, y, z \in \mathbb{R}$).

Exercise 2.18 Show:

1. $(\Phi \Leftrightarrow \Psi) \equiv (\neg\Phi \Leftrightarrow \neg\Psi)$,
2. $(\neg\Phi \Leftrightarrow \Psi) \equiv (\Phi \Leftrightarrow \neg\Psi)$.

Exercise 2.19 Show that $\Phi \equiv \Psi$ is true iff $\Phi \Leftrightarrow \Psi$ is logically valid.

Exercise 2.20 Determine (either using truth tables or Theorem 2.10) which of the following are equivalent, next check your answer by computer:

1. $\neg P \Rightarrow Q$ and $P \Rightarrow \neg Q$,
2. $\neg P \Rightarrow Q$ and $Q \Rightarrow \neg P$,
3. $\neg P \Rightarrow Q$ and $\neg Q \Rightarrow P$,
4. $P \Rightarrow (Q \Rightarrow R)$ and $Q \Rightarrow (P \Rightarrow R)$,
5. $P \Rightarrow (Q \Rightarrow R)$ and $(P \Rightarrow Q) \Rightarrow R$,
6. $(P \Rightarrow Q) \Rightarrow P$ and P,
7. $P \vee Q \Rightarrow R$ and $(P \Rightarrow R) \wedge (Q \Rightarrow R)$.

Exercise 2.21 Answer as many of the following questions as you can:

1. Construct a formula Φ involving the letters P and Q that has the following truth table.

P	Q	Φ
t	t	t
t	f	t
f	t	f
f	f	t

2. How many truth tables are there for 2-letter formulas altogether?

3. Can you find formulas for all of them?

4. Is there a general method for finding these formulas?

5. What about 3-letter formulas and more?

2.3 Making Symbolic Form Explicit

In a sense, propositional reasoning is not immediately relevant for mathematics. Few mathematicians will ever feel the urge to write down a disjunction of two statements like $3 < 1 \vee 1 < 3$. In cases like this it is clearly "better" to only write down the right-most disjunct.

Fortunately, once variables enter the scene, propositional reasoning suddenly becomes a very useful tool: the connectives turn out to be quite useful for combining *open* formulas. An open formula is a formula with one or more unbound variables in it. Variable binding will be explained below, but here is a first example of a formula with an unbound variable x. A disjunction like $3 < x \vee x < 3$ is (in some cases) a useful way of expressing that $x \neq 3$.

Example. Consider the following (true) sentence:

$$\text{Between every two rational numbers there is a third one.} \quad (2.1)$$

The property expressed in (2.1) is usually referred to as *density* of the rationals. We will take a systematic look at proving such statements in Chapter 3.

Exercise 2.22 Can you think of an argument showing that statement (2.1) is true?

A Pattern. There is a *logical pattern* underlying sentence (2.1). To make it visible, look at the following, more explicit, formulation. It uses *variables* x, y and z for arbitrary rationals, and refers to the *ordering* $<$ of the set $\mathbb{Q}$ of rational numbers.

$$\text{For all rational numbers } x \text{ and } z, \text{ if } x < z, \text{ then some} \quad (2.2)$$
$$\text{rational number } y \text{ exists such that } x < y \text{ and } y < z.$$

You will often find '$x < y$ and $y < z$' shortened to: $x < y < z$.

Quantifiers Note the words *all* (or: *for all*), *some* (or: *for some, some... exists, there exists... such that*, etc.). They are called *quantifiers*, and we use the symbols ∀ and ∃ as shorthands for them.

With these shorthands, plus the shorthands for the connectives that we saw above, and the shorthand ... ∈ ℚ for the property of being a rational, we arrive at the following compact symbolic formulation:

$$\forall x \in \mathbb{Q} \, \forall z \in \mathbb{Q} \, (x < z \;\Rightarrow\; \exists y \in \mathbb{Q} (x < y \,\wedge\, y < z)). \qquad (2.3)$$

We will use example (2.3) to make a few points about the proper use of the vocabulary of logical symbols. An expression like (2.3) is called a sentence or a formula. Note that the example formula (2.3) is composite: we can think of it as constructed out of simpler parts. We can picture its structure as in Figure 2.2.

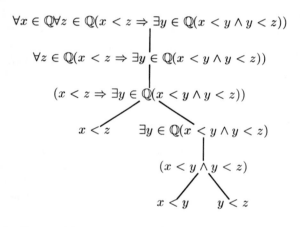

Figure 2.2: Composition of Example Formula from its Sub-formulas.

As the figure shows, the example formula is formed by putting the quantifier prefix $\forall x \in \mathbb{Q}$ in front of the result of putting quantifier prefix $\forall z \in \mathbb{Q}$ in front of a simpler formula, and so on.

The two consecutive universal quantifier prefixes can also be combined into $\forall x, z \in \mathbb{Q}$. This gives the phrasing

$$\forall x, z \in \mathbb{Q} (x < z \;\Rightarrow\; \exists y \in \mathbb{Q} (x < y \,\wedge\, y < z)).$$

Putting an ∧ between the two quantifiers is incorrect, however. In other words, the expression $\forall x \in \mathbb{Q} \,\wedge\, \forall z \in \mathbb{Q}(x < z \Rightarrow \exists y \in \mathbb{Q}(x < y \,\wedge\, y < z))$ is considered *un*grammatical. The reason is that the formula part $\forall x \in \mathbb{Q}$ is

2.3. MAKING SYMBOLIC FORM EXPLICIT

itself not a formula, but a prefix that turns a formula into a more complex formula. The connective ∧ can only be used to construct a new formula out of two simpler formulas, so ∧ cannot serve to construct a formula from $\forall x \in \mathbb{Q}$ and another formula.

The symbolic version of the density statement uses parentheses. Their function is to indicate the way the expression has been formed and thereby to show the *scope* of operators. The scope of a quantifier-expression is the formula that it combines with to form a more complex formula. The scopes of quantifier-expressions and connectives in a formula are illustrated in the structure tree of that formula. Figure 2.2 shows that the scope of the quantifier-expression $\forall x \in \mathbb{Q}$ is the formula

$$\forall z \in \mathbb{Q}(x < z \Rightarrow \exists y \in \mathbb{Q} \ (x < y \land y < z)),$$

the scope of $\forall z \in \mathbb{Q}$ is the formula

$$(x < z \Rightarrow \exists y \in \mathbb{Q}(x < y \land y < z)),$$

and the scope of $\exists y \in \mathbb{Q}$ is the formula $(x < y \land y < z)$.

Exercise 2.23 Give structure trees of the following formulas (we use shorthand notation, and write $A(x)$ as Ax for readability).

1. $\forall x(Ax \Rightarrow (Bx \Rightarrow Cx))$.

2. $\exists x(Ax \land Bx)$.

3. $\exists x Ax \land \exists x Bx$.

The expression *for all* (and similar ones) and its shorthand, the symbol ∀, is called the *universal quantifier*; the expression *there exists* (and similar ones) and its shorthand, the symbol ∃, is called the *existential quantifier*. The letters x, y and z that have been used in combination with them are *variables*. Note that 'for some' is equivalent to 'for at least one'.

Unrestricted and Restricted Quantifiers, Domain of Quantification
Quantifiers can occur *unrestricted*: $\forall x(x \geq 0)$, $\exists y \forall x(y > x)$, and *restricted*: $\forall x \in A(x \geq 0)$, $\exists y \in B(y < a)$ (where A and B are sets).

In the unrestricted case, there should be some *domain of quantification* that often is implicit in the context. E.g., if the context is real analysis, $\forall x$ may mean *for all reals x...*, and $\forall f$ may mean *for all real-valued functions f...*.

Example 2.24 $\mathbb{R}$ is the set of real numbers. The fact that the $\mathbb{R}$ has no greatest element can be expressed with restricted quantifiers as:

$$\forall x \in \mathbb{R} \exists y \in \mathbb{R}(x < y).$$

If we specify that all quantifiers range over the reals (i.e., if we say that $\mathbb{R}$ is the domain of quantification) then we can drop the explicit restrictions, and we get by with $\forall x \exists y (x < y)$.

The use of restricted quantifiers allows for greater flexibility, for it permits one to indicate different domains for different quantifiers.

Example 2.25

$$\forall x \in \mathbb{R} \forall y \in \mathbb{R}(x < y \Rightarrow \exists z \in \mathbb{Q} \ (x < z < y)).$$

Instead of $\exists x(Ax \wedge \ldots)$ one can write $\exists x \in A(\ldots)$. The advantage when all quantifiers are thus restricted is that it becomes immediately clear that the domain is subdivided into different sub domains or types. This can make the logical translation much easier to comprehend.

Remark. We will use standard names for the following domains: $\mathbb{N}$ for the natural numbers, $\mathbb{Z}$ for the integer numbers, $\mathbb{Q}$ for the rational numbers, and $\mathbb{R}$ for the real numbers. More information about these domains can be found in Chapter 8.

Exercise 2.26 Write as formulas with restricted quantifiers:

1. $\exists x \exists y (x \in \mathbb{Q} \wedge y \in \mathbb{Q} \wedge x < y)$.

2. $\forall x (x \in \mathbb{R} \Rightarrow \exists y (y \in \mathbb{R} \wedge x < y))$.

3. $\forall x (x \in \mathbb{Z} \Rightarrow \exists m, n (m \in \mathbb{N} \wedge n \in \mathbb{N} \wedge x = m - n))$.

Exercise 2.27 Write as formulas without restricted quantifiers:

1. $\forall x \in \mathbb{Q} \exists m, n \in \mathbb{Z}(n \neq 0 \wedge x = m/n)$.

2. $\forall x \in F \forall y \in D(Oxy \Rightarrow Bxy)$.

2.3. MAKING SYMBOLIC FORM EXPLICIT

Bound Variables. Quantifier expressions $\forall x$, $\exists y, \ldots$ (and their restricted companions) are said to *bind* every occurrence of $x, y, \ldots$ in their scope. If a variable occurs *bound* in a certain expression then the meaning of that expression does not change when all bound occurrences of that variable are replaced by another one.

Example 2.28 $\exists y \in \mathbb{Q}(x < y)$ has the same meaning as $\exists z \in \mathbb{Q}(x < z)$. This indicates that y is bound in $\exists y \in \mathbb{Q}(x < y)$. But $\exists y \in \mathbb{Q}(x < y)$ and $\exists y \in \mathbb{Q}(z < y)$ have different meanings, for the first asserts that there exists a rational number greater than some given number x, and the second that there exists a rational number greater than some given z.

Universal and existential quantifiers are not the only variable binding operators used by mathematicians. There are several other constructs that you are probably familiar with which can bind variables.

Example 2.29 (Summation, Integration.) The expression $\sum_{i=1}^{5} i$ is nothing but a way to describe the number 15 ($15 = 1 + 2 + 3 + 4 + 5$), and clearly, 15 does in no way depend on i. Use of a different variable does not change the meaning: $\sum_{k=1}^{5} k = 15$. Here are the Haskell versions:

```
Prelude> sum [ i | i <- [1..5] ]
15
Prelude> sum [ k | k <- [1..5] ]
15
```

Similarly, the expression $\int_0^1 x dx$ denotes the number $\frac{1}{2}$ and does not depend on x.

Example 2.30 (Abstraction.) Another way to bind a variable occurs in the *abstraction notation* $\{x \in A \mid P\}$, cf. (4.1), p. 116. The Haskell counterpart to this is list comprehension:

```
[ x | x <- list, property x ]
```

The choice of variable does not matter. The same list is specified by:

```
[ y | y <- list, property y ]
```

The way set comprehension is used to define sets is similar to the way list comprehension is used to define lists, and this is similar again to the way lambda abstraction is used to define functions. See Section 2.4.

Bad Habits. It is not unusual to encounter our example-statement (2.1) displayed as follows.

> For all rationals x and y, if $x < y$, then both $x < z$ and $z < y$ hold for some rational z.

Note that the meaning of this is not completely clear. With this expression the true statement that $\forall x, y \in \mathbb{Q}\, \exists z \in \mathbb{Q}\, (x < y \Rightarrow (x < z \wedge z < y))$ could be meant, but what also could be meant is the false statement that $\exists z \in \mathbb{Q}\, \forall x, y \in \mathbb{Q}\, (x < y \Rightarrow (x < z \wedge z < y))$.

Putting quantifiers both at the front and at the back of a formula results in ambiguity, for it becomes difficult to determine their scopes. In the worst case the result is an ambiguity between statements that mean entirely different things.

It does not look too well to let a quantifier bind an expression that is not a variable, such as in:

> for all numbers $n^2 + 1$, ...

Although this habit does not always lead to unclarity, it is better to avoid it, as the result is often rather hard to comprehend. If you insist on quantifying over complex terms, then the following phrasing is suggested: for all numbers *of the form* $n^2 + 1, \ldots$

Of course, in the implementation language, terms like n + 1 are important for *pattern matching*.

Translation Problems. It is easy to find examples of English sentences that are hard to translate into the logical vernacular. E.g., in *between two rationals is a third one* it is difficult to discover a universal quantifier and an implication.

Also, note that indefinites in natural language may be used to express universal statements. Consider the sentence *a well-behaved child is a quiet child*. The indefinite articles here may suggest *existential* quantifiers; however, the reading that is clearly meant has the form

$$\forall x \in C\, (\text{Well-behaved}(x) \Rightarrow \text{Quiet}(x)).$$

A famous example from philosophy of language is: *if a farmer owns a*

2.3. MAKING SYMBOLIC FORM EXPLICIT

donkey, he beats it. Again, in spite of the indefinite articles, the meaning is universal:

$$\forall x \forall y ((\text{Farmer}(x) \wedge \text{Donkey}(y) \wedge \text{Own}(x,y)) \Rightarrow \text{Beat}(x,y)).$$

In cases like this, translation into a formula reveals the logical meaning that remained hidden in the original phrasing.

In mathematical texts it also occurs quite often that the indefinite article *a* is used to make universal statements. Compare Example 2.43 below, where the following universal statement is made: *A real function is continuous if it satisfies the ε-δ-definition.*

Exercise 2.31 Translate into formulas, taking care to express the intended meaning:

1. The equation $x^2 + 1 = 0$ has a solution.

2. A largest natural number does not exist.

3. The number 13 is prime (use $d|n$ for '*d* divides *n*').

4. The number n is prime.

5. There are infinitely many primes.

Exercise 2.32 Translate into formulas:

1. Everyone loved Diana. (Use the expression $L(x,y)$ for: x loved y, and the name d for Diana.)

2. Diana loved everyone.

3. Man is mortal. (Use $M(x)$ for 'x is a man', and $M'(x)$ for 'x is mortal'.)

4. Some birds do not fly. (Use $B(x)$ for 'x is a bird' and $F(x)$ for 'x can fly'.)

Exercise 2.33 Translate into formulas, using appropriate expressions for the predicates:

1. Dogs that bark do not bite.

2. All that glitters is not gold.

3. Friends of Diana's friends are her friends.

4.*The limit of $\frac{1}{n}$ as n approaches infinity is zero.

Expressing Uniqueness. If we combine quantifiers with the relation $=$ of identity, we can make definite statements like 'there is precisely one real number x with the property that for any real number y, $xy = y$'. The logical rendering is (assuming that the domain of discussion is $\mathbb{R}$):

$$\exists x(\forall y(x \cdot y = y) \wedge \forall z(\forall y(z \cdot y = y) \Rightarrow z = x)).$$

The first part of this formula expresses that at least one x satisfies the property $\forall y(x \cdot y = y)$, and the second part states that any z satisfying the same property is identical to that x.

The logical structure becomes more transparent if we write P for the property. This gives the following translation for 'precisely one object has property P':

$$\exists x(Px \wedge \forall z(Pz \Rightarrow z = x)).$$

Exercise 2.34 Use the identity symbol $=$ to translate the following sentences:

1. Everyone loved Diana except Charles.

2. Every man adores at least two women.

3. No man is married to more than one woman.

Long ago the philosopher Bertrand Russell has proposed this logical format for the translation of the English definite article. According to his theory of description, the translation of *The King is raging* becomes:

$$\exists x(\text{King}(x) \wedge \forall y(\text{King}(y) \Rightarrow y = x) \wedge \text{Raging}(x)).$$

Exercise 2.35 Use Russell's recipe to translate the following sentences:

1. The King is not raging.

2. The King is loved by all his subjects. (use $K(x)$ for 'x is a King', and $S(x,y)$ for 'x is a subject of y').

Exercise 2.36 Translate the following logical statements back into English.

1. $\exists x \in \mathbb{R}(x^2 = 5)$.

2. $\forall n \in \mathbb{N} \exists m \in \mathbb{N}(n < m)$.

3. $\forall n \in \mathbb{N} \neg \exists d \in \mathbb{N}(1 < d < (2^n + 1) \wedge d|(2^n + 1))$.

4. $\forall n \in \mathbb{N} \exists m \in \mathbb{N}(n < m \wedge \forall p \in \mathbb{N}(p \leqslant n \vee m \leqslant p))$.

5. $\forall \varepsilon \in \mathbb{R}^+ \exists n \in \mathbb{N} \forall m \in \mathbb{N}(m \geqslant n \Rightarrow (|a - a_m| \leqslant \varepsilon))$. ($\mathbb{R}^+$ is the set of positive reals; $a, a_0, a_1, \ldots$ refer to real numbers.)

Remark. Note that translating back and forth between formulas and plain English involves making decisions about a domain of quantification and about the predicates to use. This is often a matter of taste. For instance, how does one choose between $P(n)$ for 'n is prime' and the spelled out

$$n > 1 \land \neg \exists d \in \mathbb{N}(1 < d < n \land d|n),$$

which expands the definition of being prime? Expanding the definitions of mathematical concepts is not always a good idea. The purpose of introducing the word *prime* was precisely to hide the details of the definition, so that they do not burden the mind. The art of finding the right mathematical phrasing is to introduce precisely the amount and the kind of complexity that are needed to handle a given problem. ∎

Before we will start looking at the language of mathematics and its conventions in a more systematic way, we will make the link between mathematical definitions and implementations of those definitions.

2.4 Lambda Abstraction

The following description defines a specific function that does not depend at all on x:

The function that sends x to x^2.

Often used notations are $x \mapsto x^2$ and $\lambda x.x^2$. The expression $\lambda x.x^2$ is called a *lambda term*.

If t is an expression of type b and x is a variable of type a then $\lambda x.t$ is an expression of type $a \to b$, i.e., $\lambda x.t$ denotes a function. This way of defining functions is called lambda abstraction.

Note that *the function that sends y to y^2* (notation $y \mapsto y^2$, or $\lambda y.y^2$) describes the same function as $\lambda x.x^2$.

In Haskell, function definition by lambda abstraction is available. Compare the following two definitions:

```
square1 :: Integer -> Integer
square1 x = x^2

square2 :: Integer -> Integer
square2 = \ x -> x^2
```

In the first of these, the function is defined by means of an unguarded equation. In the second, the function is defined as a lambda abstract. The Haskell way of lambda abstraction goes like this. The syntax is: \ v -> body, where v is a variable of the argument type and body an expression of the result type. It is allowed to abbreviate \ v -> \ w -> body to \ v w -> body. And so on, for more than two variables. E.g., both of the following are correct:

```
m1 :: Integer -> Integer -> Integer
m1 = \ x -> \ y -> x*y

m2 :: Integer -> Integer -> Integer
m2 = \ x y -> x*y
```

And again, the choice of variables does not matter.

Also, it is possible to abstract over tuples. Compare the following definition of a function that solves quadratic equations by means of the well-known 'abc'-formula

$$x = \frac{-b \pm \sqrt{b^2 - 4ac}}{2a}.$$

```
solveQdr :: (Float,Float,Float) -> (Float,Float)
solveQdr = \ (a,b,c) -> if a == 0 then error "not quadratic"
                       else let d = b^2 - 4*a*c in
                       if d < 0 then error "no real solutions"
                       else
                         ((- b + sqrt d) / 2*a,
                          (- b - sqrt d) / 2*a)
```

To solve the equation $x^2 - x - 1 = 0$, use solveQdr (1,-1,-1), and you will get the (approximately correct) answer (1.61803,-0.618034). Approximately correct, for 1.61803 is an approximation of the golden ratio, $\frac{1+\sqrt{5}}{2}$, and -0.618034 is an approximation of $\frac{1-\sqrt{5}}{2}$.

One way to think about quantified expressions like $\forall x P x$ and $\exists y P y$ is as combinations of a quantifier expression $\forall$ or $\exists$ and a lambda term $\lambda x.Px$ or $\lambda y.Py$. The lambda abstract $\lambda x.Px$ denotes the property of being a P.

The quantifier ∀ is a function that maps properties to truth values according to the recipe: if the property holds of the whole domain then **t**, else **f**. The quantifier ∃ is a function that maps properties to truth values according to the recipe: if the property holds of anything at all then **t**, else **f**. This perspective on quantification is the basis of the Haskell implementation of quantifiers in Section 2.8.

2.5 Definitions and Implementations

Here is an example of a definition in mathematics. A natural number n is **prime** if $n > 1$ and no number m with $1 < m < n$ divides n.

We can capture this definition of being prime in a formula, using $m|n$ for 'm divides n', as follows (we assume the natural numbers as our domain of discourse):

$$n > 1 \land \neg \exists m (1 < m < n \land m|n). \tag{2.4}$$

Another way of expressing this is the following:

$$n > 1 \land \forall m ((1 < m < n) \Rightarrow \neg m|n). \tag{2.5}$$

If you have trouble seeing that formulas (2.4) and (2.5) mean the same, don't worry. We will study such *equivalences* between formulas in the course of this chapter.

If we take the domain of discourse to be the domain of the natural numbers $\mathbb{N} = \{0, 1, 2, \ldots\}$, then formula (2.5) expresses that n is a prime number.

We can make the fact that the formula is meant as a definition explicit by introducing a predicate name P and linking that to the formula:[4]

$$P(n) :\equiv n > 1 \land \forall m((1 < m < n) \Rightarrow \neg m|n). \tag{2.6}$$

One way to think about this definition is as a **procedure** for testing whether a natural number is prime. Is 83 a prime? Yes, because none of $2, 3, 4, \ldots, 9$ divides 83. Note that there is no reason to check $10, \ldots$, for since $10 \times 10 > 83$ any factor m of 83 with $m \geqslant 10$ will not be the smallest factor of 83, and a smaller factor should have turned up before.

The example shows that we can make the prime procedure more efficient. We only have to try and find the *smallest* factor of n, and any b with $b^2 > n$ cannot be the smallest factor. For suppose that a number b with

---

[4]:≡ means: 'is by definition equivalent to'.

$b^2 \geqslant n$ divides n. Then there is a number a with $a \times b = n$, and therefore $a^2 \leqslant n$, and a divides n. Our definition can therefore run:

$$P(n) \; :\equiv \; n > 1 \land \forall m((1 < m \land m^2 \leqslant n) \Rightarrow \neg m|n). \tag{2.7}$$

In Chapter 1 we have seen that this definition is equivalent to the following:

$$P(n) \; :\equiv \; n > 1 \land \mathrm{LD}(n) = n. \tag{2.8}$$

The Haskell implementation of the primality test was given in Chapter 1.

2.6 Abstract Formulas and Concrete Structures

The formulas of Section 2.1 are "handled" using truth values and tables. Quantificational formulas need a **structure** to become meaningful. Logical sentences involving variables can be interpreted in quite different structures. A structure is a domain of quantification, together with a meaning for the abstract symbols that occur. A meaningful statement is the result of interpreting a logical formula in a certain structure. It may well occur that interpreting a given formula in one structure yields a true statement, while interpreting the same formula in a different structure yields a false statement. This illustrates the fact that we can use one logical formula for many different purposes.

Look again at the example formula (2.3), now displayed without reference to $\mathbb{Q}$ and using a neutral symbol $\mathbf{R}$. This gives:

$$\forall x \, \forall y \, (\, x\mathbf{R}y \; \Longrightarrow \; \exists z \, (\, x\mathbf{R}z \; \land \; z\mathbf{R}y \,)\,). \tag{2.9}$$

It is only possible to read this as a meaningful statement if

1. it is understood which is the underlying domain of quantification, and

2. what the symbol $\mathbf{R}$ stands for.

Earlier, the set of rationals $\mathbb{Q}$ was used as the domain, and the ordering $<$ was employed instead of the —in itself meaningless— symbol $\mathbf{R}$. In the context of $\mathbb{Q}$ and $<$, the quantifiers $\forall x$ and $\exists z$ in (2.9) should be read as: *for all rationals x ...*, resp., *for some rational z ...*, whereas $\mathbf{R}$ should be viewed as standing for $<$. In that particular case, the formula expresses the *true* statement that, between every two rationals, there is a third one.

2.6. ABSTRACT FORMULAS AND CONCRETE STRUCTURES

However, one can also choose the set $\mathbb{N} = \{0, 1, 2, \ldots\}$ of natural numbers as domain and the corresponding ordering $<$ as the meaning of $\mathbf{R}$. In that case, the formula expresses the *false* statement that between every two natural numbers there is a third one.

A specification of (i) a domain of quantification, to make an unrestricted use of the quantifiers meaningful, and (ii) a meaning for the unspecified symbols that may occur (here: $\mathbf{R}$), will be called a *context* or a *structure* for a given formula.

As you have seen here: given such a context, the formula can be "read" as a meaningful assertion about this context that can be either true or false.

Open Formulas, Free Variables, and Satisfaction. If one deletes the first quantifier expression $\forall x$ from the example formula (2.9), then the following remains:

$$\forall y \, (\, x\mathbf{R}y \implies \exists z \, (\, x\mathbf{R}z \land z\mathbf{R}y \,) \,). \tag{2.10}$$

Although this expression does have the *form* of a statement, it in fact is not such a thing. Reason: statements are either *true* or *false*; and, even if a quantifier domain and a meaning for $\mathbf{R}$ were specified, what results cannot be said to be true or false, as long as we do not know what it is that the variable x (which no longer is bound by the quantifier $\forall x$) stands for.

However, the expression can be turned into a statement again by replacing the variable x by (the name of) some object in the domain, or —what amounts to the same— by agreeing that x denotes this object.

For instance, if the domain consists of the set $\mathbb{N} \cup \{q \in \mathbb{Q} \mid 0 < q < 1\}$ of natural numbers together with all rationals between 0 and 1, and the meaning of $\mathbf{R}$ is the usual ordering relation $<$ for these objects, then the expression turns into a truth upon replacing x by 0.5 or by assigning x this value. We say that 0.5 *satisfies* the formula in the given domain.

However, (2.10) turns into a falsity when we assign 2 to x; in other words, 2 *does not* satisfy the formula.

Of course, one can delete a next quantifier as well, obtaining:

$$x\mathbf{R}y \implies \exists z \, (\, x\mathbf{R}z \land z\mathbf{R}y \,).$$

Now, both x and y have become free, and, next to a context, values have to be assigned to both these variables in order to determine a truth value.

An occurrence of a variable in an expression that is not (any more) in the scope of a quantifier is said to be *free* in that expression. Formulas that contain free variables are called *open*.

An open formula can be turned into a statement in two ways: (i) adding quantifiers that bind the free variables; (ii) replacing the free variables by (names of) objects in the domain (or stipulating that they have such objects as values).

Exercise 2.37 Consider the following formulas.

1. $\forall x \forall y(x\mathbf{R}y)$,
2. $\forall x \exists y(x\mathbf{R}y)$.
3. $\exists x \forall y(x\mathbf{R}y)$.
4. $\exists x \forall y(x = y \lor x\mathbf{R}y)$.
5. $\forall x \exists y(x\mathbf{R}y \land \neg \exists z(x\mathbf{R}z \land z\mathbf{R}y))$.

Are these formulas *true* or *false* in the following contexts?:

a. Domain: $\mathbb{N} = \{0, 1, 2, \ldots\}$; meaning of $\mathbf{R}$: $<$,

b. Domain: $\mathbb{N}$; meaning of $\mathbf{R}$: $>$,

c. Domain: $\mathbb{Q}$ (the set of rationals); meaning of $\mathbf{R}$: $<$,

d. Domain: $\mathbb{R}$ (the set of reals); meaning of $x\mathbf{R}y$: $y^2 = x$,

e. Domain: set of all human beings; meaning of $\mathbf{R}$: father-of,

f. Domain: set of all human beings; meaning of $x\mathbf{R}y$: x loves y.

Exercise 2.38 In Exercise 2.37, delete the first quantifier on x in formulas 1–5. Determine for which values of x the resulting open formulas are satisfied in each of the structures a–f.

2.7 Logical Handling of the Quantifiers

Goal To learn how to recognize simple logical equivalents involving quantifiers, and how to manipulate negations in quantified contexts.

2.7. LOGICAL HANDLING OF THE QUANTIFIERS

Validities and Equivalents. Compare the corresponding definitions in Section 2.2.

1. A logical formula is called (logically) *valid* if it turns out to be true in *every* structure.

2. Formulas are (logically) *equivalent* if they obtain the same truth value in *every* structure (i.e., if there is no structure in which one of them is true and the other one is false).

   Notation: $\Phi \equiv \Psi$ expresses that the quantificational formulas Φ and Ψ are equivalent.

Exercise 2.39 (The propositional version of this is in Exercise 2.19 p. 48.) Argue that Φ and Ψ are equivalent iff $\Phi \Leftrightarrow \Psi$ is valid.

Because of the reference to *every possible* structure (of which there are infinitely many), these are quite complicated definitions, and it is nowhere suggested that you will be expected to decide on validity or equivalence in every case that you may encounter. In fact, in 1936 it was proved rigorously, by Alonzo Church (1903–1995) and Alan Turing (1912–1954) that no one can! This illustrates that the complexity of quantifiers exceeds that of the logic of connectives, where truth tables allow you to decide on such things in a mechanical way, as is witnessed by the Haskell functions that implement the equivalence checks for propositional logic.

Nevertheless: the next theorem already shows that it is sometimes very well possible to recognize whether formulas are valid or equivalent — if only these formulas are sufficiently simple.

Only a few useful equivalents are listed next. Here, $\Psi(x)$, $\Phi(x,y)$ and the like denote logical formulas that may contain variables x (or x, y) free.

Theorem 2.40

1. $\forall x \forall y \Phi(x,y) \equiv \forall y \forall x \Phi(x,y);$
   $\exists x \exists y \Phi(x,y) \equiv \exists y \exists x \Phi(x,y),$

2. $\neg \forall x \Phi(x) \equiv \exists x \neg \Phi(x);$
   $\neg \exists x \Phi(x) \equiv \forall x \neg \Phi(x);$
   $\neg \forall x \neg \Phi(x) \equiv \exists x \Phi(x);$
   $\neg \exists x \neg \Phi(x) \equiv \forall x \Phi(x),$

3. $\forall x (\Phi(x) \wedge \Psi(x)) \equiv (\forall x \Phi(x) \wedge \forall x \Psi(x));$
   $\exists x (\Phi(x) \vee \Psi(x)) \equiv (\exists x \Phi(x) \vee \exists x \Psi(x)).$

Proof. There is no neat truth table method for quantification, and there is no neat proof here. You just have to follow common sense. For instance (part 2, first item) common sense dictates that not every x satisfies Φ if, and only if, some x does not satisfy Φ. ∎

Of course, common sense may turn out not a good adviser when things get less simple. Chapter 3 hopefully will (partly) resolve this problem for you.

Exercise 2.41 For every sentence Φ in Exercise 2.36 (p. 56), consider its negation $\neg\Phi$, and produce a more positive equivalent for $\neg\Phi$ by working the negation symbol through the quantifiers.

Order of Quantifiers. Theorem 2.40.1 says that the order of similar quantifiers (all universal or all existential) is irrelevant. But note that this is not the case for quantifiers of different kind.

On the one hand, if you know that $\exists y \forall x \Phi(x,y)$ (which states that there is one y such that for all x, $\Phi(x,y)$ holds) is true in a certain structure, then *a fortiori* $\forall x \exists y \Phi(x,y)$ will be true as well (for each x, take this *same* y). However, if $\forall x \exists y \Phi(x,y)$ holds, it is far from sure that $\exists y \forall x \Phi(x,y)$ holds as well.

Example 2.42 The statement that $\forall x \exists y (x < y)$ is true in $\mathbb{N}$, but the statement $\exists y \forall x (x < y)$ in this structure wrongly asserts that there exists a greatest natural number.

Restricted Quantification. You have met the use of restricted quantifiers, where the restriction on the quantified variable is membership in some domain. But there are also other types of restriction.

Example 2.43 (Continuity) According to the "ε-δ-definition" of continuity, a real function f is *continuous* if (domain $\mathbb{R}$):

$$\forall x \, \forall \varepsilon > 0 \, \exists \delta > 0 \, \forall y \, (\, |x - y| < \delta \implies |f(x) - f(y)| < \varepsilon \,).$$

This formula uses the restricted quantifiers $\forall \varepsilon > 0$ and $\exists \delta > 0$ that enable a more compact formulation here.

2.7. LOGICAL HANDLING OF THE QUANTIFIERS

Example 2.44 Consider our example statement (2.3). Here it is again:

$$\forall y \forall x (x < y \implies \exists z (x < z \land z < y))$$

This can also be given as

$$\forall y \forall x < y \exists z < y (x < z),$$

but this reformulation stretches the use of this type of restricted quantification probably a bit too much.

Remark. If A is a subset of the domain of quantification, then

$$\forall x \in A\ \Phi(x) \text{ means the same as } \forall x(x \in A \Rightarrow \Phi(x)),$$

whereas

$$\exists x \in A\ \Phi(x) \text{ is tantamount with } \exists x(x \in A \land \Phi(x)).$$

∎

Warning: The restricted *universal* quantifier is explained using $\Rightarrow$, whereas the *existential* quantifier is explained using $\land$!

Example 2.45 'Some Mersenne numbers are prime' is correctly translated as $\exists x(Mx \land Px)$. The translation $\exists x(Mx \Rightarrow Px)$ is wrong. It is much too weak, for it expresses (in the domain $\mathbb{N}$) that there is a natural number x which is either not a Mersenne number or it is a prime. Any prime will do as an example of this, and so will any number which is not a Mersenne number.

In the same way, 'all prime numbers have irrational square roots' is translated as $\forall x \in \mathbb{R}(Px \Rightarrow \sqrt{x} \notin \mathbb{Q})$. The translation $\forall x \in \mathbb{R}(Px \land \sqrt{x} \notin \mathbb{Q})$ is wrong. This time we end up with something which is too strong, for this expresses that every real number is a prime number with an irrational square root.

Restricted Quantifiers Explained. There is a version of Theorem 2.40 that employs restricted quantification. This version states, for instance, that $\neg \forall x \in A\ \Phi$ is equivalent to $\exists x \in A\ \neg\Phi$, and so on. The equivalence follows immediately from the remark above. We now have, e.g., that $\neg \forall x \in A \Phi(x)$ is equivalent to $\neg \forall x(x \in A \Rightarrow \Phi(x))$, which in turn is equivalent to (Theorem 2.40) $\exists x \neg(x \in A \Rightarrow \Phi(x))$, hence to (and *here* the implication turns into a conjunction — cf. Theorem 2.10) $\exists x(x \in A \land \neg\Phi(x))$, and, finally, to $\exists x \in A \neg\Phi(x)$.

Exercise 2.46 Does it hold that $\neg \exists x \in A\, \Phi(x)$ is equivalent to $\exists x \notin A\, \Phi(x)$? If your answer is 'yes', give a proof, if 'no', then you should show this by giving a simple refutation (an example of formulas and structures where the two formulas have different truth values).

Exercise 2.47 Is $\exists x \notin A\, \neg\Phi(x)$ equivalent to $\exists x \in A\, \neg\Phi(x)$? Give a proof if your answer is 'yes', and a refutation otherwise.

Exercise 2.48 Produce the version of Theorem 2.40 (p. 63) that employs restricted quantification. Argue that your version is correct.

Example 2.49 (Discontinuity Explained) The following formula describes what it means for a real function f to be discontinuous in x:

$$\neg \forall \varepsilon > 0\, \exists \delta > 0\, \forall y\, (\, |x-y| < \delta \implies |f(x)-f(y)| < \varepsilon\,).$$

Using Theorem 2.40, this can be transformed in three steps, moving the negation over the quantifiers, into:

$$\exists \varepsilon > 0\, \forall \delta > 0\, \exists y\, \neg\, (\, |x-y| < \delta \implies |f(x)-f(y)| < \varepsilon\,).$$

According to Theorem 2.10 this is equivalent to

$$\exists \varepsilon > 0\, \forall \delta > 0\, \exists y\, (\, |x-y| < \delta \,\wedge\, \neg |f(x)-f(y)| < \varepsilon\,),$$

i.e., to

$$\exists \varepsilon > 0\, \forall \delta > 0\, \exists y\, (\, |x-y| < \delta \,\wedge\, |f(x)-f(y)| \geqslant \varepsilon\,).$$

What has emerged now is a clearer "picture" of what it means to be discontinuous in x: there must be an $\varepsilon > 0$ such that for every $\delta > 0$ ("no matter how small") a y can be found with $|x-y| < \delta$, whereas $|f(x)-f(y)| \geqslant \varepsilon$; i.e., there are numbers y "arbitrarily close to x" such that the values $f(x)$ and $f(y)$ remain at least ε apart.

Different Sorts. Several *sorts* of objects, may occur in one and the same context. (For instance, sometimes a problem involves vectors as well as reals.) In such a situation, one often uses different variable naming conventions to keep track of the differences between the sorts. In fact, sorts are just like the basic types in a functional programming language.

Just as good naming conventions can make a program easier to understand, naming conventions can be helpful in mathematical writing. For instance: the letters $n, m, k, \ldots$ are often used for natural numbers, $f, g, h, \ldots$ usually indicate that functions are meant, etc.

2.8. QUANTIFIERS AS PROCEDURES

The interpretation of quantifiers in such a case requires that not one, but several domains are specified: one for every sort or type. Again, this is similar to providing explicit typing information in a functional program for easier human digestion.

Exercise 2.50 That the sequence $a_0, a_1, a_2, \ldots \in \mathbb{R}$ converges to a, i.e., that $\lim_{n \to \infty} a_n = a$, means that $\forall \delta > 0 \exists n \forall m \geqslant n (|a - a_m| < \delta)$. Give a positive equivalent for the statement that the sequence $a_0, a_1, a_2, \ldots \in \mathbb{R}$ does not converge.

2.8 Quantifiers as Procedures

One way to look at the meaning of the universal quantifier ∀ is as a procedure to test whether a set has a certain property. The test yields **t** if the set equals the whole domain of discourse, and **f** otherwise. This means that ∀ is a procedure that maps the domain of discourse to **t** and all other sets to **f**. Similarly for restricted universal quantification. A restricted universal quantifier can be viewed as a procedure that takes a set A and a property P, and yields **t** just in case the set of members of A that satisfy P equals A itself.

In the same way, the meaning of the unrestricted existential quantifier ∃ can be specified as a procedure. ∃ takes a set as argument, and yields **t** just in case the argument set is non-empty. A restricted existential quantifier can be viewed as a procedure that takes a set A and a property P, and yields **t** just in case the set of members of A that satisfy P is non-empty.

If we implement sets as lists, it is straightforward to implement these quantifier procedures. In Haskell, they are predefined as `all` and `any` (these definitions will be explained below):

```
any, all     :: (a -> Bool) -> [a] -> Bool
any p        = or  . map p
all p        = and . map p
```

The typing we can understand right away. The functions `any` and `all` take as their first argument a function with type `a` inputs and type `Bool` outputs (i.e., a test for a property), as their second argument a list over type `a`, and return a truth value. Note that the list representing the restriction is the *second* argument.

To understand the implementations of `all` and `any`, one has to know that `or` and `and` are the generalizations of (inclusive) disjunction and conjunction to lists. (We have already encountered `and` in Section 2.2.) They have type `[Bool] -> Bool`.

Saying that all elements of a list `xs` satisfy a property `p` boils down to: the list `map p xs` contains only `True` (see Section 1.8). Similarly, saying that some element of a list `xs` satisfies a property `p` boils down to: the list `map p xs` contains at least one `True`. This explains the implementation of `all`: first apply `map p`, next apply `and`. In the case of `any`: first apply `map p`, next apply `or`.

The action of applying a function `g :: b -> c` after a function `f :: a -> b` is performed by the function `g . f :: a -> c`, the composition of `f` and `g`. See Section 6.3 below.

The definitions of `all` and `any` are used as follows:

```
Prelude> any (<3) [0..]
True
Prelude> all (<3) [0..]
False
Prelude>
```

The functions `every` and `some` get us even closer to standard logical notation. These functions are like `all` and `any`, but they first take the restriction argument, next the body:

```
every, some :: [a] -> (a -> Bool) -> Bool
every xs p = all p xs
some  xs p = any p xs
```

Now, e.g., the formula $\forall x \in \{1,4,9\} \exists y \in \{1,2,3\}\ x = y^2$ can be implemented as a test, as follows:

```
TAMO> every [1,4,9] (\ x -> some [1,2,3] (\ y -> x == y^2))
True
```

But caution: the implementations of the quantifiers are procedures, not algorithms. A call to `all` or `any` (or `every` or `some`) need not terminate. The call

```
every [0..] (>=0)
```

2.9. FURTHER READING

will run forever. This illustrates once more that the quantifiers are in essence more complex than the propositional connectives. It also motivates the development of the method of proof, in the next chapter.

Exercise 2.51 Define a function unique :: (a -> Bool) -> [a] -> Bool that gives True for unique p xs just in case there is exactly one object among xs that satisfies p.

Exercise 2.52 Define a function parity :: [Bool] -> Bool that gives True for parity xs just in case an even number of the xss equals True.

Exercise 2.53 Define a function evenNR :: (a -> Bool) -> [a] -> Bool that gives True for evenNR p xs just in case an even number of the xss have property p. (Use the parity function from the previous exercise.)

2.9 Further Reading

If you find that the pace of the introduction to logic in this chapter is too fast for you, you might wish to have a look at the more leisurely paced [NK04]. Excellent books about computer science applications of logic are [Bur98] and [HR00]. A good introduction to mathematical logic is Ebbinghaus, Flum and Thomas [EFT94].

Chapter 3

The Use of Logic: Proof

Preview

This chapter describes how to write simple proofs. Section 3.1 is about style of presentation, while Section 3.2 gives the general format of the proof rules. Sections 3.3 and 3.4 describe the rules that govern the use of the connectives and the quantifiers in the proof process. The recipes are summarized in Section 3.5. Section 3.6 gives some strategic hints for handling proof problems. Section 3.7 applies the proof recipes to reasoning about prime numbers; this section also illustrates how the computer can be used (sometimes) as an instrument for checking particular cases, and therefore as a tool for refuting general claims.

Representation and proof are two sides of the same coin. In order to handle the stuff of mathematics, we start out from definitions and try to find meaningful relations. To check our intuitions and sometimes just to make sure that our definitions accomplish what we had in mind we have to provide proofs for our conjectures. To handle abstract objects in an implementation we have to represent them in a concrete way. Again, we have to check that the representation is faithful to the original intention.

It turns out that proofs and implementations have many things in common. In the first place, *variables* are a key ingredient in both. Variables are used to denote members of certain sets. In our example proofs, we will use x, y for rational numbers and real numbers, and m, n for integers. Similarly, the variables used in the implementation of a definition range over certain sets, indicated by means of typings for the variables.

The main purpose of this chapter is to spur you on to develop good habits in setting up mathematical proofs. Once acquired, these habits will help you in many ways. Being able to 'see structure' in a mathematical proof will enable you to easily read and digest the proofs you find in papers and textbooks. When you have learned to see the patterns of proof, you will discover that they turn up again and again in any kind of formal reasoning.

Because it takes time to acquire new habits, it is not possible to digest the contents of this chapter in one or two readings. Once the recommended habits are acquired, but only then, the contents of the chapter have been properly digested, and the chapter will have served its purpose.

The module containing the code of this chapter depends on the module containing the code of the previous chapter. In the module declaration we take care to *import* that module. This is done with the reserved keyword import.

```
module TUOLP

where

import TAMO
```

3.1 Proof Style

The *objects* of mathematics are strange creatures. They do not exist in physical space. No one ever saw the number 1. One can consistently argue that mathematics has no subject at all, or maybe that the subject matter of mathematics is in the mind. But what does that mean? For sure, different mathematicians have the same objects in mind when they investigate prime numbers. But how can that be, if mathematics has no subject matter?

As to the *method* of mathematics, there is no disagreement: the heart of the matter is the notion of *proof*. A proof is an argument aimed at convincing yourself and others of the truth of an assertion. Some proofs are simple, but others can be pieces of art with aesthetic and intellectual qualities.

In daily life, people argue a lot, but these efforts are not always capable of convincing others. If you have a row with your partner, more often than not it is extremely difficult to assess who is right. (If this were an easy matter, why have a row in the first place?) By contrast, in mathematical matters it usually is very clear who is right and who is wrong. Doing

3.1. PROOF STYLE

mathematics is a sports where the rules of the game are very clear. Indeed, mathematical proofs go undisputed most of the time. This remarkable phenomenon is probably due to the idealized character of mathematical objects. Idealization is a means by which mathematics provides access to the essence of things, and grasping the essence of something is a hallmark of the creative mind. It remains a source of wonder that the results are so often applicable to the real world.

The mathematical content of most of the proofs discussed in this chapter is minute; later on you will call such proofs "routine". For the time being, the attention is focused on the *form* of proofs. However, to begin with, here follows a list of stylistic commandments.

A proof is made up of sentences. Therefore:

$\boxed{1}$ *Write correct English, try to express yourself clearly.*

If you are not a native speaker you are in the same league as the authors of this book. Still, English is the *lingua franca* of the exact sciences, so there is no way around learning enough of it for clear communication. Especially when not a native speaker, keep your style simple.

Write with a reader in mind, and do not leave your reader in the dark concerning your proof plans.

$\boxed{2}$ *Make sure the reader knows exactly what you are up to.*

Applying the rules of proof of the next section in the correct way will usually take care of this. In particular, when you feel the need to start writing symbols —often, variables— inform the reader what they stand for and do not let them fall out of the blue.

$\boxed{3}$ *Say what you mean when introducing a variable.*

Section 3.2 again shows you the way.

Use the symbolic shorthands for connectives and quantifiers sparingly. Something like the continuity definition on p. 64 with its impressive sequence of four quantifiers is about the maximum a well-educated mathematician can digest.

$\boxed{4}$ *Don't start a sentence with symbols; don't write formulas only.*

It is best to view symbolic expressions as *objects* in their own right. E.g., write: *The formula* Φ *is true*, or: *It holds that* Φ. Note that a proof that consists of formulas only is not a suitable way to inform the reader about what has been going on in your brain.

⑤ *Use words or phrases like* 'thus', 'therefore', 'hence', 'it follows that',*etc. to link up your formulas. Be relevant and succinct.*

Do not use the implication symbol $\Rightarrow$ in cases that really require "thus" or "therefore". Beginner's proofs can often be recognized by their excessive length. Of course, it will be difficult at first to keep to the proper middle road between commandments (4) and (5).

When embarking on a proof, it is a good idea to write down exactly (i) what can be used in the proof (the *Given*), (ii) and what is to be proved. It is very helpful to use a schema for this.

⑥ *When constructing proofs, use the following schema:*

> **Given:** ...
> **To be proved:** ...
> **Proof:** ...

N.B.: A proof using *Mathematical Induction* offers some extra structure; cf. Section 7.1 below and p. 390.

In the course of this chapter you will discover that proofs are highly structured pieces of texts. The following guideline will help you keep track of that structure.

⑦ *Use layout (in particular, indentation) to identify subproofs and to keep track of the scopes of assumptions.*

The general shape of a proof containing subproofs will be discussed in Section 3.2.

⑧ *Look up definitions of defined notions, and use these definitions to re-write both* Given *and* To be proved.

In particular, elimination of defined notions from the sentence to be proved will show you clearly what it is that you have to prove.

Do not expect that you will be able to write down a correct proof at the first try. Do not expect to succeed at the second one. You will have to accept the fact that, for the time being, your efforts will not result in faultless proofs. Nevertheless: keep trying!

⑨ *Make sure you have a sufficient supply of scrap paper, make a fair copy of the end-product — whether you think it to be flawless or not.*

Finally: before handing in what you wrote, try to let it rest for at least one night. Then, and only then,

3.2. PROOF RECIPES 75

$\boxed{10}$ *Ask yourself two things: Is this correct? Can others read it?*

The honest answers usually will be negative at first... Fortunately, you did not finish your supply of scrap paper. Surely, if you have understood properly, you must be able to get it down correctly eventually. (Apply to this the law of contraposition in Theorem 2.10, p. 45.)

3.2 Proof Recipes

Goal Develop the ability to apply the proof rules in simple contexts. The proof rules are recipes that will allow you to cook up your own proofs. Try to distill a recipe from every rule, and and make sure you remember how to apply these recipes in the appropriate situation.

It is completely normal that you get stuck in your first efforts to prove things. Often, you will not even know how to make a first move. This section then may come to your help. It provides you with rules that govern the behaviour of the logical phrases in mathematical proof, and thus with recipes to use while constructing a proof.

In fact, constructing proofs has a lot in common with writing computer programs. In structured programming, layout can be used to reveal the building blocks of a program more clearly. We will also use layout as a tool to reveal structure.

The most important structure principle is that a proof can contain subproofs, just like a program may contain procedures which have their own sub-procedures, and so on. We will indicate subproofs by means of indentation. The general structure of a proof containing a subproof is as follows:

```
┌─────────────────────────────┐
│ Given: A, B, ...            │
│ To be proved: P             │
│ Proof:                      │
│   ...                       │
│   ┌───────────────────────┐ │
│   │ Suppose C ...         │ │
│   │ To be proved: Q       │ │
│   │ Proof: ...            │ │
│   │   ...                 │ │
│   │ Thus Q                │ │
│   └───────────────────────┘ │
│   ...                       │
│ Thus P                      │
└─────────────────────────────┘
```

To be sure, a subproof may itself contain subproofs:

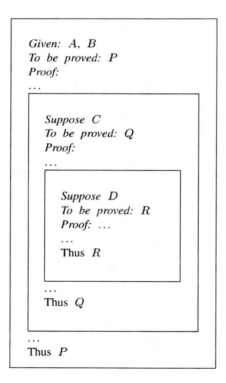

3.2. PROOF RECIPES

The purpose of 'Suppose' is to add a new given to the list of assumptions that may be used, but only for the duration of the subproof of which 'Suppose' is the head. If the current list of givens is $P_1, \ldots, P_n$ then 'Suppose Q' extends this list to $P_1, \ldots, P_n, Q$. In general, inside a box, you can use all the givens and assumptions of all the including boxes. Thus, in the innermost box of the example, the givens are A, B, C, D. This illustrates the importance of indentation for keeping track of the 'current box'.

There are some 15 rules discussed here for all seven notions of logic. At a first encounter, this may seem an overwhelming number. However, only some of these are really enlightening; several are so trivial that you won't even notice using them. An example is:

Given: P, Q
Thus $P \wedge Q$.

What is really remarkable is this: together these 15 rules are sufficient to tackle every possible *proof problem.*

Of course, this does not mean that the process of proving mathematical facts boils down in the end to mastery of a few simple rules. Think of it as chess: the rules of the game are extremely simple, and these rules can be used to play very complicated and beautiful games. To be able to play you must know the rules. But if your knowledge does not extend beyond the rules, you will only be capable of playing games that are rather dull. Learning to play chess does not stop with learning the rules, and what is beautiful in a game of chess cannot be found in the rules. Similarly, the rules of proof cannot teach you more than how to prove the simplest of things. (In the beginning, that is difficult enough.) It is ideas that make a proof interesting or beautiful, but the rules keep silent about these things.

Classification of Rules.

Every logical symbol has its own rules of use. There are basically two ways in which you can encounter a logical symbol: it can either appear in the given or in the statement that is to be proved. In the first case the rule to use is an *elimination* rule, in the second case an *introduction* rule. Elimination rules enable you to reduce a proof problem to a new, hopefully simpler, one. Introduction rules make clear how to prove a goal of a certain given shape. All rules are summarized in Section 3.5.

Safety. As we go along, we will provide arguments that try to explain why a certain rule is *safe*. A rule is safe if it will never allow you to prove something false on the basis of statements that are true. Obviously, this is a requirement that proofs should fulfill. That the rules for connectives are safe is due to truth tables. Safety of two of the quantifier rules is obvious; the remaining ones are tough nuts.

Don't worry when these explanations in some cases appear mystifying. Eventually, you'll understand!

3.3 Rules for the Connectives

Implication

Here come a complicated but important introduction rule and a trivial one for elimination.

Introduction The introduction rule for implication is the *Deduction Rule*. It enables you to reduce the problem of proving an implication $\Phi \Rightarrow \Psi$. Instead, it prescribes to assume Φ as an additional new *Given*, and asks you to derive that Ψ.

Given: ...
To be proved: $\Phi \Rightarrow \Psi$
Proof:
    Suppose Φ
    To be proved: Ψ
    Proof: ...
Thus $\Phi \Rightarrow \Psi$.

Safety. In case Φ is false, the implication $\Phi \Rightarrow \Psi$ will be true anyway. Thus, the case that is left for you to consider is when Φ is true, meaning that you can assume it as given. But then of course you should show that Ψ is true as well (otherwise, $\Phi \Rightarrow \Psi$ would be false).

Example 3.1 We show that the implication $P \Rightarrow R$ is provable on the basis of the given $P \Rightarrow Q$ and $Q \Rightarrow R$. Thus, employing the schema in the 7th commandment (p. 74) for the first time:

Given: $P \Rightarrow Q$, $Q \Rightarrow R$
To be proved: $P \Rightarrow R$
Proof:
    Given: $P \Rightarrow Q$, $Q \Rightarrow R$ (old), P (new)
    To be proved: R
    Proof: From $P \Rightarrow Q$ and P, conclude Q.
    Next, from $Q \Rightarrow R$ and Q, conclude R.
Thus, according to the deduction rule, $P \Rightarrow R$

3.3. RULES FOR THE CONNECTIVES

Here is a slightly more concise version:

Given: $P \Rightarrow Q$, $Q \Rightarrow R$
To be proved: $P \Rightarrow R$
Proof:
>   *Suppose* P
>   *To be proved:* R
>   *Proof:* From $P \Rightarrow Q$ and P, conclude Q.
>   Next, from $Q \Rightarrow R$ and Q, conclude R.

Thus $P \Rightarrow R$

Detailed vs. Concise Proofs. The proof just given explains painstakingly how the Deduction Rule has been applied in order to get the required result. However, *in practice*, you should write this in a still more concise way:

Given: $P \Rightarrow Q$, $Q \Rightarrow R$
To be proved: $P \Rightarrow R$
Proof:
>   *Suppose* P
>   From $P \Rightarrow Q$ and P, conclude Q.
>   Next, from $Q \Rightarrow R$ and Q, conclude R.

Note that the final concise version does not mention the Deduction Rule at all. The actual application of this rule should follow as a last line, but that is left for the reader to fill in. This is not considered an incomplete proof: from the situation it is perfectly clear that this application is understood.

Several other examples of detailed proofs and their concise versions are given in what follows. Here is the key to the proof of an implication:

If the 'to be proved' is an implication $\Phi \Rightarrow \Psi$, then your proof should start with the following Obligatory Sentence:

Suppose that Φ holds.

The obligatory first sentence accomplishes the following things (cf. the 2nd commandment on p. 73).

- It informs the reader that you are going to apply the Deduction Rule in order to establish that $\Phi \Rightarrow \Psi$.

- The reader also understands that it is now Ψ that you are going to derive (instead of $\Phi \Rightarrow \Psi$).

- Thus, starting with the obligatory sentence informs the reader in an efficient way about your plans.

Reductive Character. In Example 3.1 the problem of showing that from the givens $P \Rightarrow Q, Q \Rightarrow R$ it can be proved that $P \Rightarrow R$ was *reduced*, using the Deduction Rule, to the problem of showing that from the givens $P \Rightarrow Q, Q \Rightarrow R, P$ it can be proved that R. This requires a subproof.

Marking Subproofs with Indentation. Note that *only during that subproof* you are allowed to use the new given P. Some logic texts recommend the use of markers to indicate beginning and end of such a subproof, between which the extra given is available. In our example proofs in this chapter, we indicate the beginnings and ends of subproofs by means of indentation. In more colloquial versions of the proofs, identifying the beginning and end of a subproof is left to the reader.

In this particular case, the reduced proof problem turned out to be easy: P and $P \Rightarrow Q$ together produce Q; whereas Q and $Q \Rightarrow R$ together produce R, as desired. (Of course, in general, such a subproof may require new reductions.)

The last two steps in this argument in fact apply the *Elimination Rule* of implication, which is almost too self-evident to write down:

Elimination. This rule is also called *Modus Ponens*. In words: from $\Phi \Rightarrow \Psi$ and Φ you can conclude that Ψ.

In a schema:

Given: $\Phi \Rightarrow \Psi$, Φ
Thus Ψ.

Safety. Immediate from truth tables: if $\Phi \Rightarrow \Psi$ and Φ are both true, then so is Ψ.

The two proof rules for implication enable you to handle implications in all types of proof problems. You may want to use more obvious properties of implication, but fact is that they usually can be derived from the two given ones. Example 3.1 is a case in point, as is Exercise 3.2. (We will not bother you with the exceptions; one is in Exercise 3.9.)

Exercise 3.2 Apply both implication rules to prove $P \Rightarrow R$ from the givens $P \Rightarrow Q$, $P \Rightarrow (Q \Rightarrow R)$.

3.3. RULES FOR THE CONNECTIVES

Conjunction

The conjunction rules are almost too obvious to write down.

Introduction. A conjunction follows from its two conjuncts taken together.

In a schema:

Given: Φ, Ψ
Thus $\Phi \wedge \Psi$.

Elimination. From a conjunction, both conjuncts can be concluded.
   Schematically:

Given: $\Phi \wedge \Psi$
Thus Φ.

Given: $\Phi \wedge \Psi$
Thus Ψ.

Suppose we want to check whether the sum of even natural numbers always is even. It is not difficult to implement this check in Haskell, using the built-in function `even` and the list of even numbers.

```
evens = [ x | x <- [0..], even x ]
```

Formulating the check is easy, but of course we won't get an answer, as the check takes an infinite amount of time to compute.

```
TUOLP> forall [ m + n | m <- evens, n <- evens ] even
```

If we want to check a statement about an infinite number of cases, we can either look for a counterexample, or we can attempt to give a proof. In the present case, the latter is easy.

Example 3.3 Assume that $n, m \in \mathbb{N}$.
   To show: (m is even $\wedge$ n is even) $\Rightarrow$ $m + n$ is even.

Detailed proof:
   Assume that (m even $\wedge$ n even).
   Then ($\wedge$-elimination) m and n are both even.
   For instance, $p, q \in \mathbb{N}$ exist such that $m = 2p$, $n = 2q$.
   Then $m + n = 2p + 2q = 2(p + q)$ is even.

The result follows using the Deduction Rule.

Concise proof:
>Assume that m and n are even.
>For instance, $m = 2p$, $n = 2q$, $p, q \in \mathbb{N}$.
>Then $m + n = 2p + 2q = 2(p + q)$ is even.

Exercise 3.4 Assume that $n, m \in \mathbb{N}$.
Show: (m is odd $\wedge$ n is odd) $\Rightarrow$ $m + n$ is even.

Equivalence

An equivalence can be thought of as the conjunction of two implications. Thus, the rules follow from those for $\Rightarrow$ and $\wedge$.

Introduction. In order to prove $\Phi \Leftrightarrow \Psi$, you have to accomplish two things (cf. Example 2.3):

($\Rightarrow$) add Φ as a new given, and show that Ψ;

($\Leftarrow$) add Ψ as a new given, and show that Φ.

If you can do this, $\Phi \Leftrightarrow \Psi$ has been proved.

Concise Proof Schema.

Given: ...
To be proved: $\Phi \Leftrightarrow \Psi$
Proof:
>*Suppose* Φ
>*To be proved:* Ψ
>*Proof:* ...
>
>*Suppose* Ψ
>*To be proved:* Φ
>*Proof:* ...

Thus $\Phi \Leftrightarrow \Psi$.

Instead of $\Leftrightarrow$ you may also encounter 'iff' or 'if and only if'. A proof of a statement of the form 'Φ iff Ψ' consists of two subproofs: the proof of the 'only if' part and the proof of the 'if' part. Caution: the 'only if' part is the proof of $\Phi \Rightarrow \Psi$, and the 'if' part is the proof of $\Phi \Leftarrow \Psi$. This is because 'Φ only if Ψ' means '$\Phi \Rightarrow \Psi$', and 'Φ if Ψ' means '$\Phi \Leftarrow \Psi$'. Thus, we get:

3.3. RULES FOR THE CONNECTIVES

Given: ...
To be proved: Φ iff Ψ
Proof:
Only if: *Suppose* Φ
        To be proved: Ψ
        Proof: ...
If: *Suppose* Ψ
        To be proved: Φ
        Proof: ...
Thus Φ iff Ψ.

Elimination. Schematically:

Given: Φ ⇔ Ψ, Φ, ...
Thus Ψ

Given: Φ ⇔ Ψ, Ψ, ...
Thus Φ

Exercise 3.5 Show:

1. From $P \Leftrightarrow Q$ it follows that $(P \Rightarrow R) \Leftrightarrow (Q \Rightarrow R)$,

2. From $P \Leftrightarrow Q$ it follows that $(R \Rightarrow P) \Leftrightarrow (R \Rightarrow Q)$.

Negation

General Advice. In no matter what concrete mathematical situation, before applying any of the negation rules given below: whether you want to prove or use a negated sentence, you should first attempt to convert into something *positive*. If this succeeds, you can turn to the other rules.

Theorems 2.10 (p. 45) and 2.40 (p. 63) contain some tools for doing this: you can move the negation symbol inward, across quantifiers and connectives. If this process terminates, the "mathematics" of the situation often allows you to eliminate the negation symbol altogether. E.g., $\neg \forall x (x < a \lor b \leqslant x)$ can be rewritten as $\exists x (a \leqslant x \land x < b)$. For a more complicated example, cf. (2.49) on p. 66.

Remark. The general advice given above *does not apply* to the exercises in the present chapter. Firstly, many of the exercises below are purely logical: there is no mathematical context at all. Secondly, all exercises here are designed to be solved using the rules only; possible shortcuts via the results of Chapter 2 will often trivialize them.

Introduction. If $\neg\Phi$ is to be proved, do the following. Assume Φ as a new given, and attempt to prove something (depending on the context) that is evidently false.

This strategy clearly belongs to the kind of rule that reduces the proof problem.

Schematically:

Given: ...
To be proved: $\neg\Phi$
Proof:
    Suppose Φ
    To be proved: $\bot$
    Proof: ...
Thus $\neg\Phi$.

Evidently False. Here, $\bot$ stands for the evidently false statement.

In a mathematical context, this can be anything untrue, such as $1 = 0$. For a more complicated falsehood, cf. the proof that limits are unique on p. 310.

In the logical examples below, $\bot$ may consist of the occurrence of a statement together with its negation. In that case, one statement *contradicts* the other. For instance, you might derive a sentence $\neg\Psi$, thereby contradicting the occurrence of Ψ among the given Γ. Cf. Examples 3.8 and 3.30.

Example 3.6 The proof of Theorem 3.33 (the number of primes is not finite) is an example of the method of negation introduction. Another example can be found in the proof of Theorem 8.14 (the square root of 2 is not rational).

Exercise 3.7 Produce proofs for:

1. Given: $P \Rightarrow Q$. To show: $\neg Q \Rightarrow \neg P$,

2. Given $P \Leftrightarrow Q$. To show: $\neg P \Leftrightarrow \neg Q$.

Safety. Suppose that from Γ, Φ it follows that $\bot$, and that all given Γ are satisfied. Then Φ cannot be true. (Otherwise, your proof would show the evidently false $\bot$ to be true as well.) Thus, $\neg\Phi$ must be true.

Elimination. When you intend to use the given $\neg\Phi$, you can attempt to prove, on the basis of the other given, that Φ must hold. In that case, the elimination rule declares the proof problem to be solved, *no matter what* the statement To be proved!

Schematically:

3.3. RULES FOR THE CONNECTIVES

Given: Φ, $\neg\Phi$
Thus Ψ.

Safety. The rule cannot help to be safe, since you will never find yourself in a situation where Φ and $\neg\Phi$ are both true. (Remarkably, *this nevertheless is a useful rule!*)

There is one extra negation rule that can be used in every situation: *Proof by Contradiction*, or *Reductio ad Absurdum*.

Proof by Contradiction. In order to prove Φ, add $\neg\Phi$ as a new given, and attempt to deduce an evidently false statement.

In a schema:

Given: ...
To be proved: Φ
Proof:
    Suppose $\neg\Phi$
    To be proved: $\bot$
    Proof: ...
Thus Φ.

Safety. The argument is similar to that of the introduction rule.

Advice. Beginners are often lured into using this rule. The given $\neg\Phi$ that comes in free looks so inviting! However, many times it must be considered poisoned, making for a cluttered bunch of confused givens that you will not be able to disentangle. It is a killer rule that often will turn itself against its user, especially when that is a beginner. Proof by Contradiction should be considered your last way out. Some proof problems do need it, but if possible you should proceed without: you won't get hurt and a simpler and more informative proof will result.

Comparison. Proof by Contradiction looks very similar to the $\neg$ introduction rule (both in form and in spirit), and the two are often confused. Indeed, in ordinary mathematical contexts, it is usually "better" to move negation inside instead of applying $\neg$-introduction.

Example 3.8 From $\neg Q \Rightarrow \neg P$ it follows that $P \Rightarrow Q$.

Given: $\neg Q \Rightarrow \neg P$
To be proved: $P \Rightarrow Q$
Detailed proof:

>   *Suppose P*
>   *To be proved: Q*
>   *Proof:*
>   > *Suppose* ¬Q
>   > *To be proved:* ⊥
>   > *Proof:*
>   > From ¬Q and ¬Q ⇒ ¬P derive ¬P.
>   > From P and ¬P derive ⊥.
>
>   Thus, Q, by contradiction.

Thus, P ⇒ Q, by the Deduction Rule.

Concise proof:
> Assume that P.
> If ¬Q, then (by ¬Q ⇒ ¬P) it follows that ¬P.
> Contradiction.

Exercise 3.9* Show that from $(P \Rightarrow Q) \Rightarrow P$ it follows that P.

Hint. Apply Proof by Contradiction. (The implication rules do not suffice for this admittedly exotic example.)

Disjunction

Introduction. A disjunction follows from each of its disjuncts.

Schematically:

Given: Φ
Thus Φ ∨ Ψ.

Given: Ψ
Thus Φ ∨ Ψ.

Safety is immediate from the truth tables.

Elimination. You can use a given Φ ∨ Ψ by giving *two* proofs: one employing Φ, and one employing Ψ.

In a proof schema:

Given: Φ ∨ Ψ, ...
To be proved: Λ
Proof:
> *Suppose* Φ
> *To be proved:* Λ
> *Proof:* ...

3.3. RULES FOR THE CONNECTIVES

    Suppose Ψ
    To be proved: Λ
    Proof: ...
Thus Λ.

Example 3.10 We show that from $P \vee Q, \neg P$ it follows that Q.

Given: $P \vee Q, \neg P$.
To be proved: Q.
Proof:
    Suppose P. Then from P and $\neg P$ we get Q.
    Suppose Q. Then Q holds by assumption.
Therefore Q.

Exercise 3.11 Assume that A, B, C and D are statements.

1. From the given $A \Rightarrow B \vee C$ and $B \Rightarrow \neg A$, derive that $A \Rightarrow C$. (Hint: use the previous example.)

2. From the given $A \vee B \Rightarrow C \vee D$, $C \Rightarrow A$, and $B \Rightarrow \neg A$, derive that $B \Rightarrow D$.

Example 3.12 Here is a proof of the second DeMorgan law (Theorem 2.10):

Given: $\neg(P \vee Q)$.
To be proved: $\neg P \wedge \neg Q$.
Detailed proof:
    Assume P. By $\vee$-introduction, it follows that $P \vee Q$.
    This contradicts the given (i.e., we have an evident falsity here).
By $\neg$-introduction, we get $\neg P$.
In a similar way, $\neg Q$ can be derived.
By $\wedge$-introduction, it follows that $\neg P \wedge \neg Q$.

The concise version (no rules mentioned):
    Assume P. Then *a fortiori* $P \vee Q$ holds, contradicting the given.
Thus, $\neg P$. Similarly, $\neg Q$ is derivable.
Thus, $\neg P \wedge \neg Q$ follows.

Example 3.13 The following example from the list of equivalences in Theorem 2.10 is so strange that we give it just to prevent you from getting entangled.

Given: $\neg(P \Rightarrow Q)$
To be proved: $P \wedge \neg Q$
Proof: By $\wedge$-introduction, it suffices to prove both P and $\neg Q$.

> *To be proved:* P
> *Proof:* (by contradiction)
> > Suppose that $\neg P$.
> > Then if P holds, Q follows by $\neg$-elimination.
> > Thus (Deduction Rule), we get that $P \Rightarrow Q$.
> > However, this contradicts the given.
>
> *To be proved:* $\neg Q$
> *Proof:* (by $\neg$ introduction)
> > Assume that Q.
> > Then, by a trivial application of the Deduction Rule, $P \Rightarrow Q$ follows.
> > (Trivial, since we do not need P at all to conclude Q.)
> > Again, this contradicts the given.

Note that the rule for implication introduction can be used for reasoning by cases, as follows. Because $P \vee \neg P$ is a logical truth, it can always be added to the list of givens. If the two sub cases P and $\neg P$ both yield conclusion Q, then this proves Q. Here is the schema:

Given: ...
To be proved: Q
Proof:
> Suppose P.
> *To be proved:* Q.
> *Proof:* ...
>
> Suppose $\neg P$.
> *To be proved:* Q.
> *Proof:* ...

Thus Q.

This pattern of reasoning is used in the following examples.

Example 3.14

Let $n \in \mathbb{N}$. *To be proved:* $n^2 - n$ is even.
Proof:
> Assume n even.
> Then $n = 2m$, so $n^2 - n = (n-1)n = (2m-1)2m$,

3.4. RULES FOR THE QUANTIFIERS

and therefore $n^2 - n$ is even.
Assume n odd.
Then $n = 2m + 1$, so $n^2 - n = (n-1)n = 2m(2m+1)$,
and therefore $n^2 - n$ is even.

Thus $n^2 - n$ is even.

Exercise 3.15 Show that for any $n \in \mathbb{N}$, division of n^2 by 4 gives remainder 0 or 1.

Example 3.16 Let $\mathbb{R}$ be the universe of discourse, and let $P(x)$ be the following property:
$$x \notin \mathbb{Q} \wedge x^{\sqrt{2}} \in \mathbb{Q}.$$
In other words, x has property P iff x is irrational and $x^{\sqrt{2}}$ is rational. We will show that either $\sqrt{2}$ or $\sqrt{2}^{\sqrt{2}}$ has property P.

Suppose $\sqrt{2}^{\sqrt{2}} \in \mathbb{Q}$.
Then, since $\sqrt{2} \notin \mathbb{Q}$ (Theorem 8.14), we know that $\sqrt{2}$ has P.
Thus, $P(\sqrt{2}) \vee P(\sqrt{2}^{\sqrt{2}})$.
Suppose $\sqrt{2}^{\sqrt{2}} \notin \mathbb{Q}$.
Then, since $(\sqrt{2}^{\sqrt{2}})^{\sqrt{2}} = \sqrt{2}^{\sqrt{2} \cdot \sqrt{2}} = \sqrt{2}^2 = 2 \in \mathbb{Q}$, we know that $\sqrt{2}^{\sqrt{2}}$ has P.
Thus, $P(\sqrt{2}) \vee P(\sqrt{2}^{\sqrt{2}})$.

Therefore $P(\sqrt{2}) \vee P(\sqrt{2}^{\sqrt{2}})$.

Exercise 3.17 Prove the remaining items of Theorem 2.10 (p. 45). To prove $\Phi \equiv \Psi$ means (i) to derive Ψ from the given Φ and (ii) to derive Φ from the given Ψ.

3.4 Rules for the Quantifiers

The rules for the quantifiers come in two types: for the unrestricted, and for the restricted versions. Those for the restricted quantifiers can be derived from the others: see Exercise 3.32 p. 100.

Universal Quantifier

Introduction.

> When asked to prove that $\forall x\, E(x)$, you should start a proof by writing the Obligatory Sentence:
>
> **Suppose that c is an arbitrary object.**

You then proceed to show that *this* object (about which you are not supposed to assume extra information; in particular, it should *not* occur earlier in the argument) has the property E in question.
Schematic form:

Given: ...
To be proved: $\forall x E(x)$
Proof:
>   Suppose c is an arbitrary object
>   To be proved: $E(c)$
>   Proof: ...

Thus $\forall x E(x)$

Here is the Modification suitable for the restricted universal quantifier:

> If $\forall x \in A\, E(x)$ is to be proved, you should start proof this time by the, again, obligatory:
>
> **Suppose that c is any object in A.**

You proceed to show that *this* object (about which you only assume that it belongs to A) has the property E in question.
Schematic form:

Given: ...
To be proved: $\forall x \in A\ E(x)$
Proof:
>   Suppose c is any object in A
>   To be proved: $E(c)$
>   Proof: ...

Thus $\forall x \in A\ E(x)$

3.4. RULES FOR THE QUANTIFIERS

Arbitrary Objects. You may wonder what an *arbitrary object* is. For instance: what is an arbitrary natural number? Is it large? small? prime? etc. What exactly is an arbitrary object in A when this set happens to be a singleton? Are there objects that are *not* arbitrary?

Answer: the term 'arbitrary object' is only used here as an aid to the imagination; it indicates something unspecified about which no special assumptions are made.

Imagine that you allow someone else to pick an object, and that you don't care what choice is made. 'Suppose c is an arbitrary A' is the same as saying to the reader: 'Suppose you provide me with a member c from the set A; the choice is completely up to you.'

Often, a universal quantifier occurs in front of an implication. Therefore, you may find the following rule schema useful.

Given: ...
To be proved: $\forall x(P(x) \Rightarrow Q(x))$
Proof:
> *Suppose c is any object such that* $P(c)$
> *To be proved:* $Q(c)$
> *Proof:* ...

Thus $\forall x(P(x) \Rightarrow Q(x))$

This rule is derivable from the introduction rules for $\forall$ and $\Rightarrow$. Note that it is very similar to the recipe for restricted $\forall$-introduction. Cf. Exercise 3.18.

Elimination.
Schematic form:

Given: $\forall x\, E(x)$
Thus $E(t)$.

Here, t is any object of your choice.

That this rule is safe is obvious: if every thing satisfies E, then in particular so must t.

Modification suitable for the restricted universal quantifier, in schematic form:

Given: $\forall x \in A\, E(x),\ t \in A$
Thus $E(t)$.

Exercise 3.18 Show, using $\forall$-introduction and Deduction Rule: if from $\Gamma, P(c)$ it follows that $Q(c)$ (where c satisfies P, but is otherwise "arbitrary"), then from Γ it follows that $\forall x(P(x) \Rightarrow Q(x))$.

Existential Quantifier

Introduction. In order to show that $\exists x\, E(x)$, it suffices to specify one object t for which $E(t)$ holds.

Schematic form:

Given: $E(t)$
Thus, $\exists x\, E(x)$.

Here, t can be anything: any example satisfying E can be used to show that $\exists x E(x)$.

Modification suitable for the restricted existential quantifier, in schematic form:

Given: $E(t), t \in A$
Thus, $\exists x \in A\, E(x)$.

That these rules are safe goes without saying.

Example-objects. An object t such that $E(t)$ holds is an *example-object* for E. Thus, the introduction rule concludes $\exists x E(x)$ from the existence of an example-object.

However, it is not always possible or feasible to prove an existential statement by exhibiting a specific example-object, and there are (famous) proofs of existential statements that do not use this rule.

Example 3.19 A transcendent real is a real which is not a root of a polynomial with integer coefficients (a polynomial $f(x) = a_n x^n + a_{n-1} x^{n-1} + \ldots + a_1 x + a_0$, where the a_i are integers and $a_n \neq 0$).

By an argument establishing that the set of reals must be strictly larger than the set of roots of polynomials with integer coefficients, it is relatively easy to show that transcendent reals exist. (Compare the reasoning about degrees of infinity in Chapter 11.)

Still, it is not immediately clear how to get from this argument at an example-transcendent real. In particular, it is hard to see that e and π are transcendent reals.

Example 3.20 There is a proof that, in chess, either white has a winning strategy, or black has a strategy with which he cannot lose. However, the proof neither informs you which of the two cases you're in, nor describes the strategy in a usable way.

Example 3.21 For a logical example, look at the proof given in Example 3.30 (p. 99). What is proved there is $\exists x \neg \Phi(x)$, but (although $\exists$ introduction is used somewhere) no example-x for Φ is exhibited. On the basis of the given $\neg \forall x \Phi(x)$, it is unrealistic to expect such an example.

3.4. RULES FOR THE QUANTIFIERS

Sometimes, it is known only that an example-object must be present among the members of a certain (finite) set but it is impossible to pinpoint the right object. See the following example.

Example 3.22 *Given:* $P(a) \lor P(b)$
To be proved: $\exists x P(x)$.
Proof: By $\lor$-elimination, it is sufficient to prove $\exists x P(x)$ from both $P(a)$ and $P(b)$. But this is immediate, by $\exists$ introduction.

Example 3.23 To make Example 3.22 more concrete, consider the following question.

Is there an irrational number α with the property that $\alpha^{\sqrt{2}}$ is rational?

In Example 3.16 we established that either $\sqrt{2}$ or $\sqrt{2}^{\sqrt{2}}$ has this property. Thus, the answer to the question is 'yes'. But the reasoning does not provide us with an *example* of such a number.

It is a general feature of Proofs by Contradiction of existential statements that no example objects for the existential will turn up in the proof, and this is one reason to stay away from this rule as long as you can in such cases.

Elimination. (Note the similarity with the $\lor$-rule.)

When you want to use that $\exists x\, E(x)$ in an argument to prove Λ, you write the Obligatory Sentence:

Suppose that c is an object that satisfies E.

However, this is all that you are supposed to assume about c. Now, you proceed to prove Λ on the basis of this assumption.

Schematic form:

Given: $\exists x E(x)$, ...
To be proved: Λ
Proof:
    Suppose c is an object that satisfies E
    To be proved: Λ
    Proof: ...
Thus Λ

Modification suitable for the restricted existential quantifier:

> When you want to use that $\exists x \in A\ E(x)$ in an argument to prove Λ, you write
>
> **Suppose that c is an object in A that satisfies E.**

Again, this is all that you are supposed to assume about c. Subsequently, you proceed to prove Λ on the basis of this assumption.

Schematic form:

Given: $c \in A, \exists x E(x), \ldots$
To be proved: Λ
Proof:
    Suppose c is an object in A that satisfies E
    To be proved: Λ
    Proof: ...
Thus Λ

3.5 Summary of the Proof Recipes

Here is a summary of the rules, with introduction rules on the left hand side, and elimination rules on the right hand side. Proof by Contradiction has been put in the Elimination-column.

The Recipes for $\Rightarrow$ Introduction and Elimination

Given: ...
 To be proved: $P \Rightarrow Q$
 Proof:
     Suppose P
     To be proved: Q
     Proof: ...
 Thus $P \Rightarrow Q$.
	Given: $P, P \Rightarrow Q, \ldots$
	 Thus Q

The Recipes for ⇔ Introduction and Elimination

Given: ...
 To be proved: $P \Leftrightarrow Q$
 Proof:
     Suppose P
     To be proved: Q
     Proof: ...
     Suppose Q
     To be proved: P
     Proof: ...
 Thus $P \Leftrightarrow Q$.
	Given: P, $P \Leftrightarrow Q$, ...
	 Thus Q
	
	 Given: Q, $P \Leftrightarrow Q$, ...
	 Thus P

The Recipes for ¬ Introduction and Elimination

Given: ...
 To be proved: $\neg P$
 Proof:
     Suppose P
     To be proved: $\bot$
     Proof: ...
 Thus $\neg P$.
	Given: ...
	 To be proved: P
	 Proof:
	     Suppose $\neg P$
	     To be proved: $\bot$
	     Proof: ...
	 Thus P.
	
	 Given: P, $\neg P$
	 Thus Q.

The Recipes for ∧ Introduction and Elimination

Given: P, Q Thus $P \wedge Q$.	Given: $P \wedge Q$ Thus P. Given: $P \wedge Q$ Thus Q.

The Recipes for ∨ Introduction and Elimination

Given: P
 Thus $P \vee Q$.
 	
 Given: Q
 Thus $P \vee Q$.
	Given: $P \vee Q$, ...
	 To be proved: R
	 Proof:
	   Suppose P
	   To be proved: R
	   Proof: ...
	
	   Suppose Q
	   To be proved: R
	   Proof: ...
	 Thus R.

The Recipes for ∀ Introduction and Elimination

Given: ...
 To be proved: $\forall x E(x)$
 Proof:
   Suppose c is an arbitrary object
   To be proved: $E(c)$
   Proof: ...
 Thus $\forall x E(x)$
	Given: $\forall x E(x)$, ...
	 Thus $E(t)$.

Given: ...
 To be proved: $\forall x \in A \ E(x)$
 Proof:
   Suppose c is any object in A
   To be proved: $E(c)$
   Proof: ...
 Thus $\forall x \in A \ E(x)$
	Given: $\forall x \in A \ E(x)$, $t \in A$, ...
	 Thus $E(t)$.

The Recipes for ∃ Introduction and Elimination

Given: $E(t)$, ... Thus $\exists x E(x)$.	Given: $\exists x E(x)$, ... To be proved: P Proof: $\quad$ Suppose c is an object that satisfies E $\quad$ To be proved: P $\quad$ Proof: ... Thus P

Given: $t \in A$, $E(t)$, ... Thus $\exists x \in A \; E(x)$.	Given: $\exists x \in A \; E(x)$, ... To be proved: P Proof: $\quad$ Suppose c is an object in A that satisfies E $\quad$ To be proved: P $\quad$ Proof: ... Thus P

3.6 Some Strategic Guidelines

Here are the most important guidelines that enable you to solve a proof problem.

1. **Do not** concentrate on the given, by trying to transform that into what is to be proved.

2. **Instead**, concentrate on (the form of) what is to be proved.

3. A number of rules enable you to simplify the proof problem. For instance:

   - When asked to prove $P \Rightarrow Q$, add P to the givens and try to prove Q. (Deduction Rule).
   - When asked to prove $\forall x \; E(x)$, prove $E(c)$ for an arbitrary c instead (∀-introduction).

4. Only **after** you have reduced the problem as far as possible you should look at the givens in order to see which of them can be used.

   - When one of the givens is of the form $P \vee Q$, and R is to be proved, make a case distinction: first add P to the givens and prove R, next add Q to the givens and prove R.
   - When one of the givens is of the form $\exists x \, E(x)$, and P is to be proved, give the object that satisfies E a name, by adding $E(c)$ to the givens. Next, prove P.

5. It is usually a good idea to move negations inward as much as possible before attempting to apply $\neg$-introduction.

6. Stay away from Proof by Contradiction as long as possible.

Example 3.24 To show: from $\forall x(P(x) \Rightarrow Q(x)), \exists x P(x)$ it follows that $\exists x Q(x)$.

Concise proof: Using the second given, assume that x is such that $P(x)$ holds. Applying the first given to this x, it follows that $P(x) \Rightarrow Q(x)$. Thus, we have that $Q(x)$. Conclusion: $\exists x Q(x)$.

Exercise 3.25 Show:

1. from $\forall x(P(x) \Rightarrow Q(x)), \forall x P(x)$ it follows that $\forall x Q(x)$,

2. from $\exists x(P(x) \Rightarrow Q(x)), \forall x P(x)$ it follows that $\exists x Q(x)$.

What about: from $\exists x(P(x) \Rightarrow Q(x)), \exists x P(x)$ it follows that $\exists x Q(x)$?

Exercise 3.26 From the given

$$\forall x \exists y (x \mathbf{R} y), \forall x \forall y (x \mathbf{R} y \Rightarrow y \mathbf{R} x), \forall x \forall y \forall z (x \mathbf{R} y \wedge y \mathbf{R} z \Rightarrow x \mathbf{R} z),$$

derive that $\forall x (x \mathbf{R} x)$.

Exercise 3.27 Give proofs for the following:

1. From $\forall x \forall y \forall z (x \mathbf{R} y \wedge y \mathbf{R} z \Rightarrow x \mathbf{R} z), \forall x \neg x \mathbf{R} x$ it follows that $\forall x \forall y (x \mathbf{R} y \Rightarrow \neg y \mathbf{R} x)$,

2. From $\forall x \forall y (x \mathbf{R} y \Rightarrow \neg y \mathbf{R} x)$ it follows that $\forall x \neg x \mathbf{R} x$,

3. From $\forall x \forall y (x \mathbf{R} y \wedge x \neq y \Rightarrow \neg y \mathbf{R} x)$ it follows that $\forall x \forall y (x \mathbf{R} y \wedge y \mathbf{R} x \Rightarrow x = y)$,

3.6. SOME STRATEGIC GUIDELINES

4. From $\forall x \neg x\mathbf{R}x, \forall x \forall y (x\mathbf{R}y \Rightarrow y\mathbf{R}x), \forall x \forall y \forall z (x\mathbf{R}y \land y\mathbf{R}z \Rightarrow x\mathbf{R}z)$
   it follows that $\neg \exists x \exists y (x\mathbf{R}y)$.

The following exercise is an example on how to move a quantifier in a prefix of quantifiers.

Exercise 3.28 Show: from $\forall y \exists z \forall x P(x, y, z)$ it follows that $\forall x \forall y \exists z P(x, y, z)$.

Example 3.29 That a real function f is *continuous* means that (cf. p. 64), in the domain $\mathbb{R}$,

$$\forall x \, \forall \varepsilon > 0 \, \exists \delta > 0 \, \forall y \, (\, |x - y| < \delta \; \Rightarrow \; |f(x) - f(y)| < \varepsilon \,).$$

Uniform continuity of f means

$$\forall \varepsilon > 0 \, \exists \delta > 0 \, \forall x \, \forall y \, (\, |x - y| < \delta \; \Longrightarrow \; |f(x) - f(y)| < \varepsilon \,).$$

Compared with the first condition, the quantifier $\forall x$ has moved two places. According to Exercise 3.28 (where $P(x, \varepsilon, \delta)$ is $\forall y(|x - y| < \delta \Rightarrow |f(x) - f(y)| < \varepsilon)$), continuity is implied by uniform continuity. (But, as you may know, the implication in the other direction does not hold: there are continuous functions that are not uniformly so.)

Example 3.30

Given: $\neg \forall x \Phi(x)$.
To be proved: $\exists x \neg \Phi(x)$.
Proof:
We apply Proof by Contradiction.
   Assume, for a contradiction, that $\neg \exists x \neg \Phi(x)$.
   We will show that $\forall x \Phi(x)$.
      Suppose x is arbitrary.
      To show that $\Phi(x)$, we apply Proof by Contradiction again.
         Assume $\neg \Phi(x)$.
         Then $\exists x \neg \Phi(x)$, and contradiction with $\neg \exists x \neg \Phi(x)$.
      Thus $\Phi(x)$.
   Therefore $\forall x \Phi(x)$, and contradiction with $\neg \forall x \Phi(x)$.
Thus $\exists x \neg \Phi(x)$.

The concise and more common way of presenting this argument (that leaves for the reader to find out which rules have been applied and when) looks as follows.

Proof. Assume that $\neg\exists x\neg\Phi(x)$ and let x be arbitrary.
If $\neg\Phi(x)$ is true, then so is $\exists x\neg\Phi(x)$, contradicting the assumption.
Thus, $\Phi(x)$ holds; and, since x was arbitrary, we conclude that $\forall x\Phi(x)$.
However, this contradicts the first given; which proves that, in fact, $\exists x\neg\Phi(x)$ must be true.

Exercise 3.31 Prove the other equivalences of Theorem 2.40 (p. 63). (To prove $\Phi \equiv \Psi$ means (i) deriving Ψ from the given Φ and (ii) deriving Φ from the given Ψ.)

Exercise 3.32 Derive the rules for the restricted quantifiers from the others, using the facts (cf. the remark preceding Example 2.45, page 65) that $\forall x \in A\ E(x)$ is equivalent with $\forall x(x \in A \Rightarrow E(x))$, and $\exists x \in A\ E(x)$ is equivalent with $\exists x(x \in A \land E(x))$.

With practice, you will start to see that it is often possible to condense proof steps. Here are some examples of condensed proof recipes:

Given: ...
To be proved: $\forall x(A(x) \Rightarrow B(x))$.
Proof:
    Suppose c is an arbitrary object such that $A(c)$.
    To be proved: $B(c)$.
    Proof: ...
Thus $\forall x(A(x) \Rightarrow B(x))$.

Given: ...
To be proved: $\forall x \forall y\ A(x, y)$.
Proof:
    Suppose c and d are arbitrary objects.
    To be proved: $A(c, d)$.
    Proof: ...
Thus $\forall x \forall y\ A(x, y)$.

> Given: ...
> To be proved: $\forall x \in A \forall y \in B \ R(x,y)$.
> Proof:
> > Suppose c, d are arbitrary objects such that $A(c)$ and $B(d)$.
> > To be proved: $R(c,d)$.
> > Proof: ...
>
> Thus $\forall x \in A \forall y \in B \ R(x,y)$.

> Given: ...
> To be proved: $\forall x \forall y (R(x,y) \Rightarrow S(x,y))$.
> Proof:
> > Suppose c, d are arbitrary objects such that $R(c,d)$.
> > To be proved: $S(c,d)$.
> > Proof: ...
>
> Thus $\forall x \forall y (R(x,y) \Rightarrow S(x,y))$.

3.7 Reasoning and Computation with Primes

In this section we will demonstrate the use of the computer for investigating the theory of prime numbers. For this, we need the code for `prime` that was given in Chapter 1. It is repeated here:

```
prime :: Integer -> Bool
prime n | n < 1     = error "not a positive integer"
        | n == 1    = False
        | otherwise = ldp n == n where
   ldp     = ldpf primes
   ldpf (p:ps) m | rem m p == 0 = p
                 | p^2 > m      = m
                 | otherwise    = ldpf ps m
   primes = 2 : filter prime [3..]
```

Euclid (fourth century B.C.) proved the following famous theorem about prime numbers.

Theorem 3.33 There are infinitely many prime numbers.

Proof. Suppose there are only finitely many prime numbers, and $p_1, \ldots, p_n$ is a list of all primes. Consider the number $m = (p_1 p_2 \cdots p_n) + 1$. Note that m is not divisible by p_1, for dividing m by p_1 gives quotient $p_2 \cdots p_n$ and remainder 1. Similarly, division by $p_2, p_3, \ldots$ always gives a remainder 1.

Thus, we get the following:

- $\text{LD}(m)$ is prime,
- For all $i \in \{1, \ldots n\}$, $\text{LD}(m) \neq p_i$.

Thus, we have found a prime number $\text{LD}(m)$ different from all the prime numbers in our list $p_1, \ldots, p_n$, contradicting the assumption that $p_1, \ldots, p_n$ was the full list of prime numbers. Therefore, there must be infinitely many prime numbers. ∎

Exercise 3.34 Let $A = \{4n + 3 \mid n \in \mathbb{N}\}$ (See Example 5.92 below). Show that A contains infinitely many prime numbers. (Hint: any prime > 2 is odd, hence of the form $4n + 1$ or $4n + 3$. Assume that there are only finitely many primes of the form $4n + 3$, say $p_1, \ldots, p_m$. Consider the number $N = 4p_1 \cdots p_m - 1 = 4(p_1 \cdots p_m - 1) + 3$. Argue that N must contain a factor $4q + 3$, using the fact that $(4a + 1)(4b + 1)$ is of the form $4c + 1$.)

Use `filter prime [ 4*n + 3 | n <- [0..] ]` to generate the primes of this form.

Euclid's proof suggests a general recipe for finding bigger and bigger primes. Finding *examples* of very large primes is another matter, of course, for how do you know whether a particular natural number is a likely candidate for a check?

Example 3.35 A famous conjecture made in 1640 by Pierre de Fermat (1601–1665) is that all numbers of the form

$$2^{2^n} + 1$$

are prime. This holds for $n = 0, 1, 2, 3, 4$, for we have: $2^{2^0} + 1 = 2^1 + 1 = 3$, $2^{2^1} + 1 = 2^2 + 1 = 5$, $2^{2^2} + 1 = 2^4 + 1 = 17$, $2^{2^3} + 1 = 2^8 + 1 = 257$, which is prime, and $2^{2^4} + 1 = 2^{16} + 1 = 65537$, which is prime. Apparently, this is as far as Fermat got.

Our Haskell implementation of `prime` allows us to refute the conjecture for $n = 5$, using the built-in function `^` for exponentiation. We get:

```
TUOLP> prime (2^2^5 + 1)
False
```

3.7. REASONING AND COMPUTATION WITH PRIMES

This counterexample to Fermat's conjecture was discovered by the mathematician Léonard Euler (1707–1783) in 1732.

The French priest and mathematician Marin Mersenne (1588–1647; Mersenne was a pen pal of Descartes) found some large prime numbers by observing that $M_n = 2^n - 1$ sometimes is prime when n is prime.

Exercise 3.36 It is not very difficult to show that if n is composite, $M_n = 2^n - 1$ is composite too. Show this. (Hint: Assume that $n = ab$ and prove that $xy = 2^n - 1$ for the numbers $x = 2^b - 1$ and $y = 1 + 2^b + 2^{2b} + \cdots + 2^{(a-1)b}$).

But when n is prime, there is a chance that $2^n - 1$ is prime too. Examples are $2^2 - 1 = 3$, $2^3 - 1 = 7$, $2^5 - 1 = 31$. Such primes are called Mersenne primes.

Example 3.37 Let us use the computer to find one more Mersenne prime. Put the procedure prime in a file and load it. Next, we use ^ for exponentiation to make a new Mersenne guess, as follows:

```
TUOLP> prime 5
True
TUOLP> prime (2^5-1)
True
TUOLP> 2^5-1
31
TUOLP> prime (2^31-1)
True
TUOLP> 2^31-1
2147483647
TUOLP>
```

It may interest you to know that the fact that $2^{31} - 1$ is a prime was discovered by Euler in 1750. Using a computer, this fact is a bit easier to check.

We have already seen how to generate prime numbers in Haskell (Examples 1.22 and 1.23). We will now present an elegant alternative: a lazy list implementation of the *Sieve of Eratosthenes*. The idea of the sieve is this. Start with the list of all natural numbers $\geqslant 2$:

$$2, 3, 4, 5, 6, 7, 8, 9, 10, 11, 12, 13, 14, 15, 16, 17, 18, 19, 20,$$
$$21, 22, 23, 24, 25, 26, 27, 28, 29, 30, 31, 32, 33, 34, 35,$$
$$36, 37, 38, 39, 40, 41, 42, 43, 44, 45, 46, 47, 48, \ldots$$

In the first round, mark 2 (the first number in the list) as prime, and mark all multiples of 2 for removal in the remainder of the list (marking for removal indicated by over-lining):

$$\boxed{2}, 3, \overline{4}, 5, \overline{6}, 7, \overline{8}, 9, \overline{10}, 11, \overline{12}, 13, \overline{14}, 15, \overline{16}, 17, \overline{18}, 19, \overline{20},$$
$$21, \overline{22}, 23, \overline{24}, 25, \overline{26}, 27, \overline{28}, 29, \overline{30}, 31, \overline{32}, 33, \overline{34}, 35,$$
$$\overline{36}, 37, \overline{38}, 39, \overline{40}, 41, \overline{42}, 43, \overline{44}, 45, \overline{46}, 47, \overline{48}, \ldots$$

In the second round, mark 3 as prime, and mark all multiples of 3 for removal in the remainder of the list:

$$\boxed{2}, \boxed{3}, \overline{4}, 5, \overline{6}, 7, \overline{8}, \overline{9}, \overline{10}, 11, \overline{12}, 13, \overline{14}, \overline{15}, \overline{16}, 17, \overline{18}, 19, \overline{20},$$
$$\overline{21}, \overline{22}, 23, \overline{24}, 25, \overline{26}, \overline{27}, \overline{28}, 29, \overline{30}, 31, \overline{32}, \overline{33}, \overline{34}, 35, \ldots$$
$$\overline{36}, 37, \overline{38}, \overline{39}, \overline{40}, 41, \overline{42}, 43, \overline{44}, \overline{45}, \overline{46}, 47, \overline{48}, \ldots$$

In the third round, mark 5 as prime, and mark all multiples of 5 for removal in the remainder of the list:

$$\boxed{2}, \boxed{3}, \overline{4}, \boxed{5}, \overline{6}, 7, \overline{8}, \overline{9}, \overline{10}, 11, \overline{12}, 13, \overline{14}, \overline{15}, \overline{16}, 17, \overline{18}, 19, \overline{20},$$
$$\overline{21}, \overline{22}, 23, \overline{24}, \overline{25}, \overline{26}, \overline{27}, \overline{28}, 29, \overline{30}, 31, \overline{32}, \overline{33}, \overline{34}, \overline{35},$$
$$\overline{36}, 37, \overline{38}, \overline{39}, \overline{40}, 41, \overline{42}, 43, \overline{44}, \overline{45}, \overline{46}, 47, \overline{48}, \ldots$$

And so on. A remarkable thing about the Sieve is that the only calculation it involves is counting. If the 3-folds are to be marked in the sequence of natural numbers starting from 3, walk through the list while counting $1, 2, 3$ and mark the number 6, next walk on while counting $1, 2, 3$ and mark the number 9, and so on. If the 5-folds are to be marked in the sequence the natural numbers starting from 5, walk on through the sequence while counting $1, 2, 3, 4, 5$ and mark the number 10, next walk on while counting $1, 2, 3, 4, 5$ and mark the number 15, and so on.

In the Haskell implementation we mark numbers in the sequence [2..] for removal by replacing them with 0. When generating the sieve, these zeros are skipped.

3.7. REASONING AND COMPUTATION WITH PRIMES

```
sieve :: [Integer] -> [Integer]
sieve (0 : xs) = sieve xs
sieve (n : xs) = n : sieve (mark xs 1 n)
  where
  mark :: [Integer] -> Integer -> Integer -> [Integer]
  mark (y:ys) k m | k == m    = 0 : (mark ys 1     m)
                  | otherwise = y : (mark ys (k+1) m)

primes :: [Integer]
primes = sieve [2..]
```

This gives:

```
TUOLP> primes
[2,3,5,7,11,13,17,19,23,29,31,37,41,43,47,53,59,61,67,71,73,79,
83,89,97,101,103,107,109,113,127,131,137,139,149,151,157,163,
167,173,179,181,191,193,197,199,211,223,227,229,233,239,241,251,
257,263,269,271,277,281,283,293,307,311,313,317,331,337,347,349,
353,359,367,373,379,383,389,397,401,409,419,421,431,433,439,443,
449,457,461,463,467,479,487,491,499,503,509,521,523,541,547,557,
563,569,571,577,587,593,599,601,607,613,617,619,631,641,643,647,
653,659,661,673,677,683,691,701,709,719,727,733,739,743,751,757,
761,769,773,787,797,809,811,821,823,827,829,839,853,857,859,863,
877,881,883,887,907,911,919,929,937,941,947,953,967,971,977,983,
991,997,1009,1013,1019,1021,1031,1033,1039,1049,1051,1061,1063,
{Interrupted!}
```

Does this stream ever dry up? We know for sure that it doesn't, because of Euclid's proof.

It is possible, by the way, to take a finite initial segment of an infinite Haskell list. This is done with the built in function `take`, as follows:

```
TUOLP> take 100 primes
[2,3,5,7,11,13,17,19,23,29,31,37,41,43,47,53,59,61,67,71,73,79,
83,89,97,101,103,107,109,113,127,131,137,139,149,151,157,163,
167,173,179,181,191,193,197,199,211,223,227,229,233,239,241,251,
257,263,269,271,277,281,283,293,307,311,313,317,331,337,347,349,
353,359,367,373,379,383,389,397,401,409,419,421,431,433,439,443,
449,457,461,463,467,479,487,491,499,503,509,521,523,541]
TUOLP>
```

Exercise 3.38 A slightly faster way to generate the primes is by starting out from the odd numbers. The stepping and marking will work as before, for if you count k positions in the odd numbers starting from any odd number $a = 2n + 1$, you will move on to number $(2n + 1) + 2k$, and if a is a multiple of k, then so is $a + 2k$. Implement a function `fasterprimes :: [Integer]` using this idea. The odd natural numbers, starting from 3, can be generated as follows:

```
oddsFrom3 :: [Integer]
oddsFrom3 = 3 : map (+2) oddsFrom3
```

Still faster is to clean up the list at every step, by *removing* multiples from the list as you go along. We will come back to this matter in Section 10.1.

Exercise 3.39 Write a Haskell program to refute the following statement about prime numbers: if $p_1, \ldots, p_k$ are all the primes $< n$, then $(p_1 \times \cdots \times p_k) + 1$ is a prime.

A computer is a useful instrument for refuting guesses or for checking particular cases. But if, instead of checking a guess for a particular case, you want to check the truth of interesting *general* statements it is of limited help. You can use the function `mersenne` to generate Mersenne primes, but the computer will not tell you whether this stream will dry up or not ...

```
mersenne = [ (p,2^p - 1) | p <- primes, prime (2^p - 1) ]
```

This is what a call to `mersenne` gives:

```
TUOLP> mersenne
[(2,3),(3,7),(5,31),(7,127),(13,8191),(17,131071),(19,524287),
(31,2147483647)
```

If you are interested in how this goes on, you should check out GIMPS ("Great Internet Mersenne Prime Search") on the Internet. To generate slightly more information, we can define:

```
notmersenne = [ (p,2^p - 1) | p <- primes, not (prime (2^p-1)) ]
```

3.7. REASONING AND COMPUTATION WITH PRIMES

This gives:

```
TUOLP> notmersenne
[(11,2047),(23,8388607),(29,536870911),(37,137438953471),
(41,2199023255551),(43,8796093022207),(47,140737488355327),
(53,9007199254740991),(59,576460752303423487)
```

The example may serve to illustrate the limits of what you can do with a computer when it comes to generating mathematical insight. If you make an interesting mathematical statement, there are three possibilities:

- You succeed in proving it. This establishes the statement as a theorem.

- You succeed in disproving it (with or without the help of a computer). This establishes the statement as a refuted conjecture.

- Neither of the above. This may indicate that you have encountered an open problem in mathematics. It may also indicate, of course, that you haven't been clever enough.

Example 3.40 Here is an example of an open problem in mathematics:

Are there infinitely many Mersenne primes?

It is easy to see that Euclid's proof strategy will not work to tackle this problem. The assumption that there is a finite list $p_1, \ldots, p_n$ of Mersenne primes does yield a larger prime, but nothing guarantees that this larger prime number is again of the form $2^m - 1$.

Mersenne primes are related to so-called perfect numbers. A perfect number is a number n with the curious property that the sum of all its divisors equals $2n$, or, in other words, the sum of all proper divisors of n equals n (we call a divisor d of n *proper* if $d < n$). The smallest perfect number is 6, for its proper divisors are $1, 2$ and 3, and $1 + 2 + 3 = 6$, and it is easy to check that $1, 2, 3, 4$ and 5 are not perfect.

Euclid proved that if $2^n - 1$ is prime, then $2^{n-1}(2^n - 1)$ is perfect. Examples of perfect numbers found by Euclid's recipe are: $2 \cdot (2^2 - 1) = 6$, $2^2 \cdot (2^3 - 1) = 28$, $2^4 \cdot (2^5 - 1) = 496$.

Exercise 3.41 How would you go about yourself to prove the fact Euclid proved? Here is a hint: if $2^n - 1$ is prime, then the proper divisors of $2^{n-1}(2^n - 1)$ are

$$1, 2, 2^2, \ldots, 2^{n-1}, 2^n - 1, 2(2^n - 1), 2^2(2^n - 1), \ldots, 2^{n-2}(2^n - 1).$$

Here is a function for generating the list of proper divisors of a natural number. This is not an efficient way to generate proper divisors, but never mind.

```
pdivisors :: Integer -> [Integer]
pdivisors n = [ d | d <- [1..(n-1)], rem n d == 0 ]
```

With this it is easy to check that 8128 is indeed a perfect number:

```
TUOLP> pdivisors 8128
[1,2,4,8,16,32,64,127,254,508,1016,2032,4064]
TUOLP> sum (pdivisors 8128)
8128
```

Even more spectacularly, we have:

```
TUOLP> prime (2^13 -1)
True
TUOLP> 2^12 * (2^13 -1)
33550336
TUOLP> pdivisors 33550336
[1,2,4,8,16,32,64,128,256,512,1024,2048,4096,8191,16382,32764,
65528,131056,262112,524224,1048448,2096896,4193792,8387584,
16775168]
TUOLP> sum [1,2,4,8,16,32,64,128,256,512,1024,2048,4096,8191,16382,
32764,65528,131056,262112,524224,1048448,2096896,4193792,8387584,
16775168]
33550336
TUOLP>
```

Prime pairs are pairs $(p, p+2)$ where both p and $p+2$ are prime. Prime pairs can be generated as follows:

```
primePairs :: [(Integer,Integer)]
primePairs = pairs primes
  where
  pairs (x:y:xys) | x + 2 == y = (x,y): pairs (y:xys)
                  | otherwise  = pairs (y:xys)
```

This gives:

3.7. REASONING AND COMPUTATION WITH PRIMES

```
TUOLP> take 50 primePairs
[(3,5),(5,7),(11,13),(17,19),(29,31),(41,43),(59,61),(71,73),
(101,103),(107,109),(137,139),(149,151),(179,181),(191,193),
(197,199),(227,229),(239,241),(269,271),(281,283),(311,313),
(347,349),(419,421),(431,433),(461,463),(521,523),(569,571),
(599,601),(617,619),(641,643),(659,661),(809,811),(821,823),
(827,829),(857,859),(881,883),(1019,1021),(1031,1033),
(1049,1051),(1061,1063),(1091,1093),(1151,1153),(1229,1231),
(1277,1279),(1289,1291),(1301,1303),(1319,1321),(1427,1429),
(1451,1453),(1481,1483),(1487,1489)]
```

Does this stream ever dry up? We don't know, for the question whether there are infinitely many prime pairs is another open problem of mathematics.

Exercise 3.42 A prime triple is a triple $(p, p+2, p+4)$ with $p, p+2, p+4$ all prime. The first prime triple is $(3, 5, 7)$. Are there any more? Note that instructing the computer to generate them is no help:

```
primeTriples :: [(Integer,Integer,Integer)]
primeTriples = triples primes
  where
  triples (x:y:z:xyzs)
   | x + 2 == y && y + 2 == z = (x,y,z) : triples (y:z:xyzs)
   | otherwise                = triples (y:z:xyzs)
```

We get:

```
TUOLP> primeTriples
[(3,5,7)
```

Still, we can find out the answer ...How?

Exercise 3.43 Consider the following call:

```
TUOLP> filter prime [ p^2 + 2 | p <- primes ]
[11
```

Can you prove that 11 is the only prime of the form $p^2 + 2$, with p prime?

3.8 Further Reading

The distinction between finding meaningful relationships in mathematics on one hand and proving or disproving mathematical statements on the other is drawn very clearly in Polya [Pol57]. A good introduction to mathematical reasoning is [Ecc97]. More detail on the structure of proofs is given by Velleman [Vel94]. Automated proof checking in Haskell is treated in [HO00]. An all-time classic in the presentation of mathematical proofs is [Euc56].

Chapter 4

Sets, Types and Lists

Preview

The chapter introduces sets, not by means of a definition but by explaining why 'set' is a primitive notion, discusses the process of set formation, and explains some important operations on sets.

Talking about sets can easily lead to paradox, but such paradoxes can be avoided either by always forming sets on the basis of a previously given set, or by imposing certain typing constraints on the expressions one is allowed to use. Section 4.2 explains how this relates to functional programming.

The end of the Chapter discusses various ways of implementing sets using list representations. The chapter presents ample opportunity to exercise your skills in writing implementations for set operations and for proving things about sets.

```
module STAL

where

import List
import DB
```

4.1 Let's Talk About Sets

Remarkably, it is not possible to give a satisfactory definition of the notion of a set. There are several axiomatic approaches to set theory; the standard one (that is implicitly assumed throughout mathematics) is due to Zermelo and Fraenkel and dates from the beginning of the 20th century.

Axioms vs. Theorems, Primitive vs. Defined Notions. The truth of a mathematical statement is usually demonstrated by a proof. Most proofs use results that have been proved earlier. It follows that some truths must be given outright without proof. These are called *axioms*. The choice of axioms often is rather arbitrary. Criteria can be simplicity or intuitive evidence, but what must be considered an axiom can also be dictated by circumstance (for instance, a proof cannot be given within the given context, but in another context a proof would be possible). In the present context, we shall accept some of the Zermelo-Fraenkel axioms, as well as some fundamental properties of the number systems, such as the principle of mathematical induction for $\mathbb{N}$ (see Sections 7.1 and 11.1).

Statements that have been proved are called *theorems* if they are considered to be of intrinsic value, and *lemmas* if they are often used in proving theorems.

Notions are split up in two analogous categories. Their meaning may have been explained by *definitions* in terms of other notions. However, the process of defining notions must have a beginning somewhere. Thus, the need for notions that are *primitive*, i.e., undefined.

For instance, we shall consider as undefined here the notions of set and natural number. Given the notion of a set, that of a *function* can be defined. However, in a context that is not set-theoretic, it could well be an undefined notion.

Georg Cantor (1845-1915), the founding father of set theory, gave the following description.

The Comprehension Principle. *A set is a collection into a whole of definite, distinct objects of our intuition or of our thought. The objects are called the elements (members) of the set.*

Usually, the objects that are used as elements of a set are not sets themselves. To handle sets of this kind is unproblematic. But it is not excluded at all that members are sets. Thus, in practice, you can encounter sets of sets of ... sets.

4.1. LET'S TALK ABOUT SETS

Notation. If the object a is member of the set A, this is denoted by $a \in A$, or sometimes by $A \ni a$. If a is not a member of A, we express this as $a \notin A$, or $A \not\ni a$.
Example: $0 \in \mathbb{N}$, $\frac{1}{2} \notin \mathbb{N}$, $\frac{1}{2} \in \mathbb{Q}$.

A set is completely determined by its elements: this is the content of the following

Principle of Extensionality. *Sets that have the same elements are equal.*

Symbolically, for all sets A and B, it holds that:

$$\forall x(x \in A \Leftrightarrow x \in B) \implies A = B.$$

The converse of this (that equal sets have the same elements) is trivial. The Principle of Extensionality is one of Zermelo's axioms.

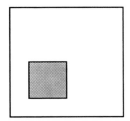

Figure 4.1: A set with a proper subset.

Subsets. The set A is called a *subset* of the set B, and B a *superset* of A; notations: $A \subseteq B$, and $B \supseteq A$, if every member of A is also a member of B. In symbols:

$$\forall x \, (x \in A \implies x \in B).$$

If $A \subseteq B$ and $A \neq B$, then A is called a *proper* subset of B.

For instance, $\{0, 2\}$ is a proper subset of $\mathbb{N}$, and the set of all multiples of 4 is a proper subset of the set of all even integers.

Note that $A = B$ iff $A \subseteq B$ and $B \subseteq A$. To show that $A \neq B$ we therefore either have to find an object c with $c \in A, c \notin B$ (in this case c is a witness of $A \not\subseteq B$), or an object c with $c \notin A, c \in B$ (in this case c is a witness of $B \not\subseteq A$). A proof of $A = B$ will in general have the following form:

> *Given:* ...
> *To be proved:* $A = B$.
> *Proof:*
> $\subseteq$: Let x be an arbitrary object in A.
>   *To be proved:* $x \in B$.
>   *Proof:*
>   ...
>   Thus $x \in B$.
> $\subseteq$: Let x be an arbitrary object in B.
>   *To be proved:* $x \in A$.
>   *Proof:*
>   ...
>   Thus $x \in A$.
> Thus $A = B$.

Warning. Sometimes, $A \subseteq B$ is written as $A \subset B$. Other authors use $A \subset B$ to indicate that A is a proper subset of B. In this book we will stick to $\subseteq$, and to express that A is properly included in B we will always use the conjunction of $A \subseteq B$ and $A \neq B$.

$\in$ **versus** $\subseteq$. Beginners often confuse $\in$ and $\subseteq$, but these relations are very different. For instance, $A \subseteq B$ implies that A and B are both sets, whereas $a \in B$ only implies that B is a set. Assuming that numbers are not sets, we have that $1 \in \{0, 1, 2\}$ and $1 \not\subseteq \{0, 1, 2\}$ (provided the latter makes sense); whereas $\{1\} \subseteq \{0, 1, 2\}$, and $\{1\} \notin \{0, 1, 2\}$.

Theorem 4.1 *For all sets A, B, C, we have that:*

1. $A \subseteq A$ *(reflexivity),*

2. $A \subseteq B \land B \subseteq A \implies A = B$ *(antisymmetry),*

3. $A \subseteq B \land B \subseteq C \implies A \subseteq C$ *(transitivity).*

Proof. 1.

To be proved: $A \subseteq A$, i.e., $\forall x(x \in A \Rightarrow x \in A)$.
Proof:
    Suppose c is any object in A. Then $c \in A$.
Therefore $\forall x(x \in A \Rightarrow x \in A)$, i.e., $A \subseteq A$.

4.1. LET'S TALK ABOUT SETS

2. This is Extensionality — be it in a somewhat different guise.
3.

To be proved: $A \subseteq B \land B \subseteq C \Rightarrow A \subseteq C$.
Proof:
> Suppose $A \subseteq B$ and $B \subseteq C$.
> To be proved: $A \subseteq C$.
> Proof:
>> Suppose c is any object in A.
>> Then by $A \subseteq B$, $c \in B$, and by $B \subseteq C$, $c \in C$.
>
> Thus $\forall x(x \in A \Rightarrow x \in C)$, i.e., $A \subseteq C$.

Thus $A \subseteq B \land B \subseteq C \Rightarrow A \subseteq C$.

∎

Remark. Note that the converse of antisymmetry also holds for $\subseteq$. In other words, if A, B are sets, then $A = B$ iff $A \subseteq B$ and $B \subseteq A$. It is because of antisymmetry of the $\subseteq$ relation that a proof that two sets A, B are equal can consist of the two subproofs mentioned above: the proof of $A \subseteq B$ and the proof of $B \subseteq A$. ∎

Exercise 4.2 Show that the superset relation also has the properties of Theorem 4.1, i.e., show that $\supseteq$ is reflexive, antisymmetric and transitive.

Enumeration. A set that has only few elements $a_1, \ldots, a_n$ can be denoted as
$$\{a_1, \ldots, a_n\}.$$
Extensionality ensures that this denotes exactly one set, for by extensionality the set is uniquely determined by the fact that it has $a_1, \ldots, a_n$ as its members.

Note that $x \in \{a_1, \ldots, a_n\}$ iff $x = a_1 \lor \cdots \lor x = a_n$.

Example 4.3 $\{0, 2, 3\}$ is the set the elements of which are 0, 2 and 3. We clearly have that $3 \in \{0, 2, 3\}$, and that $4 \notin \{0, 2, 3\}$.

Note that $\{0, 2, 3\} = \{2, 3, 0\} = \{3, 2, 2, 0\}$. Indeed, these sets have the same elements. Thus:

Order and repetition in this notation are irrelevant.

Exercise 4.4 Show, that $\{\{1, 2\}, \{0\}, \{2, 1\}\} = \{\{0\}, \{1, 2\}\}$.

An analogue to the enumeration notation is available in Haskell, where [n..m] can be used for generating a list of items from n to m. This presupposes that n and m are of the same type, and that enumeration makes sense for that type. (Technically, the type has to be of the class `Ord`; see p. 122 below.)

Sets that have many elements, in particular, infinite sets, cannot be given in this way, *unless* there is some system in the enumeration of their elements. For instance, $\mathbb{N} = \{0, 1, 2, \ldots\}$ is the set of natural numbers, and $\{0, 2, 4, \ldots\}$ is the set of even natural numbers.

Abstraction. If $P(x)$ is a certain property of objects x, the *abstraction*

$$\{ x \mid P(x) \} \tag{4.1}$$

denotes the set of things x that have property P.

Thus, for every particular object a, the expression

$$a \in \{ x \mid P(x) \}$$

is equivalent with

$$P(a).$$

By Extensionality, our talking about *the* set of x such that $P(x)$ is justified.

The abstraction notation *binds* the variable x: the set $\{x \mid P(x)\}$ in no way depends on x; $\{x \mid P(x)\} = \{y \mid P(y)\}$.

Usually, the property P will apply to the elements of a previously given set A. In that case

$$\{ x \in A \mid P(x) \}$$

denotes the set of those elements of A that have property P. For instance, the set of even natural numbers can be given by

$$\{ n \in \mathbb{N} \mid n \text{ is even } \}.$$

This way of defining an infinite set can be mimicked in functional programming by means of so-called *list comprehensions*, as follows:

```
naturals = [0..]

evens1 = [ n | n <- naturals , even n ]
```

4.1. LET'S TALK ABOUT SETS

Note the similarity between n <- naturals and $n \in \mathbb{N}$, which is of course intended by the Haskell design team. The expression even n implements the property 'n is even', so we follow the abstraction notation from set theory almost to the letter. Here is the implementation of the process that generates the odd numbers:

```
odds1 = [ n | n <- naturals , odd n ]
```

Back to the notation of set theory. A variation on the above notation for abstraction looks as follows. If f is an operation, then

$$\{ f(x) \mid P(x) \}$$

denotes the set of things of the form $f(x)$ where the object x has the property P. For instance,

$$\{ 2n \mid n \in \mathbb{N} \}$$

is yet another notation for the set of even natural numbers.

Again we have a counterpart in functional programming. Here it is:

```
evens2 = [ 2*n | n <- naturals ]
```

Still, the similarity in notation between the formal definitions and their implementations should not blind us to some annoying divergences between theory and practice. The two notations

$$\{n^2 \mid n \in \{0, \ldots, 999\}\}$$

and

$$\{n^2 \mid n \in \mathbb{N} \land n < 1000\}$$

are equivalent. They are merely two ways of specifying the set of the first 1000 square numbers. But the Haskell counterparts behave very differently:

```
small_squares1 = [ n^2 | n <- [0..999] ]
```

A call to the function `small_squares1` indeed produces a list of the first thousand square numbers, and then terminates. Note the use of `[0..999]` to enumerate a finite list. Not so with the following:

```
small_squares2 = [ n^2 | n <- naturals , n < 1000 ]
```

The way this is implemented, `n <- naturals` generates the infinite list of natural numbers in their natural order, and `n < 1000` tests each number as it is generated for the property of being less than 1000. The numbers that satisfy the test are squared and put in the result list. Unlike the previous implementation `small_squares1`, the function `small_squares2` will never terminate.

Example 4.5 (*The Russell Paradox) It is not true that to *every* property E there corresponds a set $\{x \mid E(x)\}$ of all objects that have E. The simplest example was given by Bertrand Russell (1872–1970). Consider the property of not having yourself as a member.

Most sets that you are likely to consider have this property: the set of all even natural numbers is itself not an even natural number, the set of all integers is itself not an integer, and so on. Call such sets 'ordinary'. The corresponding abstraction is $R = \{\, x \mid x \notin x \,\}$.

It turns out that the question whether the set R itself is ordinary or not is impossible to answer. For suppose $R \in R$, that is, suppose R is an ordinary set. Ordinary sets are the sets that do not have themselves as a member, so R does not have itself as a member, i.e., $R \notin R$. Suppose, on the contrary, that $R \notin R$, that is, R is an extraordinary set. Extraordinary sets are the sets that have themselves as a member, so R has itself as a member, i.e., $R \in R$.

If R were a legitimate set, this would unavoidably lead us to the conclusion that

$$R \in R \iff R \notin R,$$

which is impossible.

You do not have to be afraid for paradoxes such as the Russell paradox of Example 4.5. Only properties that you are unlikely to consider give rise to problems. In particular, if you restrict yourself to forming sets on the basis of a previously given set A, by means of the recipe

$$\{\, x \in A \mid E(x) \,\},$$

no problems will ever arise.

Example 4.6 There is no set corresponding to the property $F(x) :\equiv$ there is no infinite sequence $x = x_0 \ni x_1 \ni x_2 \ni \ldots$. To see this, assume to the contrary that F is such a set. Assume $F \in F$. This implies $F \ni F \ni F \ni F \ldots$, so by the defining property of F, $F \notin F$. Assume $F \notin F$. Then by the defining property of F, there is an infinite sequence $F = x_0 \ni x_1 \ni x_2 \ni \ldots$. Now take the infinite sequence $x_1 \ni x_2 \ni \ldots$. By the defining property of F, $x_1 \notin F$, contradicting $F = x_0 \ni x_1$.

Exercise 4.7* Assume that A is a set of sets. Show that $\{x \in A \mid x \notin x\} \notin A$.

It follows from Exercise 4.7 that every set A has a subset $B \subseteq A$ with $B \notin A$. Take $B = \{x \in A \mid x \notin x\}$.

4.2 Paradoxes, Types and Type Classes

It is a well-known fact from the theory of computation that there is no general test for checking whether a given procedure terminates for a particular input. The *halting problem* is undecidable. Intuitively, the reason for this is that the existence of an algorithm (a procedure which always terminates) for the halting problem would lead to a paradox very similar to the Russell paradox.

Here is a simple example of a program for which no proof of termination exists:

```
run :: Integer -> [Integer]
run n | n < 1 = error "argument not positive"
      | n == 1 = [1]
      | even n = n: run (div n 2)
      | odd n  = n: run (3*n+1)
```

This gives, e.g.:

STAL> run 5
[5,16,8,4,2,1]
STAL> run 6
[6,3,10,5,16,8,4,2,1]
STAL> run 7
[7,22,11,34,17,52,26,13,40,20,10,5,16,8,4,2,1]

We say that a procedure *diverges* when it does not terminate or when it aborts with an error. Stipulating divergence in Haskell is done by means of the predeclared function `undefined`, which causes an error abortion, just like `error`. In fact, Haskell has no way to distinguish between divergence and error abortion.

Now suppose `halts` can be defined. Then define the procedure `funny`, in terms of `halts`, as follows (the part of a line after `--` is a comment; in this case a warning that `funny` is no good as Haskell code):

```
funny x | halts x x = undefined    -- Caution: this
        | otherwise = True         -- will not work
```

What about the call `funny funny`? Does this diverge or halt?

Suppose `funny funny` does not halt. Then by the definition of `funny`, we are in the first case. This is the case where the argument of `funny`, when applied to itself, halts. But the argument of `funny` is `funny`. Therefore, `funny funny` does halt, and contradiction.

Suppose `funny funny` does halt. Then by the definition of `funny`, we are in the second case. This is the case where the argument of `funny`, when applied to itself, does not halt. But the argument of `funny` is `funny`. Therefore, `funny funny` does not halt, and contradiction.

Thus, there is something wrong with the definition of `funny`. The only peculiarity of the definition is the use of the `halts` predicate. This shows that such a `halts` predicate cannot be implemented.

It should be clear that `funny` is a rather close analogue to the Russell set $\{x \mid x \notin x\}$. Such paradoxical definitions are avoided in functional programming by keeping track of the *types* of all objects and operations.

As we have seen, new types can be constructed from old. Derived types are pairs of integers, lists of characters (or strings), lists of reals, and so on. How does this type discipline avoid the halting paradox? Consider the definition of `funny`. It makes a call to `halts`. What is the type of `halts`? The procedure `halts` takes as first argument a procedure, say `proc`, and as second argument an argument to that procedure, say `arg`. This means that the two arguments of `halts` have types `a -> b` and `a`, respectively. and that (`proc arg`), the result of applying `proc` to `arg`, has type `b`. But this means that the application `halts x x` in the definition of `funny` is ill-formed, for as we have seen the types of the two arguments to `halts` must be different, so the arguments themselves must be different.

For another example, take the built-in `elem` operation of Haskell which checks whether an object is element of a list. This operation is as close

4.2. PARADOXES, TYPES AND TYPE CLASSES

as you can get in Haskell to the '∈' relation of set theory. The operation expects that if its first argument is of certain type a, then its second argument is of type 'list over a'.

Thus, in Haskell, the question whether $R \in R$ does not make sense, for any R: witness the following interaction. In the transcript, Prelude> is the Haskell prompt when no user-defined files are loaded.

```
Prelude> elem 'R' "Russell"
True
Prelude> elem 'R' "Cantor"
False
Prelude> elem "Russell" "Cantor"
ERROR: Type error in application
*** expression     : "Russell" `elem` "Cantor"
*** term           : "Russell"
*** type           : String
*** does not match : Char

Prelude>
```

You would expect from this that elem has type a -> [a] -> Bool, for it takes an object of any type a as its first argument, then a list over type a, and it returns a verdict 'true' or 'false', i.e., an object of type Bool.

Almost, but not quite. The snag is that in order to check if some thing x is an element of some list of things l, one has to be able to *identify* things of the type of x. The objects that can be identified are the objects of the kinds for which equality and inequality are defined. Texts, potentially infinite streams of characters, are not of this kind. Also, the Haskell operations themselves are not of this kind, for the Haskell operations denote computation procedures, and there is no principled way to check whether two procedures perform the same task.

For suppose there were a test for equality on procedures (implemented functions). Then the following would be a test for whether a procedure f halts on input x (here /= denotes inequality):

```
halts f x =  f /= g
   where g y | y == x   = undefined    -- Caution: this
             | otherwise = f y         -- will not work
```

The where construction is used to define an auxiliary function g by stipulating that g diverges on input x and on all other inputs behaves the same as `f`. If `g` is not equal to `f`, then the difference must be in the behaviour for `x`. Since we have stipulated that g diverges for this input, we know that `f` halts on `x`. If, on the other hand, `g` and `f` are equal, then in particular `f` and `g` behave the same on input `x`, which means that `f` diverges on that input.

The types of object for which the question 'equal or not' makes sense are grouped into a collection of types called a *class*. This class is called Eq. Haskell uses == for equality and /= for inequality of objects of types in the Eq class.

Using the *hugs* command :t to ask for the type of a defined operation, we get for `elem`:

```
Prelude> :t elem
elem :: Eq a => a -> [a] -> Bool
Prelude>
```

In the present case, the type judgment means the following. If `a` is a type for which equality is defined (or, if `a` is in the Eq class), then `a -> [a] -> Bool` is an appropriate type for `elem`. In other words: for all types `a` in the Eq class it holds that `elem` is of type `a -> [a] -> Bool`.

This says that `elem` can be used to check whether an integer is a member of a list of integers, a character is a member of a string of characters, a string of characters is a member of a list of strings, and so on. But not whether an operation is a member of a list of operations, a text a member of a list of texts, and so on.

Ord is the class of types of things which not only can be tested for equality and inequality, but also for *order*: in addition to == and /=, the relations < and <= are defined. Also, it has functions min for the minimal element and max for the maximal element. The class Ord is a subclass of the class Eq.

Classes are useful, because they allow objects (and operations on those objects) to be instances of several types at once. The numeral '1' can be used as an integer, as a rational, as a real, and so on. This is reflected in Haskell by the typing:

```
Prelude> :t 1
1 :: Num a => a
Prelude>
```

All of the types integer, rational number, real number, complex number, and so on, are instances of the same class, called Num in Haskell. The class Num is a subclass of the class Eq (because it also has equality and inequality). For all types in the class Num certain basic operations, such as + and *, are defined. As we will see in Chapter 8, addition has different implementations, depending on whether we operate on $\mathbb{N}, \mathbb{Z}, \mathbb{Q}, \ldots$ and depending on the representations we choose. Still, instead of distinguishing between add, add1, add2, and so on, one could use the same *name* for all these different operations. This is standard practice in programming language design, and it is called *operator overloading*.

Exercise 4.8 Explain the following error message:

```
Prelude> elem 1 1
ERROR: [a] is not an instance of class "Num"
Prelude>
```

4.3 Special Sets

Singletons. Sets that have exactly one element are called *singletons*. The set whose only element is a is $\{a\}$; this is called the *singleton of a*. Note that it follows from the definition that $x \in \{a\}$ iff $x = a$.

Warning. Do not confuse a singleton $\{a\}$ with its element a.

In most cases you will have that $a \neq \{a\}$. For instance, in the case that $a = \{0, 1\}$, we have that $\{0, 1\} \neq \{\{0, 1\}\}$. For, $\{0, 1\}$ has two elements: the numbers 0 and 1. On the other hand, $\{\{0, 1\}\}$ has only one element: the set $\{0, 1\}$.

Remark. The question whether the equation $a = \{a\}$ has solutions (or, more generally, whether sets a exist such that $a \in a$) is answered differently by different axiomatizations of set theory. A set satisfying this equation has the shape $a = \{\{\{\cdots\cdots\}\}\}$, but of course this is unofficial notation. For the mathematical content of set theory this problem is largely irrelevant.

An example of a case where it would be useful to let sets have themselves as members would be infinite streams, like an infinite list of '1's. Such an object is easily programmed in Haskell:

```
ones = 1 : ones
```

If you load and run this, and endless stream of 1's will cover the screen, and you will have to kill the process from outside. Still, the process specified by 'first generate a '1' and then run the same process again' is well-defined, and it can plausibly be said to have itself as its second member. To be sure, the order does matter here, but any set theory can encode ordered pairs: see Section 4.5 below.

A set of the form $\{a,b\}$ is called an (unordered) *pair*. Of course, if $a = b$, then $\{a,b\} = \{a\}$ is, in fact, a singleton.

Empty Set. Finally, there is a set *without* any elements at all: the *empty* set. This curious object can be a source of distress for the beginner, because of its unexpected properties. (A first one is Theorem 4.9.) The notation for the empty set is

$$\emptyset.$$

Note that there is exactly *one* set that is empty: this is due to Extensionality.

Theorem 4.9 *For every set A, we have that*

$$\emptyset \subseteq A.$$

Proof.

Suppose x is an arbitrary object with $x \in \emptyset$.
        Then $\bot$ (contradiction with the fact that $\emptyset$ has no members).
        Therefore $x \in A$.
Thus $\forall x (x \in \emptyset \Rightarrow x \in A)$, i.e., $\emptyset \subseteq A$.

Exercise 4.10 Show:

1. $\{a\} = \{b\}$ iff $a = b$,

2. $\{a_1, a_2\} = \{b_1, b_2\}$ iff: $a_1 = b_1 \wedge a_2 = b_2$, or $a_1 = b_2 \wedge a_2 = b_1$.

Exercise 4.11 Explain that $\emptyset \neq \{\emptyset\}$, and that $\{\emptyset\} \neq \{\{\emptyset\}\}$.

Remark. Abstraction and Restricted Quantification. Note that

$$\forall x \in A \; \Phi(x) \text{ is true iff } \{x \in A \mid \Phi(x)\} = A.$$

Similarly,

$$\exists x \in A \; \Phi(x) \text{ is true iff } \{x \in A \mid \Phi(x)\} \neq \emptyset.$$

4.4 Algebra of Sets

Definition 4.12 (Intersection, Union, Difference.) Assume that A and B are sets. Then:

1. $A \cap B = \{\, x \mid x \in A \land x \in B \,\}$ is the *intersection* of A and B,

2. $A \cup B = \{\, x \mid x \in A \lor x \in B \,\}$ is their *union*, and

3. $A - B = \{\, x \mid x \in A \land x \notin B \,\}$ their *difference*.

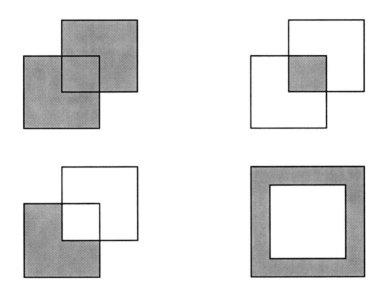

Figure 4.2: Set union, intersection, difference, and complement.

Types of Set Theoretic Expressions Often, the symbols $\cap$ and $\cup$ are confused with the connectives $\land$ and $\lor$. From Definition 4.12, their intimate connection is clear. However, their functioning is completely different: $\cap$ and $\cup$ produce, given two sets A and B, new sets $A \cap B$ resp. $A \cup B$ (thus, $\cap$ and $\cup$ can be written between sets only), whereas $\land$ and $\lor$ combine two statements Φ and Ψ into new statements $\Phi \land \Psi$ resp. $\Phi \lor \Psi$.

In the idiom of Section 4.2: the operations $\cap, \cup$ on one hand and $\wedge, \vee$ on the other have *different types*. The typing underlying the notation of set theory is much simpler than that of functional programming. To start with, we just distinguish between the types s (for set), t (for a proposition, i.e., an expression with a truth value), _ (for anything at all), and $\{s\}$ (for a family of sets, see below).

t is like `Bool` in Haskell, but s has no Haskell analogue. In set theory, sets like $\mathbb{N}, \mathbb{Z}, \mathbb{Q}, \mathbb{R}$ are all of the same type, namely s, whereas in Haskell much finer distinctions are drawn. The simplicity of the typing underlying set theory is an advantage, for it makes the language of set theory more flexible.

To find the types of the set theoretic operations we use the same recipe as in functional programming. Thus, $\cap$ and $\cup$ both have type $s \to s \to s$, for they take two set arguments and produce a new set. $\wedge$ and $\vee$ have type $t \to t \to t$, or they take two propositions and produce a new proposition. $\in$ has type $\_ \to s \to t$, for it takes anything at all as its first argument, a set as its second argument, and produces a proposition.

Exercise 4.13 What are the types of the set difference operator $-$ and of the inclusion operator $\subseteq$?

Exercise 4.14 Give the types of the following expressions:

1. $x \in \{x \mid E(x)\}$.

2. $\{x \mid E(x)\}$.

3. $(A \cap B) \subseteq C$.

4. $(A \cup B) \cap C$.

5. $\forall x (x \in A \Rightarrow x \in B)$.

6. $A = B$.

7. $a \in A \Leftrightarrow a \in B$.

The relationships between $\cap$ and $\wedge$ and between $\cup$ and $\vee$ become clearer if the equalities from Definition 4.12 are written in the form of equivalences:

1. $x \in A \cap B \iff x \in A \wedge x \in B$,

2. $x \in A \cup B \iff x \in A \vee x \in B$,

3. $x \in A - B \iff x \in A \wedge x \notin B$.

4.4. ALGEBRA OF SETS

Disjointness. Sets A and B are called *disjoint* if $A \cap B = \emptyset$.

Example 4.15 For $A = \{1,2,3\}$ and $B = \{3,4\}$, we have: $A \cup B = \{1,2,3,4\}$, $A \cap B = \{3\}$ and $A - B = \{1,2\}$. A and B are not disjoint, for $3 \in A \cap B$.

Theorem 4.16 *For all sets A, B and C, we have the following:*

1. $A \cap \emptyset = \emptyset$;
   $A \cup \emptyset = A$,

2. $A \cap A = A$;
   $A \cup A = A$ *(idempotence)*,

3. $A \cap B = B \cap A$;
   $A \cup B = B \cup A$ *(commutativity)*,

4. $A \cap (B \cap C) = (A \cap B) \cap C$;
   $A \cup (B \cup C) = (A \cup B) \cup C$ *(associativity)*,

5. $A \cap (B \cup C) = (A \cap B) \cup (A \cap C)$;
   $A \cup (B \cap C) = (A \cup B) \cap (A \cup C)$ *(distributivity)*.

By part 4.16.4, we can omit parentheses in intersections and unions of more than two sets.

Proof. Using the definitions of $\cap$, $\cup$ and $-$, these laws all reduce to Theorem 2.10 (p. 45). Still, it is instructive to give a direct proof. Here is one for the first distribution law:

$\subseteq$:
   Let x be any object in $A \cap (B \cup C)$.
   Then $x \in A$ and either $x \in B$ or $x \in C$.
      Suppose $x \in B$. Then $x \in A \cap B$, so $x \in (A \cap B) \cup (A \cap C)$.
      Suppose $x \in C$. Then $x \in A \cap C$, so $x \in (A \cap B) \cup (A \cap C)$.
   Thus in either case $x \in (A \cap B) \cup (A \cap C)$.
Thus $A \cap (B \cup C) \subseteq (A \cap B) \cup (A \cap C)$.

$\subseteq$:
   Let x be any object in $(A \cap B) \cup (A \cap C)$.
   Then either $x \in A \cap B$ or $x \in (A \cap C)$.
      Suppose $x \in A \cap B$, i.e., $x \in A$ and $x \in B$.
      Then $x \in A$ and either $x \in B$ or $x \in C$, i.e., $x \in A \cap (B \cup C)$.

Suppose $x \in A \cap C$, i.e., $x \in A$ and $x \in C$.
Then $x \in A$ and either $x \in B$ or $x \in C$, i.e., $x \in A \cap (B \cup C)$.
Thus in either case $x \in A \cap (B \cup C)$.
Therefore $(A \cap B) \cup (A \cap C) \subseteq A \cap (B \cup C)$.

By Extensionality, the required equality follows.

Finally, a third form of proof: using the definitions, the distribution law reduces, as in the above, to the equivalence

$$x \in A \wedge (x \in B \vee x \in C) \iff (x \in A \wedge x \in B) \vee (x \in A \wedge x \in C),$$

which involves the three statements $x \in A$, $x \in B$ and $x \in C$. That this comes out true always can be checked using an 8-line truth table in the usual way. ∎

Exercise 4.17 A, B and C are sets. Show:

1. $A \not\subseteq B \iff A - B \neq \emptyset$.
2. $A \cap B = A - (A - B)$.

Example 4.18 We show that $A - C \subseteq (A - B) \cup (B - C)$.

Given: $x \in (A - C)$.
To be proved: $x \in (A - B) \vee x \in (B - C)$.
Proof:
  Suppose $x \in B$. From $x \in (A - C)$ we get that $x \notin C$, so $x \in (B - C)$.
  Therefore $x \in (A - B) \vee x \in (B - C)$.

  Suppose $x \notin B$. From $x \in (A - C)$ we get that $x \in A$, so $x \in (A - B)$.
  Therefore $x \in (A - B) \vee x \in (B - C)$.

Thus $x \in (A - B) \vee x \in (B - C)$.

Exercise 4.19 Express $(A \cup B) \cap (C \cup D)$ as a union of four intersections.

Complement. Fix a set X, of which all sets to be considered are subsets. The *complement* A^c of a set $A \subseteq X$ is now defined by

$$A^c := X - A.$$

Clearly, we have that for all $x \in X$:

$$x \in A^c \iff x \notin A.$$

4.4. ALGEBRA OF SETS

Theorem 4.20

1. $(A^c)^c = A$;
   $X^c = \emptyset$;
   $\emptyset^c = X$,

2. $A \cup A^c = X$;
   $A \cap A^c = \emptyset$,

3. $A \subseteq B \Leftrightarrow B^c \subseteq A^c$,

4. $(A \cup B)^c = A^c \cap B^c$;
   $(A \cap B)^c = A^c \cup B^c$ \hfill (DeMorgan laws).

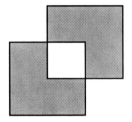

Figure 4.3: Symmetric set difference.

Symmetric Difference. The symmetric difference of two sets A and B, notation $A \oplus B$, is the set given by $\{x \mid x \in A \oplus x \in B\}$. This is the set of all objects that are either in A or in B, but not in both.

Exercise 4.21 Show that $A \oplus B = (A - B) \cup (B - A) = (A \cup B) - (A \cap B)$.

$$\left\{ \begin{array}{ccc} & \{1,2,3\} & \\ \{1,2\} & \{2,3\} & \{1,3\} \\ \{1\} & \{2\} & \{3\} \\ & \emptyset & \end{array} \right\}$$

Figure 4.4: The power set of $\{1, 2, 3\}$.

Definition 4.22 (Power Set) The *powerset* of the set X is the set $\wp(X) = \{\, A \mid A \subseteq X \,\}$ of all subsets of X.

By Theorem 4.9 and 4.1.1 we have that $\emptyset \in \wp(X)$ and $X \in \wp(X)$. So, for instance, $\wp(\{\emptyset, 1\}) = \{\emptyset, \{\emptyset\}, \{1\}, \{\emptyset, 1\}\}$. Note that $X \in \wp(A) \Leftrightarrow X \subseteq A$.

Exercise 4.23 Let X be a set with at least two elements. Then by Theorem 4.1, the relation $\subseteq$ on $\wp(X)$ has the properties of reflexivity, antisymmetry and transitivity. The relation $\leqslant$ on $\mathbb{R}$ also has these properties. The relation $\leqslant$ on $\mathbb{R}$ has the further property of *linearity*: for all $x, y \in \mathbb{R}$, either $x \leqslant y$ or $y \leqslant x$. Show that $\subseteq$ on $\wp(X)$ lacks this property.

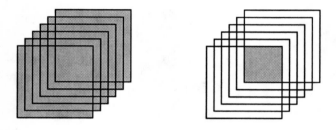

Figure 4.5: Generalized set union and intersection.

Definition 4.24 (Generalized Union and Intersection) Suppose that a set A_i has been given for every element i of a set I.

1. The *union* of the sets A_i is the set $\{x \mid \exists i \in I (x \in A_i)\}$.

   Notation: $\bigcup_{i \in I} A_i$.

2. The *intersection* of the sets A_i is the set $\{x \mid \forall i \in I (x \in A_i)\}$.

   Notation: $\bigcap_{i \in I} A_i$.

If the elements of I are sets themselves, and $A_i = i$ ($i \in I$), then $\bigcup_{i \in I} i$ is called the *union* of I;

The short notation for this set is $\bigcup I$. Similarly, $\bigcap_{i \in I} i$ is written as $\bigcap I$.

In the case that $I = \mathbb{N}$, $\bigcup_{i \in I} A_i$ and $\bigcap_{i \in I} A_i$ can also be written as $A_0 \cup A_1 \cup A_2 \cup \cdots$, resp., $A_0 \cap A_1 \cap A_2 \cap \cdots$.

4.4. ALGEBRA OF SETS

A set of sets is sometimes called a *family of sets* or a *collection of sets*. If $\mathcal{F}$ is a family of sets, $\bigcup \mathcal{F}$ and $\bigcap \mathcal{F}$ are sets. For example, if

$$\mathcal{F} = \{\{1,2,3\}, \{2,3,4\}, \{3,4,5\}\},$$

then $\bigcup \mathcal{F} = \{1,2,3,4,5\}$ and $\bigcap \mathcal{F} = \{3\}$.

Example 4.25 For $p \in \mathbb{N}$, let $A_p = \{mp \mid m \in \mathbb{N}, m \geq 1\}$. Then A_p is the set of all natural numbers that have p as a factor.
$\bigcup_{i \in \{2,3,5,7\}} A_i$ is the set of all natural numbers of the form $n \cdot 2^\alpha 3^\beta 5^\gamma 7^\delta$, with at least one of $\alpha, \beta, \gamma, \delta > 0$.
$\bigcap_{i \in \{2,3,5,7\}} A_i$ is the set of all natural numbers of the form $n \cdot 2^\alpha 3^\beta 5^\gamma 7^\delta$, with all of $\alpha, \beta, \gamma, \delta > 0$, which is the set A_{210}.

Let $\mathcal{F}$ and $\mathcal{G}$ be collections of sets. To check the truth of statements such as $\bigcup \mathcal{F} \subseteq \bigcap \mathcal{G}$ it is often useful to analyze their logical form by means of a translation in terms of quantifiers and the relation $\in$. The translation of $\bigcup \mathcal{F} \subseteq \bigcap \mathcal{G}$ becomes:

$$\forall x (\exists y (y \in \mathcal{F} \land x \in y) \Rightarrow \forall z (z \in \mathcal{G} \Rightarrow x \in z)).$$

Exercise 4.26 Give a logical translation of $\bigcap \mathcal{F} \subseteq \bigcup \mathcal{G}$ using only the relation $\in$.

Types of Generalized Union and Intersection Again, it is useful to consider the types of the operations of generalized union and intersection. These operations take families of sets and produce sets, so their type is $\{s\} \to s$, where $\{s\}$ is the type of a family of sets.

Exercise 4.27 Let $\mathcal{F}$ be a family of sets. Show that there is a set A with the following properties:

1. $\mathcal{F} \subseteq \wp(A)$,

2. For all sets B: if $\mathcal{F} \subseteq \wp(B)$ then $A \subseteq B$.

Remark. If $I = \emptyset$, then $\bigcup_{i \in I} A_i = \emptyset$, and $\bigcap_{i \in I} A_i$ is the collection of *all* sets. This last fact is an example of a trivially true implication: if $I = \emptyset$, then every statement $i \in I$ is false, hence the implication $i \in I \Rightarrow x \in A_i$ true, and everything is member of

$$\{x \mid \forall i \in I (x \in A_i)\} = \{x \mid \forall i (i \in I \Rightarrow x \in A_i)\}.$$

Therefore, the notation $\bigcap_{i \in I} A_i$ usually presupposes that $I \neq \emptyset$. ∎

Example 4.28 For $x \in X$, we have that:

$$\begin{aligned}
x \in (A \cup B)^c &\Leftrightarrow \neg(x \in A \cup B) \\
&\Leftrightarrow \neg(x \in A \vee x \in B) \\
&\stackrel{*}{\Leftrightarrow} \neg x \in A \wedge \neg x \in B \\
&\Leftrightarrow x \in A^c \wedge x \in B^c \\
&\Leftrightarrow x \in A^c \cap B^c.
\end{aligned}$$

Step ($*$) is justified by propositional reasoning. See Theorem 2.10.9. Extensionality allows us to conclude the first DeMorgan law:

$$(A \cup B)^c = A^c \cap B^c.$$

Exercise 4.29 Prove the rest of Theorem 4.20.

Exercise 4.30 Answer as many of the following questions as you can.

1. Determine: $\wp(\emptyset)$, $\wp(\wp(\emptyset))$ and $\wp(\wp(\wp(\emptyset)))$.
2. How many elements has $\wp^5(\emptyset) = \wp(\wp(\wp(\wp(\wp(\emptyset)))))$?
3. How many elements has $\wp(A)$, given that A has n elements?

Exercise 4.31 Check whether the following is true: if two sets have the same subsets, then they are equal. I.e.: if $\wp(A) = \wp(B)$, then $A = B$. Give a proof or a refutation by means of a counterexample.

Exercise 4.32 Is it true that for all sets A and B:

1. $\wp(A \cap B) = \wp(A) \cap \wp(B)$?
2. $\wp(A \cup B) = \wp(A) \cup \wp(B)$?

Provide either a proof or a refutation by counter-example.

Exercise 4.33* Show:

1. $B \cap (\bigcup_{i \in I} A_i) = \bigcup_{i \in I} (B \cap A_i)$,
2. $B \cup (\bigcap_{i \in I} A_i) = \bigcap_{i \in I} (B \cup A_i)$,
3. $(\bigcup_{i \in I} A_i)^c = \bigcap_{i \in I} A_i^c$, assuming that $\forall i \in I \; A_i \subseteq X$,
4. $(\bigcap_{i \in I} A_i)^c = \bigcup_{i \in I} A_i^c$, assuming that $\forall i \in I \; A_i \subseteq X$.

4.5. PAIRS AND PRODUCTS

Exercise 4.34* Assume that you are given a certain set A_0. Suppose you are assigned the task of finding sets $A_1, A_2, A_3, \ldots$ such that $\wp(A_1) \subseteq A_0$, $\wp(A_2) \subseteq A_1$, $\wp(A_3) \subseteq A_2, \ldots$ Show that no matter how hard you try, you will eventually fail, that is: hit a set A_n for which no A_{n+1} exists such that $\wp(A_{n+1}) \subseteq A_n$. (I.e., $\emptyset \notin A_n$.)
Hint. Suppose you can go on forever. Show this would entail $\wp(\bigcap_{i \in \mathbb{N}} A_i) \subseteq \bigcap_{i \in \mathbb{N}} A_i$. Apply Exercise 4.7.

Exercise 4.35* Suppose that the collection $\mathcal{K}$ of sets satisfies the following condition:
$$\forall A \in \mathcal{K}(A = \emptyset \vee \exists B \in \mathcal{K}(A = \wp(B))).$$
Show that every element of $\mathcal{K}$ has the form $\wp^n(\emptyset)$ for some $n \in \mathbb{N}$. (N.B.: $\wp^0(\emptyset) = \emptyset$.)

4.5 Ordered Pairs and Products

Next to the unordered pairs $\{a, b\}$ of Section 4.3, in which the order of a and b is immaterial ($\{a, b\} = \{b, a\}$), there are *ordered* pairs in which order does count. The *ordered pair* of objects a and b is denoted by

$$(a, b).$$

Here, a is the *first* and b the *second coordinate* of (a, b).

Ordered pairs behave according to the following rule:

$$(a, b) = (x, y) \implies a = x \wedge b = y. \tag{4.2}$$

This means that the ordered pair of a and b fixes the objects as well as their order. Its behaviour differs from that of the unordered pair: we always have that $\{a, b\} = \{b, a\}$, whereas $(a, b) = (b, a)$ only holds — according to (4.2) — when $a = b$.

Warning. If a and b are reals, the notation (a, b) also may denote the open interval $\{x \in \mathbb{R} \mid a < x < b\}$. The context should tell you which of the two is meant.

Defining Ordered Pairs. Defining

$$(a, b) = \{\{a\}, \{a, b\}\}$$

allows you to *prove* (4.2). Cf. Exercise 4.41.

Definition 4.36 (Products) The (Cartesian) *product* of the sets A and B is the set of all pairs (a, b) where $a \in A$ and $b \in B$. In symbols:

$$A \times B = \{\, (a, b) \mid a \in A \wedge b \in B \,\}.$$

Instead of $A \times A$ one usually writes A^2.

Example 4.37 $\{0, 1\} \times \{1, 2, 3\} = \{(0, 1), (0, 2), (0, 3), (1, 1), (1, 2), (1, 3)\}$.

When A and B are real intervals on the X resp., the Y-axis in two-dimensional space, $A \times B$ can be pictured as a rectangle.

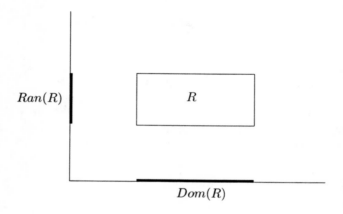

Theorem 4.38 For arbitrary sets A, B, C, D the following hold:

1. $(A \times B) \cap (C \times D) = (A \times D) \cap (C \times B)$,
2. $(A \cup B) \times C = (A \times C) \cup (B \times C)$; $(A \cap B) \times C = (A \times C) \cap (B \times C)$,
3. $(A \cap B) \times (C \cap D) = (A \times C) \cap (B \times D)$,
4. $(A \cup B) \times (C \cup D) = (A \times C) \cup (A \times D) \cup (B \times C) \cup (B \times D)$,
5. $[(A - C) \times B] \cup [A \times (B - D)] \subseteq (A \times B) - (C \times D)$.

Proof. As an example, we prove the first item of part 2.

To be proved: $(A \cup B) \times C = (A \times C) \cup (B \times C)$:
$\subseteq$:
 Suppose that $p \in (A \cup B) \times C$.
 Then $a \in A \cup B$ and $c \in C$ exist such that $p = (a, c)$.

4.5. PAIRS AND PRODUCTS

Thus (i) $a \in A$ or (ii) $a \in B$.
    (i). In this case, $p \in A \times C$, and hence $p \in (A \times C) \cup (B \times C)$.
    (ii). Now $p \in B \times C$, and hence again $p \in (A \times C) \cup (B \times C)$.
Thus $p \in (A \times C) \cup (B \times C)$.
Therefore, $(A \cup B) \times C \subseteq (A \times C) \cup (B \times C)$.

$\subseteq$:
Conversely, assume that $p \in (A \times C) \cup (B \times C)$.
Thus (i) $p \in A \times C$ or (ii) $p \in B \times C$.
    (i). In this case $a \in A$ and $c \in C$ exist such that $p = (a, c)$; a fortiori, $a \in A \cup B$ and hence $p \in (A \cup B) \times C$.
    (ii). Now $b \in B$ and $c \in C$ exist such that $p = (b, c)$; a fortiori $b \in A \cup B$ and hence, again, $p \in (A \cup B) \times C$.
Thus $p \in (A \cup B) \times C$.
Therefore, $(A \times C) \cup (B \times C) \subseteq (A \cup B) \times C$.

The required result follows using Extensionality.

∎

Exercise 4.39 Prove the other items of Theorem 4.38.

Exercise 4.40   1. Assume that A and B are non-empty and that $A \times B = B \times A$. Show that $A = B$.

2. Show by means of an example that the condition of non-emptiness in 1 is necessary. (Did you use this in your proof of 1?)

Exercise 4.41* To show that defining (a, b) as $\{\{a\}, \{a, b\}\}$ works, prove that

1. $\{a, b\} = \{a, c\} \implies b = c$,

2. $\{\{a\}, \{a, b\}\} = \{\{x\}, \{x, y\}\} \implies a = x \land b = y$.

If we assume the property of ordered pairs $(a, b) = (x, y) \implies a = x \land b = y$, we can define triples by $(a, b, c) := (a, (b, c))$. For suppose $(a, b, c) = (x, y, z)$. Then by definition, $(a, (b, c)) = (a, b, c) = (x, y, z) = (x, (y, z))$, and by the property of ordered pairs, we have that $a = x$ and $(b, c) = (y, z)$. Again, by the property of ordered pairs, $b = y$ and $c = z$. This shows that $(a, b, c) = (x, y, z) \Rightarrow (a = x \land b = y \land c = z)$.

Let us go about this more systematically, by defining ordered n-tuples over some base set A, for every $n \in \mathbb{N}$. We proceed by recursion.

Definition 4.42 (n-tuples over A)

1. $A^0 := \{\emptyset\}$,

2. $A^{n+1} := A \times A^n$.

In Haskell, ordered pairs are written as `(x1,x2)`, and there are predefined functions `fst` to get at the first member and `snd` to get at the second member. Ordered triples are written as `(x1, x2, x3)`, and so on. If `x1` has type `a` and `x2` has type `b`, then `(x1,x2)` has type `(a,b)`. Think of this type as the *product* of the types for `x1` and `x2`. Here is an example:

```
Prelude> :t  (1, 'A')
(1,'A') :: Num a => (a,Char)
Prelude>
```

4.6 Lists and List Operations

Assuming the list elements are all taken from a set A, the set of all lists over A is the set $\bigcup_{n \in \mathbb{N}} A^n$. We abbreviate this set as A^*. For every list $L \in A^*$ there is some $n \in \mathbb{N}$ with $L \in A^n$. If $L \in A^n$ we say that list L has length n.

Standard notation for the (one and only) list of length 0 is $[]$. A list of length $n > 0$ looks like $(x_0, (x_1, (\cdots, x_{n-1}) \cdots))$. This is often written as $[x_0, x_1, \ldots, x_{n-1}]$. A one element list is a list of the form $(x, [])$. In line with the above square bracket notation, this is written as $[x]$.

If one uses lists to represent sets, there is a difficulty. The list $[a, b, b]$ is different from the list $[a, b]$, for $[a, b, b]$ has length 3, and $[a, b]$ has length 2, but the sets $\{a, b, b\}$ and $\{a, b\}$ are identical. This shows that sets and lists have different identity conditions.

This is how the data type of lists is (pre-)declared in Haskell:

```
data [a] = [] | a : [a] deriving (Eq, Ord)
```

To grasp what this means, recall that in functional programming every set has a *type*. The data declaration for the type of lists over type `a`, notation `[a]` specifies that lists over type `a` are either empty or they consist of an element of type `a` put in front of a list over type `a`. Haskell uses `:` for the operation of putting an element in front of a list.

The operation `:` combines an object with a list of objects of the same type to form a new list of objects of that type. This is reflected by the type judgment:

4.6. LISTS AND LIST OPERATIONS

```
Prelude> :t (:)
(:) :: a -> [a] -> [a]
```

The data declaration for lists also tells us, by means of deriving (Eq,Ord), that if equality is defined on the objects of type a (i.e., if the type a belongs to the class Eq) then this relation carries over to lists over type a, and if the objects of type a are ordered, i.e., if the type a belongs to the class Ord (e.g., the objects of type $\mathbb{Z}$ are ordered by <), then this order carries over to lists over type a.

If we have an equality test for members of the set A, then it is easy to see how equality can be defined for lists over A (in fact, this is all taken care of by the predefined operations on lists). Lists are ordered sets, so two lists are the same if (i) they either are both empty, or (ii) they start with the same element (here is where the equality notion for elements from A comes in), and their tails are the same. In Haskell, this is implemented as follows (this is a definition from Prelude.hs):

```
instance Eq a => Eq [a] where
    []      == []      = True
    (x:xs)  == (y:ys)  = x==y && xs==ys
    _       == _       = False
```

This says that if a is an instance of class Eq, then [a] is so too. This specifies how the equality notion for objects of type a carries over to objects of type [a] (i.e., lists over that type). As we have seen, Haskell uses == for equality and = for the definition of operator values. Part (i) of the definition of list equality is taken care of by the statement that [] == [] is true, part (ii) is covered by:

```
(x:xs) == (y:ys) = x==y && xs==ys
```

This says: the truth value of (x:xs) == (y:ys) (equality of two lists that are both non-empty) is given by the conjunction of

- the first elements are the same (x == y), and
- the remainders are the same (xs == ys).

The final line _ == _ = False states that in all other cases the lists are not equal. This final line uses _ as a 'don't care pattern' or wild card.

Exercise 4.43 How does it follow from this definition that lists of different length are unequal?

A type of class Ord is a type on which the two-placed operation compare is defined, with a result of type Ordering. The type Ordering is the set

$$\{LT, EQ, GT\},$$

where LT, EQ, GT have the obvious meanings. Suppose a is a type of class Ord, i.e., compare is defined on the type a. What does a reasonable ordering of lists over type a look like? A well-known order on lists is the lexicographical order: the way words are ordered in a dictionary. In this ordering, the empty list comes first. For non-empty lists L_1, L_2, we compare their first elements, using the function compare for objects of type a. If these are the same, the order is determined by the remainders of the lists. If the first element of L_1 comes before the first element of L_2 then L_1 comes before L_2, otherwise L_2 comes before L_1. The following piece of Haskell code implements this idea, by defining the function compare for lists over type a.

```
instance Ord a => Ord [a] where
    compare []      (_:_)    = LT
    compare []      []       = EQ
    compare (_:_)   []       = GT
    compare (x:xs)  (y:ys) = primCompAux x y (compare xs ys)
```

This specifies how the ordering on type a carries over to an ordering of lists over type a. The first line says that the empty list [] is less than any non-empty list. The second line says that the empty list is equal to itself. The third line says that any non-empty list is greater than the empty list. This fully determines the relation between [] and any list. The last line uses an auxiliary function primCompAux to cover the case of two non-empty lists. This function is defined by:

```
primCompAux       :: Ord a => a -> a -> Ordering -> Ordering
primCompAux x y o =
    case compare x y of EQ -> o;
                        LT -> LT;
                        GT -> GT
```

4.6. LISTS AND LIST OPERATIONS

The type declaration of `primCompAux` says that if a is an ordered type, then `primCompAux` expects three arguments, the first one of type a, the second one of type a and the third one an element of the set (or type) Ordering, i.e., the set {LT, EQ, GT}. The result is again a member of the type Ordering.

The definition of the operation `primCompAux` uses a `case` construction, using the reserved keywords `case` and `of`, and the arrow `->` to point at the results for the various cases. It says that in case the first two arguments are equal, the function returns the element of Ordering which is its third argument, in case the first argument is less than the second argument, the function returns LT, in case the first argument is greater than the second argument, the function returns GT.

Exercise 4.44 Another ordering of lists is as follows: shorter lists come before longer ones, and for lists of the same length we compare their first elements, and if these are the same, the remainder lists. Give a formal definition of this ordering. How would you implement it in Haskell?

In list processing, fundamental operations are checking whether a list is empty, accessing the first element of a non-empty list, determining what remains after removal of the first element from a list (its `tail`). The operations for accessing the head or the tail of a list are implemented in Haskell as follows:

```
head              :: [a] -> a
head (x:_)        = x

tail              :: [a] -> [a]
tail (_:xs)       = xs
```

The type of the operation `head`: give it a list over type a and the operation will return an element of the same type a. This is specified in `head :: [a] -> a`. The type of the operation `tail` is: give it a list over type a and it will return a list over the same type. This is specified in `tail :: [a] -> [a]`.

Note that these operations are only defined on non-empty lists. `(x:_)` specifies the pattern of a non-empty list with first element x, where the nature of the tail is irrelevant. `(_:xs)` specifies the pattern of a non-empty list with tail xs, where the nature of the first element is irrelevant.

Accessing the last element of a non-empty list is done by means of recursion: the last element of a unit list $[x]$ is equal to $[x]$. The last element of a non-unit list is equal to the last element of its tail. Here is the Haskell implementation:

```
last                :: [a] -> a
last [x]            = x
last (_:xs)         = last xs
```

Note that because the list patterns [x] and (_:xs) are tried for in that order, the pattern (_:xs) in this definition matches non-empty lists that are not unit lists.

Exercise 4.45 Which operation on lists is specified by the Haskell definition in the frame below?

```
init                :: [a] -> [a]
init [x]            = []
init (x:xs)         = x : init xs
```

It is often useful to be able to test whether a list is empty or not. The following operation accomplishes this:

```
null                :: [a] -> Bool
null []             = True
null (_:_)          = False
```

Exercise 4.46 Write your own definition of a Haskell operation reverse that reverses a list.

Exercise 4.47 Write a function splitList that gives all the ways to split a list of at least two elements in two non-empty parts. The type declaration is:

```
splitList :: [a] -> [([a],[a])]
```

4.7. LIST COMPREHENSION AND DATABASE QUERY

The call `splitList [1..4]` should give:

```
STAL> splitList [1..4]
[([1],[2,3,4]),([1,2],[3,4]),([1,2,3],[4])]
```

An operation on lists that we will need in the next sections is the operation of removing duplicates. This is predefined in the Haskell module *List.hs* as nub ('nub' means essence), but here is a home-made version for illustration:

```
nub :: (Eq a) => [a] -> [a]
nub [] = []
nub (x:xs) = x : nub (remove x xs)
  where
    remove y []                   = []
    remove y (z:zs) | y == z      = remove y zs
                    | otherwise = z : remove y zs
```

What this says is, first, that if a is any type for which a relation of equality is defined, then nub operates on a list over type a and returns a list over type a.

In Haskell, strings of characters are represented as lists, and the shorthand "abc" is allowed for ['a','b','c']. Here is an example of an application of nub to a string:

```
STAL> nub "Mississippy"
"Mispy"
```

Of course, we can also use nub on lists of words:

```
STAL> nub ["Quentin Tarantino","Harrison Ford","Quentin Tarantino"]
["Quentin Tarantino","Harrison Ford"]
```

4.7 List Comprehension and Database Query

To get more familiar with list comprehensions, we will devote this section to list comprehension for database query, using the movie database module *DB.hs* given in Figure 4.6. The database that gets listed here is called db, with type DB, where DB is a synonym for the type [WordList], where WordList is again a synonym for the type [String]. The reserved keyword type is used to declare these type synonyms. Notice the difference between defining a type synonym with type and declaring a new data type with data.

```
module DB
where
type WordList = [String]
type DB       = [WordList]

db :: DB
db = [
 ["release", "Blade Runner", "1982"],
 ["release", "Alien", "1979"],
 ["release", "Titanic", "1997"],
 ["release", "Good Will Hunting", "1997"],
 ["release", "Pulp Fiction", "1994"],
 ["release", "Reservoir Dogs", "1992"],
 ["release", "Romeo and Juliet", "1996"],
 {- ... -}

 ["direct", "Brian De Palma", "The Untouchables"],
 ["direct", "James Cameron", "Titanic"],
 ["direct", "James Cameron", "Aliens"],
 ["direct", "Ridley Scott", "Alien"],
 ["direct", "Ridley Scott", "Blade Runner"],
 ["direct", "Ridley Scott", "Thelma and Louise"],
 ["direct", "Gus Van Sant", "Good Will Hunting"],
 ["direct", "Quentin Tarantino", "Pulp Fiction"],
 {- ... -}

 ["play", "Leonardo DiCaprio", "Romeo and Juliet", "Romeo"],
 ["play", "Leonardo DiCaprio", "Titanic", "Jack Dawson"],
 ["play", "Robin Williams", "Good Will Hunting", "Sean McGuire"],
 ["play", "John Travolta", "Pulp Fiction", "Vincent Vega"],
 ["play", "Harvey Keitel", "Reservoir Dogs", "Mr White"],
 ["play", "Harvey Keitel", "Pulp Fiction", "Winston Wolf"],
 ["play", "Uma Thurman", "Pulp Fiction", "Mia"],
 ["play", "Quentin Tarantino", "Pulp Fiction", "Jimmie"],
 ["play", "Quentin Tarantino", "Reservoir Dogs", "Mr Brown"],
 ["play", "Sigourney Weaver", "Alien", "Ellen Ripley"],
 {- ... -}
```

Figure 4.6: A Database Module.

4.7. LIST COMPREHENSION AND DATABASE QUERY

The database can be used to define the following lists of database objects, with list comprehension. Here db :: DB is the database list.

```
characters = nub [ x   | ["play",_,_,x] <- db ]
movies     =     [ x   | ["release",x,_] <- db ]
actors     = nub [ x   | ["play",x,_,_] <- db ]
directors  = nub [ x   | ["direct",x,_] <- db ]
dates      = nub [ x   | ["release",_,x] <- db ]
universe   = nub (characters++actors++directors++movies++dates)
```

Next, define lists of tuples, again by list comprehension:

```
direct   = [ (x,y)   | ["direct",x,y]  <- db ]
act      = [ (x,y)   | ["play",x,y,_]  <- db ]
play     = [ (x,y,z) | ["play",x,y,z]  <- db ]
release  = [ (x,y)   | ["release",x,y] <- db ]
```

Finally, define one placed, two placed and three placed predicates by means of lambda abstraction.

```
charP     = \ x       -> elem x characters
actorP    = \ x       -> elem x actors
movieP    = \ x       -> elem x movies
directorP = \ x       -> elem x directors
dateP     = \ x       -> elem x dates
actP      = \ (x,y)   -> elem (x,y) act
releaseP  = \ (x,y)   -> elem (x,y) release
directP   = \ (x,y)   -> elem (x,y) direct
playP     = \ (x,y,z) -> elem (x,y,z) play
```

We start with some conjunctive queries. 'Give me the actors that also are directors.'

```
q1 = [ x | x <- actors, directorP x ]
```

'Give me all actors that also are directors, together with the films in which they were acting.'

```
q2 = [ (x,y) | (x,y) <- act, directorP x ]
```

'Give me all directors together with their films and their release dates.' The following is *wrong*.

```
q3 = [ (x,y,z) | (x,y) <- direct, (y,z) <- release ]
```

The problem is that the two ys are unrelated. In fact, this query generates an infinite list. This can be remedied by using the equality predicate as a link:

```
q4 = [ (x,y,z) | (x,y) <- direct, (u,z) <- release, y == u ]
```

'Give me all directors of films released in 1995, together with these films.'

```
q5 = [ (x,y) | (x,y) <- direct, (u,"1995") <- release, y == u ]
```

'Give me all directors of films released after 1995, together with these films and their release dates.'

```
q6 = [ (x,y,z) | (x,y) <- direct, (u,z) <- release,
                 y == u, z > "1995"                    ]
```

'Give me the films in which Kevin Spacey acted.'

```
q7 = [ x | ("Kevin Spacey",x) <- act ]
```

4.7. LIST COMPREHENSION AND DATABASE QUERY

'Give me all films released after 1997 in which William Hurt did act.'

```
q8 = [ x | (x,y) <- release, y > "1997", actP ("William Hurt",x) ]
```

Yes/no queries based on conjunctive querying: 'Are there any films in which the director was also an actor?'

```
q9 = q1 /= []
```

'Does the database contain films directed by Woody Allen?'

```
q10 = [ x | ("Woody Allen",x) <- direct ] /= []
```

Or simply:

```
q10' = directorP "Woody Allen"
```

Disjunctive and negative queries are also easily expressed, since we have predicates and Boolean operators.

Exercise 4.48 Translate the following into a query: 'Give me the films in which Robert De Niro or Kevin Spacey acted.'

Exercise 4.49 Translate the following into a query: 'Give me all films with Quentin Tarantino as actor or director that appeared in 1994.'

Exercise 4.50 Translate the following into a query: 'Give me all films released after 1997 in which William Hurt did not act.'

4.8 Using Lists to Represent Sets

Sets are unordered, lists are ordered, but we can use lists to represent finite (or countably infinite) sets by representing sets as lists with duplicates removed, and by disregarding the order. If a finite list does not contain duplicates, its length gives the size of the finite set that it represents.

Even if we gloss over the presence of duplicates, there are limitations to the representation of sets by lists. Such representation only works if the sets to be represented are small enough. In Chapter 11 we will return to this issue.

To removing an element from a list without duplicates all we have to do is remove the *first* occurrence of the element from the list. This is done by the predefined function delete, also part of the Haskell module *List.hs*. Here is our home-made version:

```
delete :: Eq a => a -> [a] -> [a]
delete x [] = []
delete x (y:ys) | x == y    = ys
                | otherwise = y : delete x ys
```

As we have seen, the operation of elem for finding elements is built in. Here is our demo version, re-baptized elem' to avoid a clash with *Prelude.hs*.

```
elem' :: Eq a => a -> [a] -> Bool
elem' x []                   = False
elem' x (y:ys) | x == y    = True
               | otherwise = elem' x ys
```

Further operations on sets that we need to implement are union, intersection and difference. These are all built into the Haskell module *List.hs*. Our version of union:

```
union :: Eq a => [a] -> [a] -> [a]
union [] ys     = ys
union (x:xs) ys = x : union xs (delete x ys)
```

4.8. USING LISTS TO REPRESENT SETS

Note that if this function is called with arguments that themselves do not contain duplicates then it will not introduce new duplicates.

Here is an operation for taking intersections:

```
intersect :: Eq a => [a] -> [a] -> [a]
intersect []     s                = []
intersect (x:xs) s | elem x s     = x : intersect xs s
                   | otherwise    =     intersect xs s
```

Note that because the definitions of union and intersect contain calls to delete or elem they presuppose that the type a has an equality relation defined on it. This is reflected in the type declarations for the operations.

Exercise 4.51 The Haskell operation for list difference is predefined as \\ in *List.hs*. Write your own version of this.

The predefined versions of the functions elem and notElem for testing whether an object occurs in a list or not use the functions any and all:

```
elem, notElem   :: Eq a => a -> [a] -> Bool
elem            = any . (==)
notElem         = all . (/=)
```

Be cautious with this: elem 0 [1..] will run forever.

Let's turn to an operation for the list of all sublists of a given list. For this, we first need to define how one adds a new object to each of a number of lists. This can be done with the Haskell function map . Adding an element x to a list l that does not contain x is just a matter of applying the function (x:) (prefixing the element x) to l. Therefore, the following simple definition gives an operation that adds an element x to each list in a list of lists.

```
addElem :: a -> [[a]] -> [[a]]
addElem x = map (x:)
```

Note the type declaration: [[a]] denotes the type of lists over the type of lists over type a.

The operation `addElem` is used implicitly in the operation for generating the sublists of a list:

```
powerList  :: [a] -> [[a]]
powerList  [] = [[]]
powerList  (x:xs) = (powerList xs) ++ (map (x:) (powerList xs))
```

Here is the function in action:
```
STAL> powerList [1,2,3]
[[], [3], [2], [2, 3], [1], [1, 3], [1, 2], [1, 2, 3]]
```

Example 4.52 For a connection with Exercise 4.30, let us try to fool Haskell into generating the list counterparts of $\wp^2(\emptyset) = \wp(\{\emptyset\})$, $\wp^3(\emptyset)$ and $\wp^4(\emptyset)$. Some fooling is necessary, because Haskell is much less flexible than set theory about types. In set theory, we have: $\emptyset :: s$, $\{\emptyset\} :: s$, $\{\{\emptyset\}\} :: s$, for these are all sets. In Haskell, the type of the empty list is polymorphic, i.e., we have `[] :: [a]`. Thus, the list counterpart of $\wp^2(\emptyset) = \{\emptyset, \{\emptyset\}\}$, can be consistently typed:

```
STAL> :t [[],[[]]]
[[],[[]]] :: [[[a]]]
```

What happens is that the first occurrence of `[]` gets type `[[a]]`, and the second occurrence type `[a]`, so that `[[],[[]]]` is a list containing two objects of the same type `[[a]]`, and is itself of type `[[[a]]]`. However, Haskell refuses to display a list of the generic type `[a]`. The empty list will only be displayed in a setting where it is clear what the type is of the objects that are being listed:

```
STAL> [ x | x <- [1..10], x < 0 ]
[]
```

This is OK, for it is clear from the context that `[]` is an empty list of numbers.

```
STAL> []
ERROR: Cannot find "show" function for:
*** Expression : []
*** Of type    : [a]

STAL> [[],[[]]]
ERROR: Cannot find "show" function for:
*** Expression : [[],[[]]]
*** Of type    : [[[a]]]
```

4.9. A DATA TYPE FOR SETS

These are *not* OK, for the context does not make clear what the types are.

The following data type declaration introduces a data type S containing a single object Void. You should think of this as something *very* mysterious, for the Void is the unfathomable source of an infinite list-theoretic universe built from empty :: [S]. Void is used only to provide empty with a type.

```
data S = Void deriving (Eq,Show)
empty :: [S]
empty = []
```

Here are the first stages of the list universe:

```
STAL> powerList empty
[[]]
STAL> powerList (powerList empty)
[[],[[]]]
STAL> powerList (powerList (powerList empty))
[[],[[[]]],[[[]]],[[],[[]]]]
STAL> powerList (powerList (powerList (powerList empty)))
[[],[[[],[[]]],[[[]]],[[[]],[[],[[]]],[[[[]]],[[[]]],[[],[[]]]],
[[[[]]],[[[]]],[[[[]]],[[[]],[[],[[]]],[[[]],[[],[[],[[]]],[[],[[]]],
[[],[[]],[[],[[]]],[[],[[[]]],[[],[[[]]],[[],[[]]]],
[[],[[[]]],[[[]]],[[],[[[]]],[[[]]],[[],[[]]]]]]
STAL>
```

Exercise 4.53 Write functions genUnion and genIntersect for generalized list union and list intersection. The functions should be of type [[a]]-> [a]. They take a list of lists as input and produce a list as output. Note that genIntersect is undefined on the empty list of lists (compare the remark about the presupposition of generalized intersection on page 131).

4.9 A Data Type for Sets

The representation of sets as lists without duplicates has the drawback that two finite lists containing the same elements, but in a different order, e.g., [1,2,3] and [3,2,1], are unequal as lists, but equal as sets. The Haskell equality operator == gives the wrong results when we are interested in set equality. For instance, we get:

```
Prelude> [1,2,3] == [3,2,1]
False
Prelude>
```

```
module SetEq (Set(..),emptySet,isEmpty,inSet,subSet,insertSet,
              deleteSet,powerSet,takeSet,list2set,(!!!))

where

import List (delete)

infixl 9 !!!

newtype Set a = Set [a]

instance Eq a => Eq (Set a) where
  set1 == set2 = subSet set1 set2 && subSet set2 set1

instance (Show a) => Show (Set a) where
    showsPrec _ (Set s) str = showSet s str

showSet []       str = showString "{}" str
showSet (x:xs) str = showChar '{' (shows x (showl xs str))
     where showl []       str = showChar '}' str
           showl (x:xs) str = showChar ',' (shows x (showl xs str))

emptySet  :: Set a
emptySet = Set []

isEmpty  :: Set a -> Bool
isEmpty (Set []) = True
isEmpty _        = False

inSet  :: (Eq a) => a -> Set a -> Bool
inSet x (Set s) = elem x s

subSet :: (Eq a) => Set a -> Set a -> Bool
subSet (Set [])         _   = True
subSet (Set (x:xs)) set = (inSet x set) && subSet (Set xs) set

insertSet :: (Eq a) => a -> Set a -> Set a
insertSet x (Set ys) | inSet x (Set ys) = Set ys
                     | otherwise        = Set (x:ys)
```

Figure 4.7: A Module for Sets as Unordered Lists Without Duplicates.

4.9. A DATA TYPE FOR SETS

```
deleteSet :: Eq a => a -> Set a -> Set a
deleteSet x (Set xs) = Set (delete x xs)

list2set :: Eq a => [a] -> Set a
list2set [] = Set []
list2set (x:xs) = insertSet x (list2set xs)

powerSet :: Eq a => Set a -> Set (Set a)
powerSet (Set xs) = Set (map (\xs -> (Set xs)) (powerList xs))

powerList   :: [a] -> [[a]]
powerList   [] = [[]]
powerList   (x:xs) = (powerList xs) ++ (map (x:) (powerList xs))

takeSet :: Eq a => Int -> Set a -> Set a
takeSet n (Set xs) = Set (take n xs)

(!!!) :: Eq a => Set a -> Int -> a
(Set xs) !!! n = xs !! n
```

Figure 4.8: A Module for Sets as Unordered Lists Without Duplicates (ctd).

This can be remedied by defining a special data type for sets, with a matching definition of equality. All this is provided in the module *SetEq.hs* that is given in Figures 4.7 and 4.8. If a is an equality type, then Set a is the type of sets over a. The newtype declaration allows us to put Set a as a separate type in the Haskell type system.

Equality for Set a is defined in terms of the subSet relation:

```
instance Eq a => Eq (Set a) where
   set1 == set2 = subSet set1 set2 && subSet set2 set1
```

The instance declaration says that if a is in class Eq, then Set a is also in that class, with == defined as specified, in terms of the procedure subSet, that in turn is defined recursively in terms of inSet. This gives:

SetEq> Set [2,3,1] == Set [1,2,3]

```
True
SetEq> Set [2,3,3,1,1,1] == Set [1,2,3]
True
SetEq>
```

The module *SetEq.hs* gives some useful functions for the Set type. In the first place, it is convenient to be able to display sets in the usual notation.

```
instance (Show a) => Show (Set a) where
   showsPrec _ (Set s) = showSet s

showSet []     str = showString "{}" str
showSet (x:xs) str = showChar '{' ( shows x ( showl xs str))
    where showl []     str = showChar '}' str
          showl (x:xs) str = showChar ',' (shows x (showl xs str))
```

This gives:

```
SetEq> Set [1..10]
{1,2,3,4,5,6,7,8,9,10}
SetEq> powerSet (Set [1..3])
{{},{3},{2},{2,3},{1},{1,3},{1,2},{1,2,3}}
SetEq>
```

The empty set `emptySet` and the test `isEmpty` are implemented as you would expect. Useful functions for operating on sets are `insertSet` and `deleteSet`. The function `insertSet` is used to implement the translation function from lists to sets `list2set`. The function `powerSet` is implemented by lifting `powerList` to the type of sets.

Exercise 4.54 Give implementations of the operations

   `unionSet, intersectSet` and `differenceSet`,

in terms of `inSet`, `insertSet` and `deleteSet`.

Exercise 4.55 In an implementation of sets as lists without duplicates, the implementation of `insertSet` has to be changed. How?

4.9. A DATA TYPE FOR SETS

```
module Hierarchy where

import SetEq

data S = Void deriving (Eq,Show)
empty,v0,v1,v2,v3,v4,v5 :: Set S
empty = Set []
v0    = empty
v1    = powerSet v0
v2    = powerSet v1
v3    = powerSet v2
v4    = powerSet v3
v5    = powerSet v4
```

Figure 4.9: The First Five Levels of the Set Theoretic Universe.

```
Hierarchy> display 88 (take 1760 (show v5))
{{},{{{}},{{{}}},{{}},{{},{{}}}},{{{},{{{}}}},{{}}},{{{},{{{}}},{{}}},{{},{{{}}},{{}},{{
},{{{}}}},{{{}},{{{}}}},{{},{{{}}}},{{{},{{{}}},{{}}},{{},{{{}}},{{}},{{{}}},{{
},{{{}}}},{{},{{{}}},{{},{{{}}},{{}}},{{{},{{{}}},{{}},{{{}}},{{}},{{},{{{}}}}
,{{}},{{},{{{}}}},{{{},{{{}}}}},{{{},{{{}}}},{{},{{{}}},{{}}},{{},{{{}}},{{
},{{{}}},{{{}}},{{{},{{{}}}},{{},{{{}}},{{}},{{},{{{}}},{{}},{{},{{{}}}},{{
{},{{{}}},{{},{{{}}},{{{},{{{}}}},{{},{{{}}},{{}},{{{}}},{{{},{{{}}}}},{
},{{{}}}},{{},{{{}}},{{},{{{}}},{{},{{{}}},{{}}},{{{},{{{}}},{{}}},{{},{{{}}},{{
,{{{}}},{{}},{{},{{{}}}},{{}},{{},{{{}}}},{{{},{{}},{{},{{{}}}},{{{},{{}},{{},{{{}}}},{{
,{{{}}},{{}},{{},{{{}}}},{{{},{{}},{{},{{{}}}},{{{},{{}},{{},{{{}}}},{{{},{{}},{{},{{{}}}},{{
,{{{}}},{{}},{{},{{{}}}},{{}},{{},{{{}}}},{{{},{{}},{{},{{{}}}},{{{},{{}},{{},{{{}}}}},{{
{},{{}},{{},{{{}}}},{{},{{{}}},{{},{{{}}},{{},{{{}}},{{}},{{},{{{}}}},{{{},{{}},{{},{{{}}}
},{{},{{{}}}},{{},{{{}}}},{{},{{{}}},{{{}}},{{{},{{}},{{},{{{}}}},{{},{{{}}},{{},{{{}}},{{
,{{{}}},{{}},{{},{{{}}}},{{}},{{},{{{}}}},{{{},{{}},{{},{{{}}}},{{{}}},{{{},{{}},{{
},{{{}}}},{{}}},{{{},{{}},{{},{{{}}}},{{{},{{{}}}},{{{},{{{}}}},{{}}},{{{},{{{}}}},{{},{{
{}},{{}}},{{{},{{}},{{},{{{}}}},{{{},{{{}}}},{{{},{{{}}}},{{}}},{{{},{{{}}},{{},{{{}}}}
},{{{},{{},{{},{{{}}}},{{{},{{{}}}},{{{},{{{}}}},{{},{{{}}}},{{{},{{}},{{},{{{}}}},{{{}
}},{{},{{{}}}},{{{}}},{{},{{{}}}},{{},{{},{{{}}}},{{{},{{}}},{{{},{{}}},{{},{{{}}},{{
},{{{}}}},{{},{{{}}},{{{},{{{}}},{{{}}},{{{},{{{}}}},{{{},{{}}}},{{{}}},{{{},{{{}}},{{
}}}},{{},{{{}}}},{{{}}},{{},{{{}}}},{{}},{{},{{{}}}},{{{},{{}}}},{{{},{{}}}},{{},{{{}}}},{{}}
,{{},{{{}}}}}},{{{},{{}}}},{{},{{{}}}},{{{}}}},{{{},{{}}}},{{},{{{}}}},{{{}}}},{{},{{
```

Figure 4.10: An initial segment of $\wp^5(\emptyset)$.

Figure 4.9 gives a module for generating the first five levels of the set theoretic hierarchy: $V_0 = \emptyset$, $V_1 = \wp(\emptyset)$, $V_2 = \wp^2(\emptyset)$, $V_3 = \wp^3(\emptyset)$, $V_4 = \wp^4(\emptyset)$, $V_5 = \wp^5(\emptyset)$. Displaying V_5 in full takes some time, but here are the first few items. This uses the operator !!!, defined in the SetEq module as a left-associative infix operator by means of the reserved keyword infixl (the keyword infixr can be used to declare right associative infix operators).

```
Hierarchy> v5 !!! 0
{}
Hierarchy> v5 !!! 1
{{{},{{{}}},{{}},{{},{{}}}}}
Hierarchy> v5 !!! 2
{{{},{{{}}},{{}}}}
Hierarchy> v5 !!! 3
{{{},{{{}}},{{}}},{{},{{{}}},{{}},{{},{{}}}}}
Hierarchy>
```

Figure 4.10 displays a bit more of the initial segment of $\wp^5(\emptyset)$, with the help of the following function:

```
display :: Int -> String -> IO ()
display n str = putStrLn (display' n 0 str)
  where
  display' _ _ [] = []
  display' n m (x:xs) | n == m    = '\n': display' n 0 (x:xs)
                      | otherwise = x : display' n (m+1) xs
```

Exercise 4.56 What would have to change in the module *SetEq.hs* to get a representation of the empty set as 0?

Exercise 4.57*

1. How many pairs of curly braces { } occur in the expanded notation for $\wp^5(\emptyset)$, in the representation where $\emptyset$ appears as {}?

2. How many copies of 0 occur in the expanded notation for $\wp^5(\emptyset)$, in the representation where $\emptyset$ appears as 0 (Exercise 4.56)?

3. How many pairs of curly braces occur in the expanded notation for $\wp^5(\emptyset)$, in the representation where $\emptyset$ appears as 0?

4.10 Further Reading

Russell's paradox, as stated by Russell himself, can be found in [Rus67]. A further introduction to sets and set theory is Doets, Van Dalen and De Swart [DvDdS78]. A good introduction to type theory is given in [Hin97]. There are many books on database query an database design; a lucid introduction is [SKS01]. SQL (Standard Query Language) is explained in the documentation of all state of the art relational databases. Implementations of the relational database model are freely available on the Internet. See, e.g., www.postgresql.org or www.mysql.com. Logical foundations of database theory are given in [AHV95].

Chapter 5

Relations

Preview

The first section of this chapter explains the abstract notion of a relation. Section 5.2 discusses various properties of relations. Next, in Sections 5.3 and 5.4, we discuss various possible implementations of relations and relation processing. Sections 5.5 and 5.6 focus on an often occurring type of relation: the *equivalence*.

The following declaration turns the code in this chapter into a module that loads the List module (to be found in same directory as *Prelude.hs*, under the name *List.hs*), and the SetOrd module (see below, in Figures 5.4 and 5.5).

```
module REL

where

import List
import SetOrd
```

5.1 Relations as Sets of Ordered Pairs

Although you probably will not be able to explain the general notion of a relation, you are definitely familiar with a couple of instances, such as the usual ordering relation $<$ between natural numbers, or the subset relation $\subseteq$ between sets. Non-mathematical examples are the different family relationships that exist between humans. For instance, the *father-of* relation holds between two people if the first one is the father of the second.

For every two numbers $n, m \in \mathbb{N}$, the statement $n < m$ is either true or false. E.g., $3 < 5$ is true, whereas $5 < 2$ is false. In general: to a relation you can "input" a pair of objects, after which it "outputs" either *true* or *false*. depending on whether these objects are in the relationship given.

In set theory, there is a clever way to reduce the notion of a relation to that of a set. Consider again the ordering $\leqslant$ on $\mathbb{N}$. With it, associate the set R of ordered pairs (n, m) of natural numbers for which $n \leqslant m$ is true:

$$R_{\leqslant} = \{\, (n, m) \in \mathbb{N}^2 \mid n \leqslant m \,\}.$$

Note that a statement $n \leqslant m$ now has become tantamount with the condition, that $(n, m) \in R_{\leqslant}$. Thus, $(3, 5) \in R_{\leqslant}$, and $(5, 2) \notin R_{\leqslant}$.

This connection can be converted into a definition. That is, the ordering relation $\leqslant$ of $\mathbb{N}$ is *identified* with the set $R_{\leqslant}$. The following definition puts it bluntly.

Definition 5.1 (Relations, Domain, Range) A *relation* is a set of ordered pairs.

Instead of $(x, y) \in R$ — where R is a relation — one usually writes xRy, or $R(x, y)$, or Rxy.

The set $\text{dom}(R) = \{x \mid \exists y \, (\, xRy \,)\}$, i.e., the set consisting of all first coordinates of pairs in R, is called the *domain* of R and $\text{ran}(R) = \{y \mid \exists x \, (\, xRy \,)\}$, the set of second coordinates of pairs in R, its *range*.

Example 5.2 If A and B are sets, then $A \times B$ is a relation. The empty set $\emptyset$ trivially is a relation (for, *all* its members are ordered pairs).

$\text{dom}(\emptyset) = \text{ran}(\emptyset) = \emptyset$, $\text{dom}(A \times B) = A$ (provided B is non-empty: $A \times \emptyset = \emptyset$, thus $\text{dom}(A \times \emptyset) = \emptyset$), and $\text{ran}(A \times B) = B$ (analogously: provided A is non-empty).

Definition 5.3 (From ... to, Between, On) The relation R is a relation *from* A *to* B or *between* A and B, if $\text{dom}(R) \subseteq A$ and $\text{ran}(R) \subseteq B$.

A relation from A to A is called *on* A.

If R is a relation on A, then A is called the *underlying set* (of the structure that consists of the domain A and the relation R).

5.1. RELATIONS AS SETS OF ORDERED PAIRS

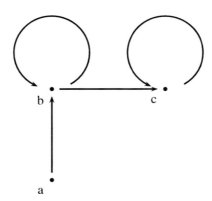

Figure 5.1: Picture of the relation $\{(a,b),(b,b),(b,c),(c,c)\}$.

Relations on a finite set A can be pictured by drawing the elements of A as dots and putting an arrow $\to$ between a and b if the pair (a,b) is in the relation. See Figure 5.1.

Example 5.4 $R = \{(1,4),(1,5),(2,5)\}$ is a relation from $\{1,2,3\}$ to $\{4,5,6\}$, and it also is a relation on $\{1,2,4,5,6\}$. Furthermore, $\text{dom}(R) = \{1,2\}$, $\text{ran}(R) = \{4,5\}$.

Example 5.5 If A is a set, then $\subseteq$ is a relation on $\wp(A)$.

Example 5.6 The relations $\leq$ and $\geq$ on $\mathbb{R}$ are subsets of the real plane $\mathbb{R}^2$, and can be pictured as in Figure 5.2.

Definition 5.7 (Identity and Inverse)

1. $\Delta_A = \{(a,b) \in A^2 \mid a = b\} = \{(a,a) \mid a \in A\}$ is a relation on A, the *identity on* A.

2. If R is a relation between A and B, then $R^{-1} = \{(b,a) \mid aRb\}$, the *inverse* of R, is a relation between B and A.

Example 5.8 The inverse of the relation 'parent of' is the relation 'child of'.

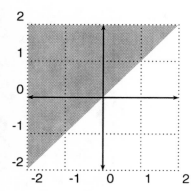 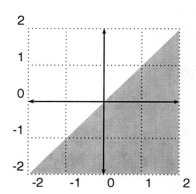

Figure 5.2: The relations $\leqslant$ and $\geqslant$ on $\mathbb{R}^2$.

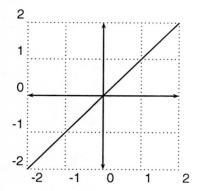 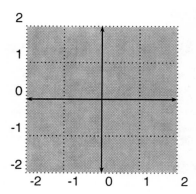

Figure 5.3: The relations $\Delta_\mathbb{R}$ and $\mathbb{R} \times \mathbb{R}$ on $\mathbb{R}^2$.

5.1. RELATIONS AS SETS OF ORDERED PAIRS

Example 5.9

1. $A \times B$ is the biggest relation from A to B.
2. $\emptyset$ is the smallest relation from A to B.
3. For the usual ordering $<$ of $\mathbb{R}$, $<^{-1} = >$.
4. $(R^{-1})^{-1} = R$; $\Delta_A^{-1} = \Delta_A$; $\emptyset^{-1} = \emptyset$ and $(A \times B)^{-1} = B \times A$.

In practice, you define a relation by means of a condition on the elements of ordered pairs. This is completely analogous to the way you define a set by giving a condition its elements should fulfill.

Example 5.10 If R is the set of pairs (n, m) $(n, m \in \mathbb{Z})$ such that (*condition*) n^2 is a divisor of m, the definition of R may look as follows:

For $n, m \in \mathbb{Z}$: $nRm :\equiv n^2 | m$.

In Haskell this is implemented as: \ n m -> rem m n^2 == 0.

Example 5.11 For all $n \in \mathbb{N}$, the set of pairs $\{(a, b) \mid a, b \in \mathbb{N}, ab = n, a \leqslant b\}$ is a relation on $\mathbb{N}$. This relation gives all the divisor pairs of n. Here is its implementation:

```
divisors :: Integer -> [(Integer,Integer)]
divisors n = [ (d, quot n d) | d <- [1..k], rem n d == 0 ]
   where k = floor (sqrt (fromInteger n))
```

Example 5.12 We can use the relation divisors of the previous example for yet another implementation of the primality test:

```
prime'' :: Integer -> Bool
prime'' = \n -> divisors n == [(1,n)]
```

Also, here are the list of divisors of a natural number, the list of all proper divisors of a natural number, and a test for being a perfect natural number:

```
divs :: Integer -> [Integer]
divs n = (fst list) ++ reverse (snd list)
   where list = unzip (divisors n)

properDivs :: Integer -> [Integer]
properDivs n = init (divs n)

perfect :: Integer -> Bool
perfect n = sum (properDivs n) == n
```

'**There is a relation between** a **and** b.' In daily life, a statement like 'there is a relation between a and b' (or: 'a and b are having a relation') may not sound unusual, but in the present context this should be avoided. Not because these are false statements, but because they are so overly true, and, therefore, uninformative. Cf. Exercise 5.13. Of course, a statement like 'a stands in relation R to b' or 'the relation R subsists between a and b' can very well be informative, when R is some specific relation.

Exercise 5.13 Show that $\forall x \, \forall y \, \exists R \, (xRy)$. ("Between every two things there exist some relation.")

In everyday situations, by saying that a and b are related we usually mean more than that it is possible to form the ordered pair of a and b. We usually mean that there is some *specific* link between a and b (for instance, you can make Mrs. a blush by mentioning Mr. b in conversation).

5.2 Properties of Relations

In this section we list some useful properties of relations on a set A. Further on, in Sections 5.3 and 5.4, we will illustrate how tests for these properties can be implemented.

A relation R is **reflexive** on A if for every $x \in A$: xRx.

Example 5.14 On any set A, the relation Δ_A is reflexive. Note that Δ_A is the smallest reflexive relation on A: it is a subset of any reflexive relation on A. In other words, a relation R is reflexive on A iff $\Delta_A \subseteq R$.

Example 5.15 The relation $\leq$ on $\mathbb{N}$ is reflexive (for every number is less than or equal to itself).

5.2. PROPERTIES OF RELATIONS

A relation R on A is **irreflexive** if for no $x \in A$: xRx.

Example 5.16 The relation $<$ on $\mathbb{N}$ is irreflexive.

Exercise 5.17 Show that a relation R on A is irreflexive iff $\Delta_A \cap R = \emptyset$.

There are relations which are neither reflexive nor irreflexive (the reader is urged to think up an example).

A relation R on A is **symmetric** if for all $x, y \in A$: if xRy then yRx.

Example 5.18 The relation 'having the same age' between people is symmetric. Unfortunately, the relation 'being in love with' between people is not symmetric.

Exercise 5.19 Show the following:

1. A relation R on a set A is symmetric iff $\forall x, y \in A (xRy \Leftrightarrow yRx)$.

2. A relation R is symmetric iff $R \subseteq R^{-1}$, iff $R = R^{-1}$.

A relation R on A is **asymmetric** if for all $x, y \in A$: if xRy then not yRx.

The relation $<$ on $\mathbb{N}$ is asymmetric. It is immediate from the definition that a relation R on A is asymmetric iff $R \cap R^{-1} = \emptyset$. Note that there are relations which are neither symmetric nor asymmetric.

Exercise 5.20 Show that every asymmetric relation is irreflexive.

A relation R on A is **antisymmetric** if for all $x, y \in A$: if xRy and yRx then $x = y$.

Example 5.21 The relation $m|n$ (m is a divisor of n) on $\mathbb{N}$ is antisymmetric. If m is a divisor of n and n is a divisor of m, then m and n are equal.

The relation in Example 5.15 is antisymmetric.

Exercise 5.22 Show from the definitions that an asymmetric relation always is antisymmetric.

The converse of the statement in Exercise 5.22 is not true: the relation $\leqslant$ on $\mathbb{N}$ provides a counterexample.

A relation R on A is **transitive** if for all $x, y, z \in A$: if xRy and yRz then xRz.

Examples of transitive relations are $<$ and $\leqslant$ on $\mathbb{N}$. If F is the relation 'friendship' on a set of people E, and every member of E endorses the laudable principle "The friends of my friends are my friends", then F is transitive.

Exercise 5.23 Show that a relation R on a set A is transitive iff

$$\forall x, z \in A (\exists y \in A (xRy \wedge yRz) \Rightarrow xRz).$$

A relation R on A is **intransitive** if for all $x, y, z \in A$: if xRy and yRz then not xRz.

Example 5.24 The relation 'father of' on the set of all human beings is intransitive.

Again: there are relations that are neither transitive nor intransitive (think up an example).

A relation R on A is a **pre-order** (or **quasi-order**) if R is transitive and reflexive.

Example 5.25 Let L be the set of all propositional formulas built from a given set of atomic propositions. Then the relation $\models$ given by

$$P \models Q \text{ iff } P \Rightarrow Q \text{ is logically valid}$$

is a pre-order on L. To check this, note that for every propositional formula P, $P \Rightarrow P$ is logically valid, so $\models$ is reflexive. Also, for all propositional formulas P, Q, R, if $P \Rightarrow Q$ and $Q \Rightarrow R$ are logically valid, then $P \Rightarrow R$ is logically valid. This can be checked as follows, by contraposition. Suppose $P \Rightarrow R$ is not logically valid. Then there is a valuation for the atomic propositions that makes P true and R false. Now there are two possibilities: this valuation makes Q true or it makes Q false. If it makes Q true, then we know that $Q \Rightarrow R$ is not logically valid. If it makes Q false, then we know that $P \Rightarrow Q$ is not logically valid. So either $P \Rightarrow Q$ or $Q \Rightarrow R$ is not logically valid. This shows that $\models$ is transitive.

Example 5.26 Let L be the set of all propositional formulas built from a given set of atomic propositions. Then the relation $\vdash$ on L given by $P \vdash Q$ iff there is a proof of Q from the given P (this relation was defined in Chapter 3) is a pre-order. See Example 3.1 for the transitivity of $\vdash$. In fact, the relation $\vdash$ coincides with the relation $\models$ of the previous example. The safety checks on the proof rules for the logical connectives that were performed in Chapter 3 guarantee that $\vdash \subseteq \models$. This fact is called the

5.2. PROPERTIES OF RELATIONS

soundness of the proof system. The inclusion in the other direction, $\models \,\subseteq\, \vdash$, is called the *completeness* of the proof system. It is outside the scope of this book, but it is proved in [EFT94].

A relation R on A is a **strict partial order** if R is transitive and irreflexive. A relation R on A is a **partial order** if R is transitive, reflexive and antisymmetric.

Note that the relation $\models$ from Example 5.25 is not antisymmetric. Take the formulas $P \wedge Q$ and $Q \wedge P$. Then we have $P \wedge Q \models Q \wedge P$ and $Q \wedge P \models P \wedge Q$, but $P \wedge Q$ and $Q \wedge P$ are different formulas. Thus, $\models$ is not a partial order.

Example 5.27 The relation $<$ on $\mathbb{N}$ is a strict partial order and the relation $\leqslant$ on $\mathbb{N}$ is a partial order. Theorem 4.1 states that for any set A, the relation $\subseteq$ on $\wp(A)$ is a partial order.

Exercise 5.28 Show that every strict partial order is asymmetric.

Exercise 5.29 Show that every relation which is transitive and asymmetric is a strict partial order.

Exercise 5.30 Show that if R is a strict partial order on A, then $R \cup \Delta_A$ is a partial order on A. (So every strict partial order is contained in a partial order.)

Exercise 5.31 Show that the inverse of a partial order is again a partial order.

A relation R on A is **linear** (or: has the **comparison property**) if for all $x, y \in A$: xRy or yRx or $x = y$. A partial order that is also linear is called a **total order**. All Haskell types in class Ord a are total orders. A set A with a total order on it is called a *chain*.

Exercise 5.32 Let S be a reflexive and symmetric relation on a set A.

A *path* is a finite sequence $a_1, \ldots, a_n$ of elements of A such that for every i, $1 \leqslant i < n$, we have that $a_i S a_{i+1}$. Such a path *connects* a_1 with a_n.

Assume that for all $a, b \in A$ there is exactly one path connecting a with b.

Fix $r \in A$. Define the relation $\leqslant$ on A by: $a \leqslant b$ iff a is one of the elements in the path connecting r with b.

Show the following:

1. $\leqslant$ is reflexive,

2. $\leqslant$ is antisymmetric.

3. $\leqslant$ is transitive,

4. for all $a \in A$, $r \leqslant a$,

5. for every $a \in A$, the set $X_a = \{x \in A \mid x \leqslant a\}$ is finite and if $b, c \in X_a$ then $b \leqslant c$ or $c \leqslant b$.

(A structure $(A, \leqslant, r)$ with these five properties is called a *tree* with *root* r. The directory-structure of a computer account is an example of a tree. Another example is the structure tree of a formula (see Section 2.3 above). These are examples of finite trees, but if A is infinite, then the tree $(A, \leqslant, r)$ will have at least one infinite branch.)

Here is a list of all the relational properties of binary relations on a set A that we discussed:

reflexivity	$\forall x\ xRx$.
irreflexivity	$\forall x\ \neg xRx$.
symmetry	$\forall xy\ (xRy \Rightarrow yRx)$.
asymmetry	$\forall xy\ (xRy \Rightarrow \neg yRx)$.
antisymmetry	$\forall xy\ (xRy \wedge yRx \Rightarrow x = y)$.
transitivity	$\forall xyz\ (xRy \wedge yRz \Rightarrow xRz)$.
intransitivity	$\forall xyz\ (xRy \wedge yRz \Rightarrow \neg xRz)$.
linearity	$\forall xy\ (xRy \vee yRx \vee x = y)$.

Because of their importance, relations that are transitive, reflexive and symmetric (equivalence relations) deserve a section of their own (see Section 5.5).

	irrefl	refl	asymm	antisymm	symm	trans	linear
pre-order		✓				✓	
strict partial order	✓		✓	✓		✓	
partial order		✓		✓		✓	
total order		✓		✓		✓	✓
equivalence		✓			✓	✓	

Exercise 5.33 Consider the following relations on the natural numbers. Check their properties (some answers can be found in the text above). The *successor* relation is the relation given by $\{(n, m) \mid n + 1 = m\}$. The *divisor* relation is $\{(n, m) \mid n \text{ divides } m\}$. The *coprime* relation C on $\mathbb{N}$ is given by $nCm :\equiv \text{GCD}(n, m) = 1$, i.e., the only factor of n that divides m is 1, and vice versa (see Section 8.2).

5.2. PROPERTIES OF RELATIONS

	<	⩽	successor	divisor	coprime
irreflexive					
reflexive					
asymmetric					
antisymmetric					
symmetric					
transitive					
linear					

Definition 5.34 If $\mathcal{O}$ is a set of properties of relations on a set A, then the $\mathcal{O}$-closure of a relation R is the smallest relation S that includes R and that has all the properties in $\mathcal{O}$.

The most important closures are the *reflexive closure*, the *symmetric closure*, the *transitive closure* and the *reflexive transitive closure* of a relation.

Remark. To show that R is the smallest relation S that has all the properties in $\mathcal{O}$, show the following:

1. R has all the properties in $\mathcal{O}$,
2. If S has all the properties in $\mathcal{O}$, then $R \subseteq S$.

∎

Exercise 5.35 Suppose that R is a relation on A.

1. Show that $R \cup \Delta_A$ is the reflexive closure of R.
2. Show that $R \cup R^{-1}$ is the symmetric closure of R.

Exercise 5.36 Let R be a transitive binary relation on A. Does it follow from the transitivity of R that its symmetric reflexive closure $R \cup R^{-1} \cup \Delta_A$ is also transitive? Give a proof if your answer is yes, a counterexample otherwise.

To define the transitive and reflexive transitive closures of a relation we need the concept of relation composition.

Composing Relations. Suppose that R and S are relations on A. The *composition* $R \circ S$ of R and S is the relation on A that is defined by

$$x(R \circ S)z :\equiv \exists y \in A(xRy \wedge ySz).$$

Furthermore, for $n \in \mathbb{N}, n \geqslant 1$ we define R^n by means of $R^1 := R$, $R^{n+1} := R^n \circ R$.

Example 5.37 In imperative programming, the meaning of a command can be viewed as a relation on a set of machine states, where a machine state is a set of pairs consisting of machine registers with their contents. If

$$s = \{(r_1, 4), (r_2, 6), \ldots\},$$

then executing $r_2 := r_1$ in state s gives a new state

$$s' = \{(r_1, 4), (r_2, 4), \ldots\}.$$

If C_1 and C_2 are commands, then we can execute them in sequential order as $C_1; C_2$ or as $C_2; C_1$. If the meaning of command C_1 is R and the meaning of command C_2 is S, then the meaning of $C_1; C_2$ is $R \circ S$ and the meaning of $C_2; C_1$ is $S \circ R$.

Exercise 5.38 Determine the composition of the relation "father of" with itself. Determine the composition of the relations "brother of" and "parent of" Give an example showing that $R \circ S = S \circ R$ can be false.

Exercise 5.39 Consider the relation

$$R = \{(0, 2), (0, 3), (1, 0), (1, 3), (2, 0), (2, 3)\}$$

on the set $A = \{0, 1, 2, 3, 4\}$.

1. Determine R^2, R^3 and R^4.

2. Give a relation S on A such that $R \cup (S \circ R) = S$.

Exercise 5.40 Verify:

1. A relation R on A is transitive iff $R \circ R \subseteq R$.

2. Give an example of a transitive relation R for which $R \circ R = R$ is false.

Exercise 5.41 Verify:

1. $Q \circ (R \circ S) = (Q \circ R) \circ S$. (Thus, the notation $Q \circ R \circ S$ is unambiguous.)

2. $(R \circ S)^{-1} = S^{-1} \circ R^{-1}$.

5.2. PROPERTIES OF RELATIONS

We will show that for any relation R on A, the relation $R^+ = \bigcup_{n \geq 1} R^n$ is the transitive closure of R.

Because $R = R^1 \subseteq \bigcup_{n \geq 1} R^n$, we know that $R \subseteq R^+$. We still have to prove that R^+ is transitive, and moreover that it is the *smallest* transitive relation that contains R.

To see that R^+ is transitive, assume xR^+y and yR^+z. From xR^+y we get that there is some $k \geq 1$ with $xR^k y$. From yR^+z we get that there is some $m \geq 1$ with $yR^m z$. It follows that $xR^{k+m} z$, and therefore xR^+z, proving that R^+ is transitive.

Proposition 5.42 R^+ is the *smallest* transitive relation that contains R.

Proof. The proposition can be restated as follows. If T is a transitive relation on A such that $R \subseteq T$, then $R^+ \subseteq T$. We prove this fact by induction (see Chapter 7).

Basis If xRy then it follows from $R \subseteq T$ that xTy.

Induction step Assume (induction hypothesis) that $R^n \subseteq T$. We have to show that $R^{n+1} \subseteq T$. Consider a pair (x, y) with $xR^{n+1}y$. Then, by the definition of R^{n+1}, there is a z such that $xR^n z$ and zRy. Because $R \subseteq T$ we have zTy, and the induction hypothesis yields xTz. By transitivity of T it now follows that xTy.

∎

Example 5.43 If $A = \{a, \{b, \{c\}\}\}$, then

$$a \in^+ A, \quad \{b, \{c\}\} \in^+ A, \quad b \in^+ A, \quad \{c\} \in^+ A \text{ and } c \in^+ A.$$

Example 5.44 On the set of human beings, the transitive closure of the 'parent of' relation is the 'ancestor of' relation, and the transitive closure of the 'child of' relation is the 'descendant of' relation.

Exercise 5.45 Show that the relation $<$ on $\mathbb{N}$ is the transitive closure of the relation $R = \{(n, n+1) \mid n \in \mathbb{N}\}$.

Exercise 5.46 Let R be a relation on A. Show that $R^+ \cup \Delta_A$ is the reflexive transitive closure of R.

The reflexive transitive closure of a relation R is often called the *ancestral* of R, notation R^*. Note that R^* is a pre-order.

Exercise 5.47 Give the reflexive transitive closure of the following relation:

$$R = \{(n, n+1) \mid n \in \mathbb{N}\}.$$

Exercise 5.48*

1. Show that an intersection of *arbitrarily many* transitive relations is transitive.

2. Suppose that R is a relation on A. Note that A^2 is one example of a transitive relation on A that extends R. Conclude that the intersection of *all* transitive relations extending R is the *least* transitive relation extending R. In other words, R^+ equals the intersection of *all* transitive relations extending R.

Exercise 5.49*

1. Show that $(R^*)^{-1} = (R^{-1})^*$.

2. Show by means of a counter-example that $(R \cup R^{-1})^* = R^* \cup R^{-1*}$ may be false.

3. Prove: if $S \circ R \subseteq R \circ S$, then $(R \circ S)^* \subseteq R^* \circ S^*$.

Exercise 5.50 Suppose that R and S are reflexive relations on the set A. Then $\Delta_A \subseteq R$ and $\Delta_A \subseteq S$, so $\Delta_A \subseteq R \cap S$, i.e., $R \cap S$ is reflexive as well. We say: reflexivity is preserved under intersection. Similarly, if R and S are reflexive, then $\Delta_A \subseteq R \cup S$, so $R \cup S$ is reflexive. Reflexivity is preserved under union. If R is reflexive, then $\Delta_A \subseteq R^{-1}$, so R^{-1} is reflexive. Reflexivity is preserved under inverse. If R and S are reflexive, $\Delta_A \subseteq R \circ S$, so $R \circ S$ is reflexive. Reflexivity is preserved under composition. Finally, if R on A is reflexive, then the complement of R, i.e., the relation $A^2 - R$, is irreflexive. So reflexivity is not preserved under complement. These closure properties of reflexive relations are listed in the table below.

We can ask the same questions for other relational properties. Suppose R and S are symmetric relations on A. Does it follow that $R \cap S$ is symmetric? That $R \cup S$ is symmetric? That R^{-1} is symmetric? That $A^2 - R$ is symmetric? That $R \circ S$ is symmetric? Similarly for the property of transitivity. These questions are summarized in the table below. Complete the table by putting 'yes' or 'no' in the appropriate places.

5.3. IMPLEMENTING RELATIONS AS SETS OF PAIRS

property	*reflexivity*	*symmetry*	*transitivity*
preserved under ∩?	yes	?	?
preserved under ∪?	yes	?	?
preserved under inverse?	yes	?	?
preserved under complement?	no	?	?
preserved under composition?	yes	?	?

5.3 Implementing Relations as Sets of Pairs

Our point of departure is a slight variation on the module *SetEq.hs* of the previous Chapter. This time, we represent sets a *ordered* lists without duplicates. See Figs. 5.4 and 5.5 for a definition of the module *SetOrd.hs*.

The definition of `deleteList` inside Figure 5.4 employs a Haskell feature that we haven't encountered before: `ys@(y:ys')` is a notation that allows us to refer to the non-empty list `(y:ys')` by means of `ys`. If the item to be deleted is greater then the first element of the list then the instruction is to do nothing. Doing nothing boils down to returning the whole list `ys`. This possibility to give a name to a pattern is just used for readability.

Next we define relations over a type a as sets of pairs of that type, i.e., `Rel a` is defined and implemented as `Set(a,a)`.

```
type Rel a = Set (a,a)
```

`domR` gives the domain of a relation.

```
domR :: Ord a => Rel a -> Set a
domR (Set r) = list2set [ x | (x,_) <- r ]
```

`ranR` gives the range of a relation.

```
ranR :: Ord a => Rel a -> Set a
ranR (Set r) = list2set [ y | (_,y) <- r ]
```

`idR` creates the identity relation Δ_A over a set A:

```
module SetOrd (Set(..),emptySet,isEmpty,inSet,subSet,insertSet,
               deleteSet,powerSet,takeSet,(!!!),list2set)

where

import List (sort)

{-- Sets implemented as ordered lists without duplicates --}

newtype Set a = Set [a] deriving (Eq,Ord)

instance (Show a) => Show (Set a) where
    showsPrec _ (Set s) str = showSet s str

showSet []     str = showString "{}" str
showSet (x:xs) str = showChar '{' ( shows x ( showl xs str))
     where showl []     str = showChar '}' str
           showl (x:xs) str = showChar ',' (shows x (showl xs str))

emptySet   :: Set a
emptySet = Set []

isEmpty  :: Set a -> Bool
isEmpty (Set []) = True
isEmpty _        = False

inSet  :: (Ord a) => a -> Set a -> Bool
inSet x (Set s) = elem x (takeWhile (<= x) s)

subSet :: (Ord a) => Set a -> Set a -> Bool
subSet (Set []) _         = True
subSet (Set (x:xs)) set = (inSet x set) && subSet (Set xs) set

insertSet :: (Ord a) => a -> Set a -> Set a
insertSet x (Set s) = Set (insertList x s)
```

Figure 5.4: A Module for Sets as Ordered Lists Without Duplicates.

```
insertList x [] = [x]
insertList x ys@(y:ys') = case compare x y of
                                GT -> y : insertList x ys'
                                EQ -> ys
                                _  -> x : ys

deleteSet :: Ord a => a -> Set a -> Set a
deleteSet x (Set s) = Set (deleteList x s)

deleteList x [] = []
deleteList x ys@(y:ys') = case compare x y of
                                GT -> y : deleteList x ys'
                                EQ -> ys'
                                _  -> ys

list2set :: Ord a => [a] -> Set a
list2set [] = Set []
list2set (x:xs) = insertSet x (list2set xs)
-- list2set xs = Set (foldr insertList [] xs)

powerSet :: Ord a => Set a -> Set (Set a)
powerSet (Set xs) =
   Set (sort (map (\xs -> (list2set xs)) (powerList xs)))

powerList :: [a] -> [[a]]
powerList [] = [[]]
powerList (x:xs) = (powerList xs)
                    ++ (map (x:) (powerList xs))

takeSet :: Eq a => Int -> Set a -> Set a
takeSet n (Set xs) = Set (take n xs)

infixl 9 !!!

(!!!) :: Eq a => Set a -> Int -> a
(Set xs) !!! n = xs !! n
```

Figure 5.5: A Module for Sets as Ordered Lists Without Duplicates (ctd).

```
idR :: Ord a => Set a -> Rel a
idR (Set xs) = Set [(x,x) | x <- xs]
```

The total relation over a set is given by:

```
totalR :: Set a -> Rel a
totalR (Set xs) = Set [(x,y) | x <- xs, y <- xs ]
```

invR inverts a relation (i.e., the function maps R to R^{-1}).

```
invR :: Ord a => Rel a -> Rel a
invR (Set []) = (Set [])
invR (Set ((x,y):r)) = insertSet (y,x) (invR (Set r))
```

inR checks whether a pair is in a relation.

```
inR :: Ord a => Rel a -> (a,a) -> Bool
inR r (x,y) = inSet (x,y) r
```

The complement of a relation $R \subseteq A \times A$ is the relation $A \times A - R$. The operation of relational complementation, relative to a set A, can be implemented as follows:

```
complR :: Ord a => Set a -> Rel a -> Rel a
complR (Set xs) r =
   Set [(x,y) | x <- xs, y <- xs, not (inR r (x,y))]
```

A check for reflexivity of R on a set A can be implemented by testing whether $\Delta_A \subseteq R$:

5.3. IMPLEMENTING RELATIONS AS SETS OF PAIRS

```
reflR :: Ord a => Set a -> Rel a -> Bool
reflR set r = subSet (idR set) r
```

A check for irreflexivity of R on A proceeds by testing whether $\Delta_A \cap R = \emptyset$:

```
irreflR :: Ord a => Set a -> Rel a -> Bool
irreflR (Set xs) r =
   all (\ pair -> not (inR r pair)) [(x,x) | x <- xs]
```

A check for symmetry of R proceeds by testing for each pair $(x, y) \in R$ whether $(y, x) \in R$:

```
symR :: Ord a => Rel a -> Bool
symR (Set []) = True
symR (Set ((x,y):pairs)) | x == y = symR (Set pairs)
                         | otherwise =
                           inSet (y,x) (Set pairs)
                           && symR (deleteSet (y,x) (Set pairs))
```

A check for transitivity of R tests for each couple of pairs $(x, y) \in R, (u, v) \in R$ whether $(x, v) \in R$ if $y = u$:

```
transR :: Ord a => Rel a -> Bool
transR (Set []) = True
transR (Set s) = and [ trans pair (Set s) | pair <- s ] where
      trans (x,y) (Set r) =
      and [ inSet (x,v) (Set r) | (u,v) <- r, u == y ]
```

Now what about relation composition? This is a more difficult matter, for how do we implement $\exists z(Rxz \land Szy)$? The key to the implementation is the following procedure for composing a single pair of objects (x, y) with a relation S, simply by forming the relation $\{(x, z) \mid (z, y) \in S\}$. This is done by:

```
composePair :: Ord a => (a,a) -> Rel a -> Rel a
composePair (x,y) (Set []) = Set []
composePair (x,y) (Set ((u,v):s))
   | y == u    = insertSet (x,v) (composePair (x,y) (Set s))
   | otherwise = composePair (x,y) (Set s)
```

For relation composition we need set union (Cf. Exercise 4.54):

```
unionSet :: (Ord a) => Set a -> Set a -> Set a
unionSet (Set [])      set2  =  set2
unionSet (Set (x:xs)) set2  =
   insertSet x (unionSet (Set xs) (deleteSet x set2))
```

Relation composition is defined in terms of composePair and unionSet:

```
compR :: Ord a => Rel a -> Rel a -> Rel a
compR (Set []) _ = (Set [])
compR (Set ((x,y):s)) r =
   unionSet (composePair (x,y) r) (compR (Set s) r)
```

Composition of a relation with itself (R^n):

```
repeatR :: Ord a => Rel a -> Int -> Rel a
repeatR r n | n < 1     = error "argument < 1"
            | n == 1    = r
            | otherwise = compR r (repeatR r (n-1))
```

Example 5.51 Let us use the implementation to illustrate Exercise 5.39.

```
r  = Set [(0,2),(0,3),(1,0),(1,3),(2,0),(2,3)]
r2 = compR r r
r3 = repeatR r 3
r4 = repeatR r 4
```

5.3. IMPLEMENTING RELATIONS AS SETS OF PAIRS

This gives:

```
REL> r
{(0,2),(0,3),(1,0),(1,3),(2,0),(2,3)}
REL> r2
{(0,0),(0,3),(1,2),(1,3),(2,2),(2,3)}
REL> r3
{(0,2),(0,3),(1,0),(1,3),(2,0),(2,3)}
REL> r4
{(0,0),(0,3),(1,2),(1,3),(2,2),(2,3)}
REL> r == r2
False
REL> r == r3
True
REL> r == r4
False
REL> r2 == r4
True
```

Also, the following test yields 'True':

```
s = Set [(0,0),(0,2),(0,3),(1,0),(1,2),(1,3),(2,0),(2,2),(2,3)]
test = (unionSet r (compR s r)) == s
```

Exercise 5.52 Extend this implementation with a function

```
restrictR :: Ord a => Set a -> Rel a -> Rel a
```

that gives the restriction of a relation to a set. In the type declaration, Set a is the restricting set.

The union of two relations R and S is the relation $R \cup S$. Since relations are sets, for this we can use unionSet.

Exercise 5.53 Use unionSet to define procedures rclosR for reflexive closure and sclosR for symmetric closure of a relation. As background set for the reflexive closure you can take the union of domain and range of the relation.

Exercise 5.54 Define a function

```
tclosR :: Ord a => Rel a -> Rel a
```

to compute the transitive closure of a relation R, for relations implemented as `Ord a => Rel a`.
Hint: compute the smallest relation S with the property that

$$S = R \cup R^2 \cup \cdots \cup R^k$$

(for some k) is transitive. Use `transR` for the transitivity test.

Exercise 5.55 If R is a relation on A, then the *indegree* of a point $a \in A$ is defined as $|\{x \in A \mid xRa\}|$ (the number of elements of A that are R-related to a). The *outdegree* of a point $a \in A$ is defined as $|\{x \in A \mid aRx\}|$ (the number of elements of A that a is R-related to). Implement these definitions as functions of the following types:

```
indegree  :: (Eq a) => Rel a -> a -> Int
outDegree :: (Eq a) => Rel a -> a -> Int
```

Exercise 5.56 If R is a relation on A, then an element a of A is a *source* of R iff a has indegree 0 and outdegree $\geqslant 1$; an element a of A is a *sink* of R iff a has outdegree 0 and indegree $\geqslant 1$. Use the results of the previous exercise to implement these definitions as functions of the following types:

```
sources :: (Eq a) => Rel a -> Set a
sinks   :: (Eq a) => Rel a -> Set a
```

5.4 Implementing Relations as Functions

A *characteristic function* is a function of type $A \to \{0, 1\}$, for some set A. Characteristic functions are so-called because they characterize subsets of a set A. The function $f : A \to \{0, 1\}$ characterizes the set $B = \{a \in A \mid f(a) = 1\} \subseteq A$. Characteristic functions implemented in Haskell have type `a -> Bool`, for some type `a`.

From the fact that a binary relation `r` is a subset of a product $A \times B$, you would expect that a binary relation is implemented in Haskell as a function of type `(a,b) -> Bool`. Given a pair of objects `(x,y)`, with `x` of type `a` and `y` of type `b`, the function proclaims the verdict `True` if `(x,y)` in the relation, `False` otherwise.

Standard relations like `==` (for identity, or equality) and `<=` (for $\leqslant$) are represented in Haskell in a slightly different way. They take their arguments one by one. Let us check their types:

5.4. IMPLEMENTING RELATIONS AS FUNCTIONS

```
Prelude> :t (==)
(==) :: Eq a => a -> a -> Bool
Prelude> :t (<=)
(<=) :: Ord a => a -> a -> Bool
Prelude>
```

What this means is: if a is a type in the class Eq, then == takes a first argument of that type and a second argument of that type and proclaims a verdict True or False, and similarly for <=, except that now the arguments have to be of a type in the class Ord.

Another example of a relation in Haskell is the following implementation divides of the relation $x|y$ ('x divides y') on the integers (x divides y if there a $q \in \mathbb{Z}$ with $y = xq$).

```
divides :: Integer -> Integer -> Bool
divides d n | d == 0    = error "divides: zero divisor"
            | otherwise = (rem n d) == 0
```

Switching back and forth between types a -> a -> Bool and (a,a) -> Bool (or, more generally, between types a -> b -> c and (a,b) -> c), can be done by means of the procedures for *currying* and *uncurrying* a function. The procedures refer to the logician H.B. Curry who helped laying the foundations for functional programming. (The initial H stands for *Haskell*; the programming language that we use in this book is also named after him.)

If f is of type (a,b) -> c, then currying f means transforming it into a function that takes its arguments one by one, i.e., a function of type a -> b -> c. The procedure curry is predefined in Haskell as follows:

```
curry       :: ((a,b) -> c) -> (a -> b -> c)
curry f x y = f (x,y)
```

If f is of type a -> b -> c, then uncurrying f means transforming it into a function that takes its arguments as a pair, i.e., a function of type (a,b) -> c. The procedure uncurry is predefined in Haskell as follows:

```
uncurry       :: (a -> b -> c) -> ((a,b) -> c)
uncurry f p   = f (fst p) (snd p)
```

As an example, here are some definitions of relations.

```
eq :: Eq a => (a,a) -> Bool
eq = uncurry (==)

lessEq :: Ord a => (a,a) -> Bool
lessEq = uncurry (<=)
```

If a relation is implemented as a procedure of type (a,b) -> Bool it is very easy to define its inverse:

```
inverse :: ((a,b) -> c) -> ((b,a) -> c)
inverse f (x,y) = f (y,x)
```

This gives:

```
REL> inverse lessEq (3,4)
False
REL> inverse lessEq (4,3)
True
REL>
```

Can we do something similar for procedures of type a -> b -> c? Yes, we can. Here is the predefined procedure flip:

```
flip        :: (a -> b -> c) -> b -> a -> c
flip f x y  = f y x
```

Here it is in action:

```
REL> flip (<=) 3 4
False
REL> flip (<=) 4 3
True
REL>
```

5.4. IMPLEMENTING RELATIONS AS FUNCTIONS

The procedure `flip` can be used to define properties from relations. Take the property of dividing the number 102. This denotes the set

$$\{d \in \mathbb{N}^+ \mid d \text{ divides } 102\} = \{1, 2, 3, 6, 17, 34, 51, 102\}.$$

It is given in Haskell by (`'divides'` 102), which in turn is shorthand for

```
flip divides 102
```

Trying this out, we get:

```
REL> filter ('divides' 102) [1..300]
[1,2,3,6,17,34,51,102]
```

We will now work out the representation of relations as characteristic functions. To keep the code compatible with the implementation given before, we define the type as `Rel'`, and similarly for the operations.

```
type Rel' a = a -> a -> Bool

emptyR' :: Rel' a
emptyR' = \ _ _ -> False

list2rel' :: Eq a => [(a,a)] -> Rel' a
list2rel' xys = \ x y -> elem (x,y) xys
```

`idR'` creates the identity relation over a list.

```
idR' :: Eq a => [a] -> Rel' a
idR' xs = \ x y -> x == y && elem x xs
```

`invR'` inverts a relation.

```
invR' :: Rel' a -> Rel' a
invR' = flip
```

`inR'` checks whether a pair is in a relation.

```
inR' :: Rel' a -> (a,a) -> Bool
inR' = uncurry
```

Checks whether a relation is reflexive, irreflexive, symmetric or transitive (on a domain given by a list):

```
reflR' :: [a] -> Rel' a -> Bool
reflR' xs r = and [ r x x | x <- xs ]

irreflR' :: [a] -> Rel' a -> Bool
irreflR' xs r = and [ not (r x x) | x <- xs ]

symR' :: [a] -> Rel' a -> Bool
symR' xs r = and [ not (r x y && not (r y x)) | x <- xs, y <- xs ]

transR' :: [a] -> Rel' a -> Bool
transR' xs r = and
               [ not (r x y && r y z && not (r x z))
                     | x <- xs, y <- xs, z <- xs ]
```

Union, intersection, reflexive and symmetric closure of relations:

```
unionR' :: Rel' a -> Rel' a -> Rel' a
unionR' r s x y = r x y || s x y

intersR' :: Rel' a -> Rel' a -> Rel' a
intersR' r s x y = r x y && s x y

reflClosure' :: Eq a => Rel' a -> Rel' a
reflClosure' r = unionR' r (==)

symClosure' :: Rel' a -> Rel' a
symClosure' r = unionR' r (invR' r)
```

Relation composition:

5.5. EQUIVALENCE RELATIONS

```
compR' :: [a] -> Rel' a -> Rel' a -> Rel' a
compR' xs r s x y = or [ r x z && s z y | z <- xs ]
```

Composition of a relation with itself:

```
repeatR' :: [a] -> Rel' a -> Int -> Rel' a
repeatR' xs r n | n < 1     = error "argument < 1"
                | n == 1    = r
                | otherwise = compR' xs r (repeatR' xs r (n-1))
```

Exercise 5.57 Use the implementation of relations `Rel'` a as characteristic functions over type a to define an example relation r with the property that `unionR r (compR r r)` is *not* the transitive closure of r.

Exercise 5.58 If a relation `r :: Rel'` a is restricted to a finite list xs, then we can calculate the transitive closure of r restricted to the list xs. Define a function

`transClosure' :: [a] -> Rel' a -> Rel' a`

for this.

Hint: compute the smallest relation S with the property that

$$S = R \cup R^2 \cup \cdots \cup R^k$$

(for some k) is transitive. Use `transR xs` for the transitivity test on domain xs.

5.5 Equivalence Relations

Definition 5.59 A relation R on A is an **equivalence relation** or **equivalence** if R is transitive, reflexive on A and symmetric.

Example 5.60 On the set of human beings the relation of having the same age is an equivalence relation.

Example 5.61 The relation $R = \{(n, m) \mid n, m \in \mathbb{N}$ and $n + m$ is even $\}$ is an equivalence relation on $\mathbb{N}$.

The equivalence test can be implemented for relations of type `Ord a => Rel a` as follows:

```
equivalenceR :: Ord a => Set a -> Rel a -> Bool
equivalenceR set r = reflR set r && symR r && transR r
```

For relations implemented as type `Rel'` a the implementation goes like this:

```
equivalenceR' :: [a] -> Rel' a -> Bool
equivalenceR' xs r = reflR' xs r && symR' xs r && transR' xs r
```

Example 5.62 The next table shows for a number of familiar relations whether they have the properties of reflexivity, symmetry and transitivity. Here:

- $\emptyset$ is the empty relation on $\mathbb{N}$,
- $\Delta = \Delta_\mathbb{N} = \{(n,n) \mid n \in \mathbb{N}\}$ is the identity on $\mathbb{N}$,
- $\mathbb{N}^2 = \mathbb{N} \times \mathbb{N}$ is the biggest relation on $\mathbb{N}$,
- $<$ and $\leqslant$ are the usual ordering relations on $\mathbb{N}$,
- Suc is the relation on $\mathbb{N}$ defined by $\text{Suc}(n,m) \equiv n+1 = m$,
- $\subseteq$ is the inclusion relation on $\wp(\mathbb{N})$.

property:	yes:	no:
reflexive (on $\mathbb{N}$ resp. $\wp(\mathbb{N})$)	Δ, $\mathbb{N}^2$, $\leqslant$, $\subseteq$	$\emptyset$, $<$, Suc
symmetric	$\emptyset$, Δ, $\mathbb{N}^2$	$<$, $\leqslant$, Suc, $\subseteq$
transitive	$\emptyset$, Δ, $\mathbb{N}^2$, $<$, $\leqslant$, $\subseteq$	Suc

The table shows that among these examples only Δ and $\mathbb{N}^2$ are equivalences on $\mathbb{N}$.

Example 5.63 The relation $\emptyset$ is (trivially) symmetric and transitive. The relation $\emptyset$ is reflexive on the set $\emptyset$, but on no other set. Thus: $\emptyset$ is an equivalence on the set $\emptyset$, but on no other set.

5.5. EQUIVALENCE RELATIONS

Example 5.64 Let A be a set. Δ_A is the smallest equivalence on A. A^2 is the biggest equivalence on A.

Example 5.65 The relation $\sim$ between vectors in 3-dimensional space $\mathbb{R}^3$ that is defined by $\vec{a} \sim \vec{b} \equiv \exists r \in \mathbb{R}^+ (\vec{a} = r\vec{b})$ is an equivalence.

Example 5.66 For any $n \in \mathbb{Z}$, $n \neq 0$, the relation $\equiv_n$ on $\mathbb{Z}$ is given by $m \equiv_n k$ iff m and k have the same remainder when divided by n. More precisely, $m \equiv_n k$ (or: $m \equiv k \pmod{n}$) iff

- $m = qn + r$, with $0 \leqslant r < n$,
- $k = q'n + r'$, with $0 \leqslant r' < n$,
- $r = r'$.

When $m \equiv_n k$ we say that m is equivalent to k modulo n. The relation $\equiv_n$ is also called the $\pmod{n}$ relation.

To show that $\equiv_n$ from Example 5.66 is an equivalence, the following proposition is useful.

Proposition 5.67 $m \equiv_n k$ iff $n \mid m - k$.

Proof. $\Rightarrow$: Suppose $m \equiv_n k$. Then $m = qn + r$ and $k = q'n + r'$ with $0 \leqslant r < n$ and $0 \leqslant r' < n$ and $r = r'$. Thus, $m - k = (q - q')n$, and it follows that $n \mid m - k$.
$\Leftarrow$: Suppose $n \mid m - k$. Then $n \mid (qn + r) - (q'n + r')$, so $n \mid r - r'$. Since $-n < r - r' < n$, this implies $r - r' = 0$, so $r = r'$. It follows that $m \equiv_n k$. ∎

From this we get that the following are all equivalent:

- $m \equiv_n k$.
- $n \mid m - k$.
- $\exists a \in \mathbb{Z} : an = m - k$.
- $\exists a \in \mathbb{Z} : m = k + an$.
- $\exists a \in \mathbb{Z} : k = m + an$.

Exercise 5.68 Show that for every $n \in \mathbb{Z}$ with $n \neq 0$ it holds that $\equiv_n$ is an equivalence on $\mathbb{Z}$.

Example 5.69 Here is a Haskell implementation of the modulo relation:

```
modulo :: Integer -> Integer -> Integer -> Bool
modulo n = \ x y -> divides n (x-y)
```

Example 5.70 The relation that applies to two finite sets in case they have the same number of elements is an equivalence on the collection of all finite sets.

The corresponding equivalence on finite lists is given by the following piece of Haskell code:

```
equalSize :: [a] -> [b] -> Bool
equalSize list1 list2 = (length list1) == (length list2)
```

Abstract equivalences are often denoted by $\sim$ or $\approx$.

Exercise 5.71 Determine whether the following relations on $\mathbb{N}$ are (i) reflexive on $\mathbb{N}$, (ii) symmetric, (iii) transitive:

1. $\{(2,3),(3,5),(5,2)\}$;

2. $\{(n,m) \mid |n-m| \geqslant 3\}$.

Exercise 5.72 $A = \{1,2,3\}$. Can you guess how many relations there are on this small set? Indeed, there must be sufficiently many to provide for the following questions.

1. Give an example of a relation on A that is reflexive, but not symmetric and not transitive.

2. Give an example of a relation on A that is symmetric, but not reflexive and not transitive.

3. Give examples (if any) of relations on A that satisfy each of the six remaining possibilities w.r.t. reflexivity, symmetry and transitivity.

5.6. EQUIVALENCE CLASSES AND PARTITIONS

Exercise 5.73 For finite sets A (0, 1, 2, 3, 4 and 5 elements and n elements generally) the following table has entries for: the number of elements in A^2, the number of elements in $\wp(A^2)$ (that is: the number of relations on A), the number of relations on A that are (i) reflexive, (ii) symmetric and (iii) transitive, and the number of equivalences on A.

A	A^2	$\wp(A^2)$	reflexive	symmetric	transitive	equivalence
0	0	1	1	1	1	1
1	1	2	1	2	2	1
2	?	?	?	?	13	—
3	?	?	?	?	—	—
4	?	?	?	?	—	—
5	?	?	?	?	—	—
n	?	?	?	?	—	—

Give all reflexive, symmetric, transitive relations and equivalences for the cases that $A = \emptyset$ (0 elements) and $A = \{0\}$ (1 element). Show there are exactly 13 transitive relations on $\{0, 1\}$, and give the 3 that are not transitive. Put numbers on the places with question marks. (You are not requested to fill in the —.)

Example 5.74 Assume relation R on A is transitive and reflexive, i.e, R is a pre-order. Then consider the relation $\sim$ on A given by: $x \sim y :\equiv xRy \wedge yRx$. The relation $\sim$ is an equivalence relation on A. Symmetry is immediate from the definition. $\sim$ is reflexive because R is. $\sim$ is transitive, for assume $x \sim y$ and $y \sim z$. Then $xRy \wedge yRx \wedge yRz \wedge zRy$, and from $xRy \wedge yRz$, by transitivity of R, xRz, and from $zRy \wedge yRx$, by transitivity of R, zRx; thus $x \sim z$.

Exercise 5.75 Suppose that R is a symmetric and transitive relation on the set A such that $\forall x \in A \exists y \in A(xRy)$. Show that R is reflexive on A.

Exercise 5.76 Let R be a relation on A. Show that R is an equivalence iff (i) $\Delta_A \subseteq R$ and (ii) $R = R \circ R^{-1}$.

5.6 Equivalence Classes and Partitions

Equivalence relations on a set A enable us to *partition* the set A into equivalence classes.

Definition 5.77 Suppose R is an equivalence relation on A and that $a \in A$. The set $|a| = |a|_R = \{\, b \in A \mid bRa \,\}$ is called the *R-equivalence class* of a, or the *equivalence class* of a *modulo* R.

Elements of an equivalence class are called *representatives* of that class.

Example 5.78 (continued from Example 5.64) The equivalence class of $a \in A$ modulo Δ_A is $\{a\}$.

The only equivalence class modulo A^2 is A itself.

Example 5.79 (continued from Example 5.65) The equivalence class of $(1,1,1) \in \mathbb{R}^3$ modulo $\sim$ is the set $\{(r,r,r) \mid r > 0\}$: half a straight line starting at the origin (not including the origin). (A "direction".)

Example 5.80 (continued from Example 5.69) The equivalence class of 2 in $\mathbb{Z}$ (mod 4) is the set
$\{\ldots, -6, -2, 2, 6, 10, 14, \ldots\} = \{2 + 4n \mid n \in \mathbb{Z}\,\}$.

The implementation yields:

```
REL> filter (modulo 4 2) [-15..15]
[-14,-10,-6,-2,2,6,10,14]
```

Example 5.81 (continued from Example 5.70) The equivalence class of $\{0,1,2\}$ modulo the equivalence of having the same number of elements is the collection of all three-element sets.

*(According to the Frege-Russell definition of natural number, this *is* the number three.)

Lemma 5.82 *Suppose that R is an equivalence on A. If $a, b \in A$, then:*

$$|a|_R = |b|_R \iff aRb.$$

Proof. $\Rightarrow$: Note that $a \in |a|_R$ (for, R is reflexive). Thus, if $|a|_R = |b|_R$, then $a \in |b|_R$, i.e., aRb.
$\Leftarrow$: Assume that aRb. Then also bRa (R is symmetric.)
$|a|_R \subseteq |b|_R$: $x \in |a|_R$ signifies xRa, and hence xRb follows (R transitive); therefore, $x \in |b|_R$.
$|b|_R \subseteq |a|_R$: similarly.
Extensionality completes the proof. ∎

Lemma 5.83 *Let R be an equivalence on A.*

1. Every equivalence class is non-empty,

2. every element of A belongs to some equivalence class,

5.6. EQUIVALENCE CLASSES AND PARTITIONS

3. *different equivalence classes are disjoint.*

Proof. 1/2. Since R is reflexive on A, we have, for every $a \in A$: $a \in |a|$.
3. Suppose that $|a|$ and $|b|$ are not disjoint. Say, $c \in |a| \cap |b|$. Then we have both cRa and cRb. Since R is symmetric, it follows that aRc. Thus, aRb (R transitive). Therefore, $|a| = |b|$ follows, using Lemma 5.82. ∎

Exercise 5.84 Use the implementation of relations Rel' a as characteristic functions over type a to implement a function rclass that takes a relation r, an object x, and a list L, and returns the list of all objects y from L such that rxy holds. The type declaration should run:

rclass :: Rel' a -> a -> [a] -> [a]

The concept of *partitioning* a set is made precise in the following definition.

Definition 5.85 A family $\mathcal{A}$ of subsets of a set A is called a **partition** of A if

- $\emptyset \notin \mathcal{A}$,
- $\bigcup \mathcal{A} = A$,
- for all $X, Y \in \mathcal{A}$: if $X \neq Y$ then $X \cap Y = \emptyset$.

The elements of a partition are called its *components*.

This definition says that every element of A is in some member of $\mathcal{A}$ and that no element of A is in more than one member of $\mathcal{A}$.

Example 5.86 $\{\{1,2\},\{3,4\}\}$ is a partition of $\{1,2,3,4\}$. $\emptyset$ (trivially) is a partition of $\emptyset$. $\mathbb{Z}$ can be partitioned into the negative integers, the positive integers, and $\{0\}$.

Exercise 5.87 Show the following: if $\{A_i \mid i \in I\}$ is a partition of A and $\{B_j \mid j \in J\}$ is a partition of B, then $\{A_i \times B_j \mid (i,j) \in I \times J\}$ is a partition of $A \times B$.

Definition 5.88 (Quotients) Assume that R is an equivalence on the set A. The collection of equivalence classes of R, $A/R = \{|a| \mid a \in A\}$, is called the *quotient* of A modulo R.

The definition of *partition* was engineered in order to ensure the following:

Theorem 5.89 *Every quotient (of a set, modulo an equivalence) is a partition (of that set).*

Proof. This is nothing but a reformulation of Lemma 5.83. ∎

Example 5.90 (continued from Examples 5.64 and 5.78)
$A/\Delta_A = \{ \{a\} \mid a \in A \}$.
$A/A^2 = \{A\}$.

Example 5.91 (continued from Examples 5.65 and 5.79)
The partition $\mathbb{R}^3/\sim$ of $\mathbb{R}^3$ has components $\{0\}$ and all half-lines.

Example 5.92 (continued from Examples 5.69 and 5.80)
The partition $\mathbb{Z}/mod(4)$ of $\mathbb{Z}$ induced by the equivalence $mod(4)$ has four components: (i) the equivalence class $\{4n \mid n \in \mathbb{Z}\}$ of 0 (that also is the equivalence class of 4, 8, 12,..., -4...), (ii) the class $\{4n+1 \mid n \in \mathbb{Z}\}$ of 1, (iii) the class $\{4n+2 \mid n \in \mathbb{Z}\}$ of 2, and (iv) $\{4n+3 \mid n \in \mathbb{Z}\}$, the class of 3. The quotient $\mathbb{Z}/mod(n)$ is usually written as $\mathbb{Z}_n$.

Example 5.93* (continued from Examples 5.70 and 5.81)
The quotient of the collection of finite sets modulo the relation "same number of elements" is —according to Frege and Russell— the set of natural numbers.

Exercise 5.94 Give the partition that the relation of Example 5.61 induces on $\mathbb{N}$.

Example 5.95 (continued from Example 5.74)
Let R be a pre-order on A. Then $\sim$ given by $x \sim y :\equiv xRy \wedge yRx$ is an equivalence relation. Consider the relation $R_\sim$ on $A/\sim$ given by $|x|R_\sim|y| :\equiv xRy$. The relation $R_\sim$ is a partial order on $A/\sim$ called the *po-set reflection* of R.

The definition is independent of the representatives, because assume $x \sim x'$ and $y \sim y'$ and xRy. Then $xRx' \wedge x'Rx$, $yRy' \wedge y'Ry$, and $x'Ry'$. From $x'Rx$, xRy and yRy' by transitivity of R, $x'Ry'$.

$R_\sim$ is reflexive, because R is. $R_\sim$ is anti-symmetric, for suppose $|x|R_\sim|y|$ and $|y|R_\sim|x|$. Then xRy and yRx, so by definition of $\sim$, $|x| = |y|$. Finally $R_\sim$ is transitive because R is.

Thus, equivalences induce partitions. But the process can be inverted.

Theorem 5.96 *Every partition (of a set) is a quotient (of that set, modulo a certain equivalence).*

5.6. EQUIVALENCE CLASSES AND PARTITIONS

Specifically: Suppose that $\mathcal{A}$ is a partition of A. Then the relation R on A defined by
$$xRy := \exists K \in \mathcal{A} \, (x, y \in K)$$
(x and y member of the same component of $\mathcal{A}$) is an equivalence on A, and $\mathcal{A}$ is the collection of equivalence classes of R.

Proof. Exercise 5.123. ∎

According to Theorems 5.89 and 5.96, equivalences and partitions are two sides of the same coin.

Example 5.97 Consider the following relation on $\{0, 1, 2, 3, 4\}$:
$$\{(0,0), (0,1), (1,0), (1,1), (2,2), (2,4), (3,3), (4,2), (4,4)\}.$$
This is an equivalence, and the corresponding partition is
$$\{\{0, 1\}, \{2, 4\}, \{3\}\}.$$

Exercise 5.98 Is the following relation an equivalence on $\{0, 1, 2, 3, 4\}$? If so, give the corresponding partition.
$$\{(0,0), (0,3), (0,4), (1,1), (1,2),$$
$$(2,1), (2,2), (3,0), (3,3), (3,4), (4,0), (4,3), (4,4)\}.$$

Example 5.99 The equivalence corresponding to the partition
$$\{\{0\}, \{1, 2, 3\}, \{4\}\}$$
of $\{0, 1, 2, 3, 4\}$ is:
$$\{(0,0), (1,1), (1,2), (1,3), (2,1), (2,2), (2,3), (3,1), (3,2), (3,3), (4,4)\}.$$

Exercise 5.100 What are the equivalences corresponding to the following partitions?:

1. $\{\{0, 3\}, \{1, 2, 4\}\}$, of $\{0, 1, 2, 3, 4\}$,

2. $\{\{n \in \mathbb{Z} \mid n < 0\}, \{0\}, \{n \in \mathbb{Z} \mid n > 0\}\}$, of $\mathbb{Z}$,

3. $\{\{\text{even numbers}\}, \{\text{odd numbers}\}\}$, of $\mathbb{N}$.

Exercise 5.101 $A = \{1, 2, 3, 4, 5\}$,
$$R = \{(1,1), (1,2), (1,4), (2,1), (2,2), (2,4),$$
$$(3,3), (3,5), (4,1), (4,2), (4,4), (5,3), (5,5)\}.$$

1. Is R an equivalence on A? If so, answer 2 and 3:

2. Determine $|2|_R$.

3. Determine A/R.

Example 5.102 If $\sim$ on $\wp(\mathbb{N})$ is given by

$$A \sim B :\equiv (A - B) \cup (B - A) \text{ is finite,}$$

then $\sim$ is reflexive. If $A \subseteq \mathbb{N}$ is arbitrary, then $(A-A)\cup(A-A) = \emptyset \cup \emptyset = \emptyset$, and $\emptyset$ is finite.

Exercise 5.103* Define the relation $\sim$ on $\wp(\mathbb{N})$ by:

$$A \sim B :\equiv (A - B) \cup (B - A) \text{ is finite.}$$

Thus, $\mathbb{N} \not\sim \emptyset$, since $(\mathbb{N} - \emptyset) \cup (\emptyset - \mathbb{N}) = \mathbb{N} \cup \emptyset = \mathbb{N}$ is infinite.

Show that $\sim$ is an equivalence (reflexivity is shown in Example 5.102).

Exercise 5.104 Define the relation R on all people by: $aRb :\equiv a$ and b have a common ancestor. Is R transitive?
Same question for the relation S defined by: $aSb :\equiv a$ and b have a common ancestor along the male line.

Example 5.105 For counting the partitions of a set A of size n, the key is to count the number of ways of partitioning a set A of size n into k non-empty classes. Let us use $\left\{{n \atop k}\right\}$ for this number, and see if we can find a recurrence for it, i.e., an equation in terms of the function for smaller arguments. Distributing n objects over 1 set can be done in only one way, so $\left\{{n \atop 1}\right\} = 1$. Similarly, distributing n objects over n non-empty sets can be done in only one way, so $\left\{{n \atop n}\right\} = 1$.

To distribute n objects over k different sets, we can either (i) put the last object into an equivalence class of its own, or (ii) put it in one of the existing classes. (i) can be done in $\left\{{n-1 \atop k-1}\right\}$ ways, for this is the number of ways to distribute $n - 1$ objects over $k - 1$ non-empty classes. (ii) can be done in $k \cdot \left\{{n-1 \atop k}\right\}$ ways, for there are k classes to choose from, and $\left\{{n-1 \atop k}\right\}$ ways to distribute $n - 1$ objects over k classes. Thus, the recurrence we need is:

$$\left\{{n \atop k}\right\} = k \cdot \left\{{n-1 \atop k}\right\} + \left\{{n-1 \atop k-1}\right\}.$$

5.6. EQUIVALENCE CLASSES AND PARTITIONS

In terms of this, the number of partitions $b(n)$ is given by:

$$b(n) = \sum_{k=1}^{n} \left\{ {n \atop k} \right\}.$$

The numbers $\left\{ {n \atop k} \right\}$ are called *Stirling set numbers*. The $b(n)$ are called *Bell numbers*.

Exercise 5.106 Implement functions `bell` and `stirling` to count the number of different partitions of a set of n elements.

Exercise 5.107 Use the result of Exercise 5.106 to fill out the last column in the table of Exercise 5.73 on p. 187.

Exercise 5.108 Show: conditions 2 and 3 of Definition 5.85 taken together are equivalent with: for every $a \in A$ there exists exactly one $K \in \mathcal{A}$ such that $a \in K$.

Exercise 5.109 Is the *intersection* $R \cap S$ of two equivalences R and S on a set A again an equivalence? What about the *union* $R \cup S$? Prove, or supply a simple counter-example. (See the table of Exercise 5.50.)

Exercise 5.110* Suppose that R and S are equivalences on A such that $R \subseteq S$. Show that every S-equivalence class is a union of R-equivalence classes.

Exercise 5.111 A list partition is the list counterpart of a partition: list partitions are of type `Eq a => [[a]]`, and a list partition `xss` of `xs` has the following properties:

- `[]` is not an element of `xss`,
- `xs` and `concat xss` have the same elements,
- if `ys` and `zs` are distinct elements of `xss`, then `ys` and `zs` have no elements in common.

Implement a function

`listPartition :: Eq a => [a] -> [[a]] -> Bool`

that maps every list `xs` to a check whether a given object of type `[[a]]` is a list partition of `xs`.

Exercise 5.112 Implement a function

```
listpart2equiv :: Ord a => [a] -> [[a]] -> Rel a
```

that generates an equivalence relation from a list partition (see Exercise 5.111). The first argument gives the domain, the second argument the list partition. Generate an error if the second argument is not a list partition on the domain given by the first argument.q

Exercise 5.113 $R = \{(0,3), (1,5), (2,0)\}$; $A = \{0, 1, 2, 3, 4, 5\}$.

1. What is the smallest (in the sense of: number of elements) equivalence $S \supseteq R$ on A?

2. Determine A/S.

3. How many equivalences exist on A that include R?

4. Give the corresponding partitions.

Exercise 5.114 Implement a function

```
equiv2listpart :: Ord a => Set a -> Rel a -> [[a]]
```

that maps an equivalence relation to the corresponding list partition (see Exercise 5.111). Generate an error if the input relation is not an equivalence.

Exercise 5.115 Use the function equiv2listpart to implement a function

```
equiv2part :: Ord a => Set a -> Rel a -> Set (Set a)
```

that maps an equivalence relation to the corresponding partition.

Exercise 5.116* R is a relation on A. One of the following is the smallest equivalence on A that includes R. Which? Prove.

1. $(\Delta_A \cup R)^+ \cup (\Delta_A \cup R^{-1})^+$.

2. $\Delta_A \cup R^+ \cup (R^{-1})^+$.

3. $\Delta_A \cup (R \cup R^{-1})^+$.

Exercise 5.117 Let R be a relation on A. Show that $S = R^* \cap R^{-1*}$ is an equivalence on A. Next, show that the relation T on the quotient A/S given by $|a|_S T |b|_S :\equiv aR^*b$ is a partial order.

5.6. EQUIVALENCE CLASSES AND PARTITIONS

Note that the reasoning of Exercise 5.117 only uses reflexivity and transitivity of R^*. So, in general, if R is a reflexive transitive relation on A (i.e., R is a pre-order), then $S = R \cap R^{-1}$ is an equivalence on A, and the relation T on the quotient A/S given by $|a|_S T |b|_S :\equiv aRb$ is a partial order.

Example 5.118 Consider the pre-order $\models$ of Example 5.25. Note that the equivalence relation $\models \cap \models^{-1}$ is nothing other than the relation $\equiv$ of logical equivalence.

Remark. In constructions with quotients, it often happens that a relation on the quotient is defined in terms of a relation on the underlying set. In such a case one should always check that the definition is proper in the sense that it is independent of the representatives that are mentioned in it.

Take again the example of Exercise 5.117, of a set A with a reflexive transitive relation R on it. If $S = R \cap R^{-1}$, then the definition of relation T on the quotient A/S given by $|a|_S T |b|_S :\equiv aRb$ is proper, because it holds that aRb, $a' \in |a|_S$ and $b' \in |b|_S$ together imply that $a'Rb'$.

To see that this is so, assume $a' \in |a|_S$ and aRb. Then $a'Sa$, so $a'Ra$, and by transitivity of R, $a'Rb$. Suppose $b' \in |b|_S$. Then $b'Sb$, so bRb'. Together with $a'Rb$ this gives, again by transitivity of R, that $a'Rb'$. ∎

Exercise 5.119 Define the relations $\sim$ and $\approx$ on $\mathbb{R}$ by $p \sim q :\equiv p \times q \in \mathbb{Z}$, $p \approx q :\equiv p - q \in \mathbb{Z}$. Are these equivalences? If so, describe the corresponding partition(s).

Exercise 5.120 Define the relation R on $\mathbb{R} \times \mathbb{R}$ by $(x, y) R (u, v)$ iff $3x - y = 3u - v$.

1. Show that R is an equivalence.

2. Describe the equivalence classes of $(0, 0)$ and $(1, 1)$.

3. Describe $\mathbb{R} \times \mathbb{R}/R$ in geometrical terms.

Exercise 5.121 Define an equivalence on $\mathbb{R} \times \mathbb{R}$ that partitions the plane in concentric circles with $(0, 0)$ as centre.

Exercise 5.122 $Q = \{(0,0), (0,1), (0,5), (2,4), (5,0)\}$. For an equivalence R on $\{0, 1, 2, 3, 4, 5\}$ it is given, that $Q \subseteq R$ and $(0, 2) \notin R$.

1. Show that $(1, 5) \in R$ and $(4, 5) \notin R$.

2. Give the partition corresponding to the smallest (in the sense of number of elements) equivalence $\supseteq Q$.

3. How many equivalences S are there on $\{0, 1, 2, 3, 4, 5\}$ such that $Q \subseteq S$ and $(0, 2) \notin S$? Give the corresponding partitions.

Exercise 5.123 Prove Theorem 5.96.
Hint. Make sure that you use all properties of partitions.

Exercise 5.124* On a certain planet there are 20 countries. Every two of them either are at war with each other or have a peace treaty. Every two countries with a common enemy have such a peace treaty. What is the least possible number of peace treaties on this planet?

Hint: Note that the relations P and W, for being at peace and at war, respectively, exclude one another, that they are both irreflexive and symmetric, and that they satisfy the principle that there are no war-triangles. For if x is at war with both y and z, then y and z have x as common enemy, hence y and z have a peace treaty:

$$\forall xyz((xWy \wedge xWz) \Rightarrow yPz).$$

Find a recurrence (see page 208) for the maximum number of countries at war among $2n$ countries. Next, use this to derive the minimum number of peace treaties among $2n$ countries.

5.7 Integer Partitions

Integer partitions of $n \in \mathbb{N}^+$ are lists of non-zero natural numbers that add up to exactly n. For example, the four integer partitions of 4 are

$$[4], [1, 3], [2, 2], [1, 1, 2], [1, 1, 1, 1].$$

The integer partitions of n correspond to the sizes of the set partitions of a set A with $|A| = n$. Here is an algorithm for generating integer partitions, in lexicographically decreasing order:

- The first integer partition of n is $[n]$.

- Let B be the last integer partition generated. If B consists of only 1's, then done. Otherwise, there is a smallest non-1 part m. To generate the next partition, subtract 1 from m and collect all the units so as to match the new smallest part $m - 1$.

5.7. INTEGER PARTITIONS

Examples The partition after $[1,1,3,3]$ is $[1,2,2,3]$, for after subtracting 1 from 3, we should pack the three units that result in parcels of size 2. The partition after $[1,1,1,1,1,1,5]$ is $[3,4,4]$, for after subtracting 1 from 5, we should pack the seven units that result in parcels with maximum size 4, which gives three units and one parcel of size 4, which in turn gives one parcel of size 3 and one of size 4. The partition after $[3,3,5]$ is $[1,2,3,5]$. The partition after $[1,3,3,4]$ is $[2,2,3,4]$.

Implementation An integer partition is represented as a list of integers. For convenience we count the number of 1's, and remove them from the partition. This gives a compressed representation $(2,[3,3])$ of $[1,1,3,3]$. These compressed partitions have type CmprPart.

```
type Part = [Int]
type CmprPart = (Int,Part)
```

Expansion of a compressed partition (n,p) is done by generating n 1's followed by p:

```
expand :: CmprPart -> Part
expand (0,p) = p
expand (n,p) = 1:(expand ((n-1),p))
```

In generating the next partition from (k,xs) we may assume that xs is non-empty, and that its elements are listed in increasing order.

The partition that follows (k,x:xs) is generated by packing (k+x,x:xs) for maximum size $x - 1$. This assumes that x is the smallest element in x:xs.

```
nextpartition :: CmprPart -> CmprPart
nextpartition (k,(x:xs)) = pack (x-1) ((k+x),xs)
```

To pack a partition (m,xs) for size 1, there is nothing to do. To pack a partition (m,xs) for maximum size $k > 1$ and $k > m$, decrease the parcel size to $k - 1$. To pack a partition (m,xs) for size $k > 1$ and $k \leqslant m$, use

k units to generate one parcel of size k, and go on with (m-k,k:xs), for the same parcel size.

```
pack :: Int -> CmprPart -> CmprPart
pack 1 (m,xs) = (m,xs)
pack k (m,xs) = if k > m  then pack (k-1) (m,xs)
                 else           pack k     (m-k,k:xs)
```

To generate all partitions starting from a given partition (n,[]), just list the partition consisting of n units, for this is the last partition. To generate all partitions starting from a given partition (n,x:xs), list this partition, and then generate from the successor of (n,x:xs).

```
generatePs :: CmprPart -> [Part]
generatePs p@(n,[])    = [expand p]
generatePs p@(n,(x:xs)) =
     (expand p: generatePs(nextpartition p))
```

Generate all partitions starting from $[n]$. The case where $n = 1$ is special, for it is the only case where the compressed form of the first partition has a non-zero number of units.

```
part :: Int -> [Part]
part n | n < 1     = error "part: argument <= 0"
       | n == 1    = [[1]]
       | otherwise = generatePs (0,[n])
```

Here is what we get out:

```
REL> part 5
[[5],[1,4],[2,3],[1,1,3],[1,2,2],[1,1,1,2],[1,1,1,1,1]]
REL> part 6
[[6],[1,5],[2,4],[1,1,4],[3,3],[1,2,3],[1,1,1,3],[2,2,2],
 [1,1,2,2],[1,1,1,1,2],[1,1,1,1,1,1]]
REL> length (part 20)
627
```

Exercise 5.125 Write a program `change :: Int -> [Int]` that returns change in EURO coins for any positive integer, in the least number of coins. Measure the values of the EURO coins 0.01, 0.02, 0.05, 0.10, 0.20, 0.50, 1, 2 in EURO cents, as 1, 2, 5, 10, 20, 50, 100, 200. Use `pack` for inspiration.

Exercise 5.126 Modify the integer partition algorithm so that it generates all the possible ways of giving coin change for amounts of money up to 10 EURO, using all available EURO coins (0.01, 0.02, 0.05, 0.10, 0.20, 0.50, 1, 2). Measure the values of the EURO coins in EURO cents, as 1, 2, 5, 10, 20, 50, 100, 200.

Exercise 5.127 How many different ways are there to give change for one EURO? Use the program from the previous exercise.

5.8 Further Reading

More on relations in the context of set theory in [DvDdS78]. Binary relations on finite sets are studied in graph theory. See Chapters 4 and 5 of [Bal91]. Relations in the context of database theory are the topic of [AHV95].

Chapter 6

Functions

Preview

In mathematics, the concept of a function is perhaps even more important than that of a set. Also, functions are crucial in computer programming, as the functional programming paradigm demonstrates. This chapter introduces basic notions and then moves on to special functions, operations on functions, defining equivalences by means of functions, and compatibility of equivalences with operations.

Many concepts from the abstract theory of functions translate directly into computational practice. Most of the example functions mentioned in this chapter can be implemented directly in Haskell by just keying in their definitions, with domains and co-domains specified as Haskell types.

Still, we have to bear in mind that an implementation of a function as a computer program is a concrete incarnation of an abstract object. The same function may be computed by vastly different computation procedures. If you key in sum [2*k | k <- [1 .. 100]] at the *hugs* prompt you get the answer 10100, and if you key in 100 * 101 you get the same answer, but the computation steps that are performed to get at the answers are different. We have already seen that there is no mechanical test for checking whether two procedures perform the same task (Section 4.2), although in particular cases such results can be proved by mathematical induction (see Chapter 7). E.g., in Example 7.4 in Chapter 7 it is proved by mathematical induction that the computational recipes $\sum_{k=1}^{n} 2k$ and $n(n+1)$ specify the same function.

```
module FCT

where

import List
```

6.1 Basic Notions

A **function** is something that transforms an object given to it into another one. The objects that can be given to a function are called its *arguments*, and the results of the transformation are called *values*. The set of arguments is called the *domain* of the function. We say that a function is *defined on* its domain.

If f is a function and x one of its arguments, then the corresponding value is denoted by $f(x)$. A function value $y = f(x)$ is called the *image* of x under f. That $f(x) = y$ can also be indicated by $f : x \longmapsto y$. The domain of f is denoted by $\text{dom}(f)$. Its *range* is $\text{ran}(f) = \{f(x) \mid x \in \text{dom}(f)\}$.

Example 6.1 A function can be given as a rule or a prescription of how to carry out the transformation.

- *First square, next add one* is the function that transforms a real $x \in \mathbb{R}$ into $x^2 + 1$.

  Letting f stand for this function, it is customary to describe it by the equation
  $$f(x) = x^2 + 1.$$

  The Haskell implementation uses the same equation:

  ```
  f x = x^2 + 1
  ```

- The function described by
  $$|x| = \begin{cases} x & \text{if } x \geqslant 0 \\ -x & \text{if } x < 0 \end{cases}$$

  transforms a real into its absolute value. The Haskell implementation given in *Prelude.hs* follows this definition to the letter:

6.1. BASIC NOTIONS

```
absReal x | x >= 0   = x
          | otherwise = -x
```

- The identity function 1_A defined on A does not "transform" at all in the usual sense of the word: given an argument $x \in A$, it outputs x itself. A polymorphic identity function is predefined in Haskell as follows:

```
id :: a -> a
id x = x
```

Set theory has the following simple definition of the concept of a function.

Definition 6.2 A **function** is a relation f that satisfies the following condition.
$$(x, y) \in f \wedge (x, z) \in f \implies y = z.$$
That is: for every $x \in \text{dom}(f)$ there is *exactly one* $y \in \text{ran}(f)$ such that $(x, y) \in f$.

If $x \in \text{dom}(f)$, then $f(x)$ is by definition the unique object $y \in \text{ran}(f)$ for which $(x, y) \in f$.

Note that we use dom here in the sense defined for relations, but that the relation and the function-sense of the notion coincide: the *domain* $\text{dom}(f)$ of the function f is $\{x \mid \exists y \, ((x, y) \in f)\}$; exactly as in the relation-case (cf. Definition 5.1 p. 158).

Similarly, the range of f coincides with the range of f as a relation.

The set-theoretic and the computational view on functions are worlds apart, for computationally a function is an *algorithm* for computing values. However, in cases of functions with finite domains it is easy to switch back and forth between the two perspectives, as the following conversions demonstrate.

```
list2fct :: Eq a => [(a,b)] -> a -> b
list2fct [] _ = error "function not total"
list2fct ((u,v):uvs) x | x == u    = v
                       | otherwise = list2fct uvs x

fct2list :: (a -> b) -> [a] -> [(a,b)]
fct2list f xs = [ (x, f x) | x <- xs ]
```

The range of a function, implemented as a list of pairs, is given by:

```
ranPairs :: Eq b => [(a,b)] -> [b]
ranPairs f = nub [ y | (_,y) <- f ]
```

If a function is defined on an enumerable domain, we can list its (finite or infinite) range starting from a given element.

```
listValues  :: Enum a => (a -> b) -> a -> [b]
listValues f i = (f i) : listValues f (succ i)
```

If we also know that the domain is bounded, we can generate the whole range as a finite list.

```
listRange :: (Bounded a, Enum a) => (a -> b) -> [b]
listRange f = [ f i | i <- [minBound..maxBound] ]
```

Example 6.3

- The function $x \mapsto x^2 + 1$ defined on $\mathbb{R}$ is identified with the relation $\{(x,y) \mid x \in \mathbb{R} \land y = x^2 + 1\} = \{(x, x^2 + 1) \mid x \in \mathbb{R}\}$.

- $f = \{(1,4), (2,4), (3,6)\}$ is a function. $\text{dom}(f) = \{1, 2, 3\}$, $\text{ran}(f) = \{4, 6\}$.

- $\{(1,4), (2,4), (2,6)\}$ is *not* a function.

6.1. BASIC NOTIONS

- Δ_X, the identity-relation on X (Definition 5.7, p. 159), also is a function from X to X. When viewed as a function, the identity relation on X is usually written as 1_X.

- If f is a function, we clearly have that $f = \{(a, f(a)) \mid a \in \text{dom}(f)\}$, and $\text{ran}(f) = \{f(a) \mid a \in \text{dom}(f)\}$.

- The relation $\emptyset$ is a function. $\text{dom}(\emptyset) = \emptyset = \text{ran}(\emptyset)$.

As is the case for relations, functions are more often given in a context of *two* sets. Compare Definition 5.3 (p. 158).

Definition 6.4 (From ... to, On, Co-domain) Suppose that X and Y are sets. A function f is *from X to Y*; notation:

$$f : X \longrightarrow Y,$$

if $\text{dom}(f) = X$ and $\text{ran}(f) \subseteq Y$. (Note the difference in terminology compared with Definition 5.3!)

In this situation, Y is called the *co-domain* of f.

A function f is said to be defined *on* X if $\text{dom}(f) = X$.

Note that the set-theoretic way of identifying the function f with the relation $R = \{(x, f(x)) \mid x \in X\}$ has no way of dealing with this situation: it is not possible to recover the intended co-domain Y from the relation R. As far as R is concerned, the co-domain of f could be any set that extends $\text{ran}(R)$.

More Than One Argument. Functions as introduced here are *unary*, i.e., they apply to only one argument. But of course, functions with more than one argument (*binary*, *ternary*...) do exist. E.g., addition and multiplication on $\mathbb{N}$ are of this kind. However, such a binary (ternary...) function can be viewed as a unary one that applies to ordered pairs (resp., triples...).

As we have seen, Haskell has predefined operations `curry` and `uncurry` to switch back and forth between functions of types `(a,b) -> c` and `a -> b -> c`. We can extend this to cases of functions that take triples, quadruples, etc. as arguments. As an example, here is the case for currying functions that take triples to functions that take three arguments, and for uncurrying functions that take three arguments to functions that take triples.

```
curry3   :: ((a,b,c) -> d) -> a -> b -> c -> d
curry3 f x y z = f (x,y,z)

uncurry3 :: (a -> b -> c -> d) -> (a,b,c) -> d
uncurry3 f (x,y,z) = f x y z
```

Fact 6.5 (Function Equality) If f and g are functions that share their domain ($\text{dom}(f) = \text{dom}(g)$) and, on it, carry out the same transformation (i.e., $\forall x \in \text{dom}(f)(f(x) = g(x))$), then — according to Extensionality 4.1, p. 113 — we have that $f = g$.

To establish that two functions $f, g : X \to Y$ are different we have to find an $x \in X$ with $f(x) \neq g(x)$. As we have seen in Section 4.2, there is no generic algorithm for checking function equality. Therefore, to establish that two functions $f, g : X \to Y$ are equal, we need a proof. The general form, spelled out in full, of such a proof is:

> Given: $f, g : X \to Y$.
> To be proved: $f = g$.
> Proof:
> > Let x be an arbitrary object in X.
> > To be proved: $f(x) = g(x)$.
> > Proof:
> > ...
> > Thus $f(x) = g(x)$.
>
> Thus $f = g$.

Example 6.6 If f and g are defined on $\mathbb{R}$ by $f(x) = x^2 + 2x + 1$, resp., $g(x) = (x+1)^2$, then $f = g$. Thus, functions are not distinguished by the details of how they actually transform (in this respect, f and g differ), but only with respect to their output-behaviour.

Warning. If functions $f : X \to Y$ and $g : X \to Z$ are given in the *domain-co-domain*-context, and $\forall x \in X(f(x) = g(x))$, then f and g count as equal only if we also have that $Y = Z$. The co-domains are taken as an integral part of the functions.

6.1. BASIC NOTIONS

Function Definitions. If $t(x)$ is an expression that describes an element of Y in terms of an element $x \in X$, then we can define a function $f : X \longrightarrow Y$ by writing:

Let the function $f : X \to Y$ be defined by $f(x) = t(x)$.

For completely specifying a function f three things are sufficient:

- Specify dom (f),
- Specify the co-domain of f,
- Give an instruction for how to construct $f(x)$ from x.

Examples of such instructions are $x \mapsto x + 1$, $x \mapsto x^2$, and in fact all the Haskell definitions of functions that we have encountered in this book.

A very convenient notation for function construction is by means of lambda abstraction (page 57). In this notation, $\lambda x.x + 1$ encodes the specification $x \mapsto x + 1$. The lambda operator is a variable binder, so $\lambda x.x + 1$ and $\lambda y.y + 1$ denote the same function. In fact, every time we specify a function foo in Haskell by means of

```
foo x y z = t
```

we can think of this as a specification $\lambda xyz.t$. If the types of x, y, z, t are known, this also specifies a domain and a co-domain. For if x :: a, y :: b, z :: c, t :: d, then $\lambda xyz.t$ has type a -> b -> c -> d.

Example 6.7

1. $t(x)$ is the expression $2x^2 + 3$:

    Let the function $g : \mathbb{R} \longrightarrow \mathbb{R}$ be defined by $g(x) = 2x^2 + 3$.

2. $t(x)$ is the expression $\int_0^x y \sin(y) dy$:

    Let the function $h : \mathbb{R} \longrightarrow \mathbb{R}$ be defined by $h(x) = \int_0^x y \sin(y)\, dy$.

Example 6.8 The Haskell way of defining functions is very close to standard mathematical practice, witness the following examples (note that all these equations define the same function):

```
f1 x = x^2 + 2 * x + 1
g1 x = (x + 1)^2
f1' = \x -> x^2 + 2 * x + 1
g1' = \x -> (x + 1)^2
```

Recurrences versus Closed Forms. A definition for a function $f : \mathbb{N} \to A$ in terms of algebraic operations is called a **closed form** definition. A function definition for f in terms of the values of f for smaller arguments is called a **recurrence** for f. The advantage of a closed form definition over a recurrence is that it allows for more efficient computation, since (in general) the computation time of a closed form does not grow exponentially with the size of the argument.

Example 6.9 Consider the following recurrence.

```
g 0 = 0
g n = g (n-1) + n
```

A closed form definition of the same function is:

```
g' n = ((n + 1) * n ) / 2
```

Exercise 6.10 Give a closed form implementation of the following function:

```
h 0 = 0
h n = h (n-1) + (2*n)
```

Exercise 6.11 Give a closed form implementation of the following function:

```
k 0 = 0
k n = k (n-1) + (2*n-1)
```

6.1. BASIC NOTIONS

To show that a particular closed form defines the same function as a given recurrence, we need a proof by induction: see Chapter 7. It is not always possible to find useful definitions in closed form, and recurrences are in general much easier to find than closed forms. E.g., a closed form for the factorial function $n!$ would be an expression that allows us to compute $n!$ with at most a fixed number of 'standard' operations. The number of operations should be independent of n, so

$$n! = \prod_{k=1}^{n} k = 1 \times \cdots \times (n-1) \times n$$

does not count, for $\cdots$ hides a number of product operations that *does* depend on n. No closed form for $n!$ is known, and $n! = \prod_{k=1}^{n} k$ performs essentially the same calculation as $n! = (n-1)! \times n$. Thus, computationally, there is nothing to choose between the following two implementations of $n!$.

```
fac 0 = 1
fac n = fac (n-1) * n

fac' n = product [ k | k <- [1..n] ]
```

Note that there is no need to add `fac' 0 = 1` to the second definition, because of the convention that `product []` gives the value 1.

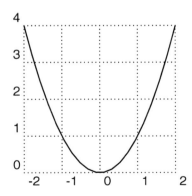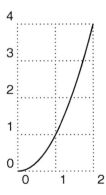

Figure 6.1: Restricting the function $\lambda x.x^2$.

A simple example of defining a function in terms of another function is the following (see Figure 6.1).

Definition 6.12 (Restrictions) Suppose that $f : X \longrightarrow Y$ and $A \subseteq X$. The *restriction* of f to A is the function $h : A \to Y$ defined by $h(a) = f(a)$. The notation for this function is $f {\upharpoonright} A$.

Here is the implementation, for functions implemented as type a -> b:

```
restrict :: Eq a => (a -> b) -> [a] -> a -> b
restrict f xs x | elem x xs = f x
                | otherwise = error "argument not in domain"
```

Here is the implementation for functions implemented as lists of pairs:

```
restrictPairs :: Eq a => [(a,b)] -> [a] -> [(a,b)]
restrictPairs xys xs = [ (x,y) |  (x,y) <- xys, elem x xs ]
```

Definition 6.13 (Image, Co-image) Suppose that $f : X \longrightarrow Y$, $A \subseteq X$ and $B \subseteq Y$.

1. $f[A] = \{f(x) \mid x \in A\}$ is called the *image* of A under f;

2. $f^{-1}[B] = \{x \in X \mid f(x) \in B\}$ is called the *co-image* of B under f.

From this definition we get:

1. $f[X] = \mathrm{ran}(f)$,

2. $f^{-1}[Y] = \mathrm{dom}\,(f)$,

3. $y \in f[A] \Leftrightarrow \exists x \in A(y = f(x))$,

4. $x \in f^{-1}[B] \Leftrightarrow f(x) \in B$.

From 3. it follows that $x \in A \Rightarrow f(x) \in f[A]$. But note that we do not necessarily have that $f(x) \in f[A] \Rightarrow x \in A$. (Example: $f = \{(0,2),(1,2)\}$, $x = 0$, $A = \{1\}$.)

6.1. BASIC NOTIONS

Two Types of Brackets. Distinguish $f(a)$ and $f[A]$. The notation $f(a)$ presupposes that $a \in \text{dom}(f)$. Then, $f(a)$ is the f-value of a. The notation $f[A]$ presupposes $A \subseteq \text{dom}(f)$. Then $f[A]$ is the *set* of values $f(x)$ where $x \in A$.

Remark. Many texts do not employ $f[\ ]$ but use $f(\ )$ throughout. In that case, you have to figure out from the context what is meant.

In the expression $f^{-1}[B]$, the part f^{-1} has no meaning when taken by itself. The notation f^{-1} will be used later on for the inverse function corresponding to f. Such an inverse only exists if f happens to be a *bijection*. However, the notation $f^{-1}[B]$ is always meaningful and does not presuppose bijectivity. ∎

Here are the implementations of *image* and *co-image*:

```
image :: Eq b => (a -> b) -> [a] -> [b]
image f xs = nub [ f x | x <- xs ]

coImage :: Eq b => (a -> b) -> [a] -> [b] -> [a]
coImage f xs ys = [ x | x <- xs, elem (f x) ys ]
```

This gives:

```
FCT> image (*2) [1,2,3]
[2,4,6]
FCT> coImage (*2) [1,2,3] [2,3,4]
[1,2]
```

Here are the versions for functions represented as lists of pairs:

```
imagePairs :: (Eq a, Eq b) => [(a,b)] -> [a] -> [b]
imagePairs f xs = image (list2fct f) xs

coImagePairs :: (Eq a, Eq b) => [(a,b)] -> [a] -> [b] -> [a]
coImagePairs f xs ys = coImage (list2fct f) xs ys
```

Exercise 6.14 Consider the relation $R = \{(0,4), (1,2), (1,3)\}$.

1. Is R a function?
   If so, determine $\text{dom}(R)$ and $\text{ran}(R)$.

2. Remember: $R^{-1} = \{(b,a) \mid (a,b) \in R\}$ is the inverse of the relation R. (Definition 5.7, p. 159).

   Is R^{-1} a function?
   If so, determine $\text{dom}(R^{-1})$ and $\text{ran}(R^{-1})$.

Exercise 6.15 Suppose that $f : X \longrightarrow Y$ and $A \subseteq X$. Verify:

1. $f[A] = \text{ran}(f{\restriction}A)$,

2. $f[\text{dom}(f)] = \text{ran}(f)$,

3. $f^{-1}[B] = \text{dom}(f \cap (X \times B))$,

4. $f^{-1}[\text{ran}(f)] = \text{dom}(f)$,

5. $f{\restriction}A = f \cap (A \times Y)$.

Exercise 6.16 Let $X = \{0,1,2,3\}$, $Y = \{2,3,4,5\}$,

$$f = \{(0,3),(1,2),(2,4),(3,2)\}.$$

Determine $f{\restriction}\{0,3\}$, $f[\{1,2,3\}]$ and $f^{-1}[\{2,4,5\}]$. Next, check your answers with the implementation.

Exercise 6.17 Suppose that $f : A \to Y$, $g : B \to Y$, and $A \cap B = \emptyset$. Show that
$$f \cup g : A \cup B \to Y.$$
What if $A \cap B \neq \emptyset$?

Exercise 6.18* Let $\mathcal{A}$ be a partition of X. For every component $A \in \mathcal{A}$ a function $f_A : A \to Y$ is given. Show, that $\bigcup_{A \in \mathcal{A}} f_A : X \to Y$.

Example 6.19 Suppose that $f : X \to Y$ and $A, B \subseteq X$. We show that $f[A - B] \supseteq f[A] - f[B]$.

Assume that $y \in f[A] - f[B]$. Then $y \in f[A]$ and $y \notin f[B]$. From the first we obtain $x \in A$ such that $y = f(x)$. From the second we see that $x \notin B$ (otherwise, $y = f(x) \in f[B]$). So, $x \in A - B$, and $y = f(x) \in f[A - B]$.

To see that the inclusion cannot be replaced by an equality, take $X = Y = A = \{0,1\}$, $B = \{1\}$, and let f be given by $f(x) = 0$. Then $f[A - B] = f[\{0\}] = \{0\}$ and $f[A] - f[B] = \{0\} - \{0\} = \emptyset$.

6.2. SURJECTIONS, INJECTIONS, BIJECTIONS

Next, suppose that $f : X \to Y$ and $C, D \subseteq Y$. We show that $f^{-1}[C - D] = f^{-1}[C] - f^{-1}[D]$:

$$\begin{aligned} x \in f^{-1}[C - D] &\iff f(x) \in C - D \\ &\iff (f(x) \in C) \wedge (f(x) \notin D) \\ &\iff x \in f^{-1}[C] \wedge x \notin f^{-1}[D] \\ &\iff x \in f^{-1}[C] - f^{-1}[D]. \end{aligned}$$

The required equality follows using Extensionality.

Exercise 6.20 Suppose that $f : X \to Y$, $A, B \subseteq X$, and $C, D \subseteq Y$. See Example 6.19. Show:

1. $A \subseteq B \Rightarrow f[A] \subseteq f[B]$;
   $C \subseteq D \Rightarrow f^{-1}[C] \subseteq f^{-1}[D]$,

2. $f[A \cup B] = f[A] \cup f[B]$;
   $f[A \cap B] \subseteq f[A] \cap f[B]$,

3. $f^{-1}[C \cup D] = f^{-1}[C] \cup f^{-1}[D]$;
   $f^{-1}[C \cap D] = f^{-1}[C] \cap f^{-1}[D]$,

4. $f[f^{-1}[C]] \subseteq C$;
   $f^{-1}[f[A]] \supseteq A$.

Give simple examples to show that the inclusions in 2 and 4 cannot be replaced by equalities.

6.2 Surjections, Injections, Bijections

If X is the domain of a function f, then for each $x \in X$ there is only one y with $(x, y) \in f$. However, there may be other elements $z \in X$ with $(z, y) \in f$. Functions for which this does not happen warrant a special name.

If f is a function from X to Y, then there may be elements of Y that are not in $f[X]$. Again, functions for which this does not happen warrant a special name.

Definition 6.21 (Surjections, Injections, Bijections)
A function $f : X \longrightarrow Y$ is called

1. *surjective*, or a *surjection*, or *onto Y*, if every element $b \in Y$ occurs as a function value of *at least* one $a \in X$, i.e., if $f[X] = Y$;

2. *injective*, an *injection*, or *one-to-one*, if every $b \in Y$ is value of *at most* one $a \in X$;

3. *bijective* or a *bijection* if it is both injective and surjective.

Example 6.22 Most functions are neither surjective, nor injective. For instance,

- $\sin : \mathbb{R} \to \mathbb{R}$ is not surjective (e.g., $2 \in \mathbb{R}$ is not a value) and not injective ($\sin 0 = \sin \pi$).

- The identity function $1_X : X \to X$ is a bijection, whatever the set X.

- Let A be a set. According to Theorems 5.89 and 5.96, the function that transforms an equivalence R on A into its quotient A/R is a bijection between the set of equivalences and the set of partitions on A.

- Consider $f = \{(0,1), (1,0), (2,1)\}$. Thus, $\mathrm{dom}\,(f) = \{0, 1, 2\}$. The function
$$f : \{0, 1, 2\} \longrightarrow \{0, 1\}$$
is surjective, but
$$f : \{0, 1, 2\} \longrightarrow \{0, 1, 2\}$$
is not surjective. The concept of surjectivity presupposes that of a co-domain. Cf. Fact 6.5 (p. 206). However, whatever this co-domain, f clearly is not injective, since 0 and 2 have the same image.

If the domain of a function is represented as a list, the injectivity test can be implemented as follows:

```
injective :: Eq b => (a -> b) -> [a] -> Bool
injective f [] = True
injective f (x:xs) =
   notElem (f x) (image f xs) && injective f xs
```

Similarly, if the domain and co-domain of a function are represented as lists, the surjectivity test can be implemented as follows:

6.2. SURJECTIONS, INJECTIONS, BIJECTIONS

```
surjective :: Eq b => (a -> b) -> [a] -> [b] -> Bool
surjective f xs [] = True
surjective f xs (y:ys) =
   elem y (image f xs) && surjective f xs ys
```

Exercise 6.23 Implement a test for bijectivity.

Exercise 6.24 Implement tests

injectivePairs, surjectivePairs, bijectivePairs

for functions represented as lists of pairs.

Proving that a Function is Injective/Surjective. The following implication is a useful way of expressing that f is injective:

$$f(x) = f(y) \implies x = y.$$

The proof schema becomes:

> *To be proved:* f is injective.
> *Proof:*
> Let x, y be arbitrary, and suppose $f(x) = f(y)$.
> $\vdots$
> Thus $x = y$.

The contraposition of $f(x) = f(y) \implies x = y$, i.e.,

$$x \neq y \implies f(x) \neq f(y),$$

of course says the same thing differently, so an equivalent proof schema is:

> *To be proved:* f is injective.
> *Proof:*
> Let x, y be arbitrary, and suppose $x \neq y$.
> $\vdots$
> Thus $f(x) \neq f(y)$.

That $f : X \to Y$ is surjective is expressed by:

$$\forall b \in Y \; \exists a \in X \; f(a) = b.$$

This gives the following pattern for a proof of surjectivity:

> To be proved: $f : X \to Y$ is surjective.
> Proof:
> Let b be an arbitrary element of Y.
> $\vdots$
> Thus there is an $a \in X$ with $f(a) = b$.

Exercise 6.25 Are the following functions injective? surjective?

1. $\sin : \mathbb{R}^+ \to \mathbb{R}$ (N.B.: $\mathbb{R}^+ = \{x \in \mathbb{R} \mid 0 < x\}$),
2. $\sin : \mathbb{R} \to [-1, +1]$,
3. $\sin : [-1, +1] \to [-1, +1]$,
4. $e^x : \mathbb{R} \to \mathbb{R}$,
5. $\tan : \mathbb{R} \to \mathbb{R}$,
6. $\log : \mathbb{R}^+ \to \mathbb{R}$,
7. $\sqrt{\ } : \mathbb{R}^+ \to \mathbb{R}^+$.

Remark. The functions of Exercise 6.25 are all predefined in Haskell: `sin`, `exp`, `tan`, `log`, `sqrt`. The base of the natural logarithm, Napier's number e, is given by `exp 1`.

Exercise 6.26 Give a formula for the number of injections from an n-element set A to a k-element set B.

Exercise 6.27 Implement a function

```
injs :: [Int] -> [Int] -> [[(Int,Int)]]
```

that takes a finite domain and a finite codomain of type `Int` and produces the list of all injections from domain to codomain, given as lists of integer pairs.

6.3. FUNCTION COMPOSITION

Exercise 6.28 The bijections on a finite set A correspond exactly to the permutations of A. Implement a function

```
perms :: [a] -> [[a]]
```

that gives all permutations of a finite list. The call `perms [1,2,3]` should yield:

[[1,2,3],[2,1,3],[2,3,1],[1,3,2],[3,1,2],[3,2,1]]

Hint: to get the permutations of (x:xs), take all the possible ways of inserting x in a permutation of xs.

6.3 Function Composition

Definition 6.29 (Composition) Suppose that $f : X \longrightarrow Y$ and $g : Y \longrightarrow Z$. Thus, the *co-domain* of f coincides with the *domain* of g. The *composition* of f and g is the function $g \circ f : X \longrightarrow Z$ defined by

$$(g \circ f)(x) = g(f(x)).$$

("First, apply f, next, apply g" — thanks to the usual "prefix"-notation for functions, the f and the g are unfortunately in the reverse order in the notation $g \circ f$. To keep this reverse order in mind it is good practice to refer to $g \circ f$ as "g after f".)

N.B.: The notation $g \circ f$ presupposes that the co-domain of f and the domain of g are the same. Furthermore, $g \circ f$ has the domain of f and the co-domain of g.

Function composition is predefined in Haskell, as follows:

```
(.)         :: (b -> c) -> (a -> b) -> (a -> c)
(f . g) x   = f (g x)
```

Example 6.30 Haskell has a general procedure `negate` for negating a number. The effect of first taking the absolute value, then negating can now be got by means of (negate . abs):

```
Prelude> (negate . abs) 5
-5
Prelude> (negate . abs) (-7)
-7
```

Example 6.31 Another example from *Prelude.hs* is:

```
even n            = n 'rem' 2 == 0
odd               = not . even
```

Exercise 6.32 Implement an operation comp for composition of functions represented as lists of pairs.

Example 6.33 In analysis, compositions abound. E.g., if $f : \mathbb{R} \to \mathbb{R}^+$ is $x \mapsto x^2 + 1$, and $g : \mathbb{R}^+ \to \mathbb{R}$ is $x \mapsto \sqrt{x}$, then $g \circ f : \mathbb{R} \to \mathbb{R}$ is $x \mapsto \sqrt{x^2 + 1}$: $(g \circ f)(x) = g(f(x)) = g(x^2 + 1) = \sqrt{x^2 + 1}$.

The identity function behaves as a unit element for composition (see Example 6.1 for 1_X):

Fact 6.34 If $f : X \to Y$, then $f \circ 1_X = 1_Y \circ f = f$.

Suppose that $f : X \to Y$, $g : Y \to Z$ and $h : Z \to U$. There are now *two* ways to define a function from X to U: (i) $(h \circ g) \circ f$, and (ii) $= h \circ (g \circ f)$. The next lemma says that these functions coincide. Thus, composition is *associative*; and we can safely write $h \circ g \circ f$ if we mean either of these.

Lemma 6.35 *If* $f : X \to Y$, $g : Y \to Z$ *and* $h : Z \to U$, *then* $(h \circ g) \circ f = h \circ (g \circ f)$.

Proof. Suppose that $x \in X$. Then:
$((h \circ g) \circ f)(x) = (h \circ g)(f(x)) = h(g(f(x))) = h((g \circ f)(x)) = (h \circ (g \circ f))(x)$.
The functions $(h \circ g) \circ f$ and $h \circ (g \circ f)$ have the same domain (X) and co-domain (U) and, on this domain, show the same action. Thus, they are (Fact 6.5, p. 206) equal. ∎

Lemma 6.36 *Suppose that* $f : X \longrightarrow Y$, $g : Y \longrightarrow Z$. *Then:*

1. $g \circ f$ *injective* $\Longrightarrow f$ *injective*,

2. $g \circ f$ *surjective* $\Longrightarrow g$ *surjective*,

3. f *and* g *injective* $\Longrightarrow g \circ f$ *injective*,

4. f *and* g *surjective* $\Longrightarrow g \circ f$ *surjective*.

6.3. FUNCTION COMPOSITION

Proof. We prove 1, 2 and 4.

1. Given: $g \circ f$ injective, i.e., $(g \circ f)(a_1) = (g \circ f)(a_2) \Rightarrow a_1 = a_2$.
To be proved: f injective, i.e., $f(a_1) = f(a_2) \Rightarrow a_1 = a_2$.
Proof: Assume $f(a_1) = f(a_2)$. Then of course, $g(f(a_1)) = g(f(a_2))$. (Applying g twice to the same argument must produce the same value twice.) But then, $(g \circ f)(a_1) = g(f(a_1)) = g(f(a_2)) = (g \circ f)(a_2)$. The given now shows that $a_1 = a_2$.

2. Given: $g \circ f$ surjective.
To be proved: g surjective. I.e., $\forall c \in Z \exists b \in Y(g(b) = c)$.
Proof: Assume $c \in Z$. Wanted: $b \in Y$ such that $g(b) = c$. Since $g \circ f$ is surjective, there is $a \in X$ such that $(g \circ f)(a) = c$. But, $g(f(a)) = (g \circ f)(a)$. I.e., $b = f(a)$ is the element looked for.

4. Given: f and g surjective.
To be proved: $g \circ f$ surjective. I.e., every $c \in Z$ is a value of $g \circ f$.
Proof: Assume $c \in Z$. Since g is surjective, $b \in Y$ exists such that $g(b) = c$. Since f is surjective, $a \in X$ exists such that $f(a) = b$. It follows that $(g \circ f)(a) = g(f(a)) = g(b) = c$. So, a is the element looked for. ∎

Note that there are statements in the spirit of Lemma 6.36 that do not hold. In particular, from the fact that $g \circ f$ is injective, it does not follow that g is injective. Consider the case where f is the (empty) function from $\emptyset$ to Y, and g an arbitrary non-injective function from $Y \to Z$. Then $g \circ f$ is the (empty) function from $\emptyset$ to Z, which is surely injective. But g by assumption is not injective.

For an example in which $g \circ f$ is bijective, but g not injective and f not surjective, take $f : \mathbb{N} \to \mathbb{N}$ given by $f(n) = n+1$ and $g : \mathbb{N} \to \mathbb{N}$ given by

$$g(n) = \begin{cases} n & \text{if } n = 0 \\ n-1 & \text{if } n \geq 1. \end{cases}$$

Clearly, $g \circ f$ is the identity on $\mathbb{N}$, but f is not a surjection, and g is not an injection.

Exercise 6.37 Can you think up an example with $g \circ f$ bijective, but g not injective and f not surjective, with the domain of f finite? If yes, give the example, if no, prove that this is impossible.

Exercise 6.38 The function $f : \{0, 1, 2, 3, 4\} \to \{0, 1, 2, 3, 4\}$ is defined by the following table:

x	0	1	2	3	4
$f(x)$	1	2	0	0	3
$(ff)(x)$					
$(fff)(x)$					
$(ffff)(x)$					

1. Determine the compositions $f \circ f$, $f \circ f \circ f$ and $f \circ f \circ f \circ f$ by completing the table.

2. How many elements has the set $\{f, f \circ f, f \circ f \circ f, \ldots\}$? (N.B.: the elements are functions!)

3. Exhibit a function $g : \{0,1,2,3,4\} \to \{0,1,2,3,4\}$ such that $\{g, g \circ g, g \circ g \circ g, \ldots\}$ has 6 elements.

Exercise 6.39* Suppose that A is a finite set and $f : A \to A$ is a bijection. Then $f^1 = f$, $f^2 = f \circ f$, $f^3 = f \circ f \circ f, \ldots$ all are bijections : $A \to A$.

1. Show that, somewhere in this sequence, there occurs the bijection 1_A. I.e., a number n exists such that $f^n = 1_A$.

2. Suppose that A has k elements. Can you give an upper bound for n?

Exercise 6.40 Suppose that $h : X \to X$ satisfies $h \circ h \circ h = 1_X$. Show that h is a bijection.

Exhibit a simple example of a set X and a function $h : X \to X$ such that $h \circ h \circ h = 1_X$, whereas $h \neq 1_X$.

Exercise 6.41 Prove Lemma 6.36.3.

Exercise 6.42 Suppose that $f : X \to Y$ and $g : Y \to Z$ are such that $g \circ f$ is bijective. Show that f is surjective iff g is injective.

Exercise 6.43 Suppose that $\lim_{i \to \infty} a_i = a$, and that $f : \mathbb{N} \to \mathbb{N}$ is injective. Show that $\lim_{i \to \infty} a_{f(i)} = a$. (Cf. also Exercise 8.22, p. 311.)

6.4 Inverse Function

If we consider $f : X \to Y$ as a relation, then we can consider its relational inverse: the set of all (y, x) with $(x, y) \in f$. However, there is no guarantee that the relational inverse of a function is again a function. In case f is injective, we know that the relational inverse of f is a partial function (some elements in the domain may not have an image). If f is also surjective, we know that the relational inverse of f is a function. Thus, an inverse function of f has to satisfy some special requirements.

6.4. INVERSE FUNCTION

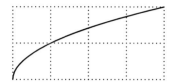

Figure 6.2: Inverse of the function $\lambda x.x^2$ (restricted to $\mathbb{R}^+$).

Definition 6.44 (Inverse Function) Suppose that $f : X \to Y$. A function $g : Y \to X$ is an *inverse* of f if both (i) $g \circ f = 1_X$, and (ii) $f \circ g = 1_Y$.

The next theorem says all there is to know about inverses. Its proof describes how to find an inverse if there is one.

Note that, by the first part of the theorem, we can safely talk about *the* inverse of a function (provided there is one).

Theorem 6.45

1. *A function has at most one inverse.*

2. *A function has an inverse iff it is bijective.*

Proof. (1) Suppose that g and h are both inverses of $f : X \to Y$. Then $g = 1_X \circ g = (h \circ f) \circ g = h \circ (f \circ g) = h \circ 1_Y = h$.

(2) (Only if.) Assume that g is inverse of f. Then since $g \circ f = 1_X$ is injective, by Lemma 6.36.1 f also is injective. Since $f \circ g = 1_Y$ is surjective, by Lemma 6.36.2 f also is surjective.

(If.) Suppose that f is a bijection. I.e., for every $y \in Y$ there is exactly one $x \in X$ such that $f(x) = y$. Thus, we can define a function $g : Y \to X$ by letting $g(y)$ be the unique x such that $f(x) = y$.

Then g is inverse of f: firstly, if $x \in X$, then $g(f(x)) = x$; secondly, if $y \in Y$, then $f(g(y)) = y$. ∎

Notation. If $f : X \to Y$ is a bijection, then its unique inverse is denoted by f^{-1}.

Remark. If $f : X \to Y$ is a bijection, then f^{-1} can denote either the inverse function $: Y \to X$, or the inverse of f considered as a relation. But from the proof of Theorem 6.45 it is clear that these are the same. ∎

Example 6.46 The real function f that is given by $f(x) = \frac{9}{5}x + 32$ allows us to convert degrees Celcius into degrees Fahrenheit. The inverse function f^{-1} is given by $f^{-1}(x) = \frac{5}{9}(x - 32)$; it converts degrees Fahrenheit back into degrees Celsius. Here are integer approximations:

```
c2f, f2c :: Int -> Int
c2f x = div (9 * x) 5 + 32
f2c x = div (5 * (x - 32)) 9
```

*Left and Right-inverse. Note that there are *two* requirements on inverse functions. If $f : X \to Y$, $g : Y \to X$, and we have that $g \circ f = 1_X$ only, then g is called *left-inverse* of f and f *right-inverse* of g.

Example 6.47 The class *Enum* is (pre-)defined in Haskell as follows:

```
class Enum a where
    succ, pred       :: a -> a
    toEnum           :: Int -> a
    fromEnum         :: a -> Int
```

fromEnum should be a *left-inverse* of toEnum:

fromEnum (toEnum x) = x

This requirement cannot be expressed in Haskell, so it is the responsibility of the programmer to make sure that it is satisfied. Examples of use of toEnum and fromEnum from *Prelude.hs*:

```
ord              :: Char -> Int
ord              = fromEnum
chr              :: Int -> Char
chr              = toEnum
```

Exercise 6.48 Show: if $f : X \to Y$ has left-inverse g and right-inverse h, then f is a bijection and $g = h = f^{-1}$.

6.4. INVERSE FUNCTION

Exercise 6.49 Suppose that $f : X \to Y$ and $g : Y \to X$. Show that the following are equivalent.

1. $g \circ f = 1_X$,
2. $\{(f(x), x) \mid x \in X\} \subseteq g$.

Exercise 6.50 $X = \{0, 1\}$, $Y = \{2, 3, 4, 5\}$, $f = \{(0, 3), (1, 4)\}$. We have that $f : X \to Y$. How many functions $g : Y \to X$ have the property, that $g \circ f = 1_X$?

Exercise 6.51 Give an example of an injection $f : X \to Y$ for which there is no $g : Y \to X$ such that $g \circ f = 1_X$.

Exercise 6.52* Show that if $f : X \to Y$ is surjective, a function $g : Y \to X$ exists such that $f \circ g = 1_Y$.

Exercise 6.53 How many right-inverses are there to the function

$$\{(0, 5), (1, 5), (2, 5), (3, 6), (4, 6)\}$$

(domain: $\{0, 1, 2, 3, 4\}$, co-domain: $\{5, 6\}$)?

Exercise 6.54   1. The surjection $f : \mathbb{R} \to \mathbb{R}^+$ is defined by $f(x) = x^2$. Give three different right-inverses for f.

2. Same question for $g : [0, \pi] \to [0, 1]$ defined by $g(x) = \sin x$.

Exercise 6.55 Suppose that $f : X \to Y$ is a surjection and $h : Y \to X$. Show that the following are equivalent.

1. h is right-inverse of f,
2. $h \subseteq \{(f(x), x) \mid x \in X\}$.

Exercise 6.56* Show:

1. Every function that has a surjective right-inverse is a bijection.
2. Every function that has an injective left-inverse is a bijection.

6.5 Partial Functions

A partial function from X to Y is a function with its domain included in X and its range included in Y. If f is a partial function from X to Y we write this as $f : X \hookrightarrow Y$. It is immediate from this definition that $f : X \hookrightarrow Y$ iff $\operatorname{dom}(f) \subseteq X$ and $f \upharpoonright \operatorname{dom}(f) : \operatorname{dom}(f) \to Y$.

A way of defining a partial function (using $\bot$ for 'undefined'):

$$f(x) = \begin{cases} \bot & \text{if ...} \\ t & \text{otherwise} \end{cases}$$

The computational importance of partial functions is in the systematic perspective they provide on exception handling. In Haskell, the crude way to deal with exceptions is by a call to the error abortion function `error`. The code below implements partial functions `succ0` and `succ1`. `succ0` is partial because the pattern (x+1) only matches *positive* integers. `succ1` has an explicit call to `error`. The disadvantage of these implementations is that they are called by another program, the execution of that other program may abort.

```
succ0 :: Integer -> Integer
succ0 (x+1) = x + 2

succ1 :: Integer -> Integer
succ1 = \ x -> if x < 0
               then error "argument out of range"
               else x+1
```

This uses the reserved keywords `if`, `then` and `else`, with the obvious meanings.

A useful technique for implementing partial functions is to represent a partial function from type `a` to type `b` as a function of type `a -> [b]`. In case of an exception, the empty list is returned. If a regular value is computed, the unit list with the computed value is returned.

```
succ2 :: Integer -> [Integer]
succ2 = \ x -> if x < 0 then [] else [x+1]
```

Composition of partial functions implemented with unit lists can be defined as follows:

6.5. PARTIAL FUNCTIONS

```
pcomp :: (b -> [c]) -> (a -> [b]) -> a -> [c]
pcomp g f = \ x -> concat [ g y | y <- f x ]
```

As an alternative to this trick with unit lists Haskell has a special data type for implementing partial functions, the data type Maybe, which is predefined as follows.

```
data Maybe a = Nothing | Just a
               deriving (Eq, Ord, Read, Show)
maybe              :: b -> (a -> b) -> Maybe a -> b
maybe n f Nothing  = n
maybe n f (Just x) = f x
```

Here is a third implementation of the partial successor function:

```
succ3 :: Integer -> Maybe Integer
succ3 = \ x -> if x < 0 then Nothing else Just (x+1)
```

The use of the predefined function maybe is demonstrated in the definition of composition for functions of type a -> Maybe b.

```
mcomp :: (b -> Maybe c) -> (a -> Maybe b) -> a -> Maybe c
mcomp g f = (maybe Nothing g) . f
```

Of course, the maybe function allows for all kinds of ways to deal with exceptions. E.g., a function of type a -> Maybe b can be turned into a function of type a -> b by the following part2error conversion.

```
part2error :: (a -> Maybe b) -> a -> b
part2error f = (maybe (error "value undefined") id) . f
```

Exercise 6.57 Define a partial function

```
stringCompare :: String -> String -> Maybe Ordering
```

for ordering strings consisting of alphabetic characters in the usual list order. If a non-alphabetic symbol occurs, the ordering function should return `Nothing`. Use `isAlpha` for the property of being an alphabetic character.

6.6 Functions as Partitions

In practice, equivalences are often defined by way of functions. *Par abus de language* functions sometimes are called partitions for that reason. Examples of such functions on the class of all people: "the gender of x" (partitions in males and females), "the color of x" (partitions in races), "the age of x" (some hundred equivalence classes). The next exercise explains how this works and asks to show that every equivalence is obtained in this way.

Exercise 6.58 Suppose that $f : A \to I$ is a surjection. Define the relation R on A by: $aRb :\equiv f(a) = f(b)$. Thus, $R = \{(a,b) \in A^2 \mid f(a) = f(b)\}$. Show:

1. R is an equivalence on A,

2. $A/R = \{f^{-1}[\{i\}] \mid i \in I\}$,

3. for every equivalence S on A there is a function g on A such that $aSb \Leftrightarrow g(a) = g(b)$.

Example 6.59 For any $n \in \mathbb{Z}$ with $n \neq 0$, let the function $\mathrm{RM}_n :: \mathbb{Z} \to \mathbb{Z}$ be given by $\mathrm{RM}_n(m) := r$ where $0 \leqslant r < n$ and there is some $a \in \mathbb{Z}$ with $m = an + r$. Then RM_n induces the equivalence $\equiv_n$ on $\mathbb{Z}$.

Here is a Haskell implementation of a procedure that maps a function to the equivalence relation inducing the partition that corresponds with the function:

```
fct2equiv :: Eq a => (b -> a) -> b -> b -> Bool
fct2equiv f x y = (f x) == (f y)
```

You can use this to test equality modulo n, as follows:

6.6. FUNCTIONS AS PARTITIONS

```
FCT> fct2equiv ('rem' 3) 2 14
True
```

Exercise 6.60* Suppose that $f : A \to B$.

1. Show: If f is an injection, then for all sets C and for every $g : A \to C$ there is a function $h : B \to C$ such that $g = h \circ f$.

2. Show: For all sets C, if to every $g : A \to C$ there is a function $h : B \to C$ such that $g = h \circ f$, then f is an injection.

Exercise 6.61* Suppose that R is an equivalence on the set A. Show: for every equivalence $S \supseteq R$ on A there exists a function $g : A/R \to A/S$ such that, for $a \in A$: $|a|_S = g(|a|_R)$.

Exercise 6.62* Suppose that $\sim$ is an equivalence on A, and that $f : A^2 \to A$ is a binary function such that for all $a, b, x, y \in A$:

$$a \sim x \wedge b \sim y \implies f(a,b) \sim f(x,y).$$

Show that a unique function $f^\sim : (A/\sim)^2 \to B$ exists such that, for $a, b \in A$: $f^\sim(|a|, |b|) = |f(a,b)|$.

Exercise 6.63* Suppose that $\sim$ is an equivalence on A, and that $R \subseteq A^2$ is a relation such that for all $a, b, x, y \in A$:

$$a \sim x \wedge b \sim y \wedge aRb \implies xRy.$$

Show that a unique relation $R^\sim \subseteq (A/\sim)^2$ exists such that for all $a, b \in A$: $|a|R^\sim|b| \Leftrightarrow aRb$.

Exercise 6.64* A and B are sets, with $B \neq \emptyset$. Define $\sim$ on $A \times B$ by: $(a,b) \sim (x,y) \equiv a = x$.

1. Show that $\sim$ is an equivalence on $A \times B$.

2. Exhibit a bijection : $(A \times B)/\sim \longrightarrow A$ from the quotient of $A \times B$ modulo $\sim$ to A.

3. Exhibit, for every equivalence class, a bijection between the class and B.

Equivalence classes (restricted to a list) for an equivalence defined by a function are generated by the following Haskell function:

```
block :: Eq b => (a -> b) -> a -> [a] -> [a]
block f x list = [ y | y <- list, f x == f y ]
```

This gives:

```
FCT> block ('rem' 3) 2 [1..20]
[2,5,8,11,14,17,20]
FCT> block ('rem' 7) 4 [1..20]
[4,11,18]
```

Exercise 6.65 Functions can be used to generate equivalences, or equivalently, partitions. In an implementation we use *list partitions*; see Exercise 5.111 for a definition. Implement an operation fct2listpart that takes a function and a domain and produces the list partition that the function generates on the domain. Some example uses of the operation are:

```
Main> fct2listpart even [1..20]
[[1,3,5,7,9,11,13,15,17,19],[2,4,6,8,10,12,14,16,18,20]]
Main> fct2listpart (\ n -> rem n 3) [1..20]
[[1,4,7,10,13,16,19],[2,5,8,11,14,17,20],[3,6,9,12,15,18]]
```

Exercise 6.66 Give an formula for the number of surjections from an n-element set A to a k-element set B. (Hint: each surjection $f : A \to B$ induces a partition. These partitions can be counted with the technique from Example 5.105.)

6.7 Products

Definition 6.67 (Product) Suppose that, for every element $i \in I$ a non-empty set X_i is given. The *product* $\prod_{i \in I} X_i$ is the set of all functions f for which $\text{dom}(f) = I$ and such that for all $i \in I$: $f(i) \in X_i$.
When $I = \{0, \ldots, n-1\}$, this product is also written as $X_0 \times \cdots \times X_{n-1}$.
If all X_i ($i \in I$) are the same, $X_i = X$, the product is written as X^I. Thus, X^I is the set of all functions $f : I \to X$.

Exercise 6.68* There are now two ways to interpret the expression $X_0 \times X_1$: (i) as $\prod_{i \in \{0,1\}} X_i$, and (ii) as $\{(x,y) \mid x \in X_0 \land y \in X_1\}$. Can you explain why there is no harm in this?

In our implementation language, product types have the form (a,b), (a,b,c), etcetera.

Exercise 6.69* Let A be any set. Exhibit two different bijections between $\wp(A)$ and $\{0,1\}^A$.

Exercise 6.70* Suppose that X and Y are sets. On the set of functions $Y^X = \{f \mid f : X \to Y\}$, the relation $\approx$ is defined by: $f \approx g \equiv$ there are bijections $i : Y \to Y$ and $j : X \to X$ such that $i \circ f = g \circ j$.

1. Show that $\approx$ is an equivalence.

2. Show: if $f, g : X \to Y$ are injective, then $f \approx g$.

3. Suppose that $Y = \{0, 1, 2\}$ and $X = \{0, 1, 2, 3\}$.

    (a) Show that $\{(0,0), (1,0), (2,1)\} \approx \{(0,1), (1,3), (2,3)\}$.

    (b) How many equivalence classes has $\approx$? For every class, produce one representative.

Exercise 6.71* Suppose that X, Y and Z are sets and that $h : Y \to Z$. Define $F : Y^X \to Z^X$ by $F(g) := h \circ g$. Show:

1. if h is injective, then F is injective,

2. if h is surjective, then F is surjective.

Exercise 6.72* Suppose that $X \neq \emptyset$, Y, and Z are sets and that $h : X \to Y$. Define $F : Z^Y \to Z^X$ by $F(g) := g \circ h$. Show:

1. if h is injective, then F is surjective,

2. if h is surjective, then F is injective.

6.8 Congruences

A function $f : X^n \to X$ is called an n-ary operation on X. Addition and multiplication are binary operations on $\mathbb{N}$ (on $\mathbb{Z}$, on $\mathbb{Q}$, on $\mathbb{R}$, on $\mathbb{C}$).

If one wants to define new structures from old, an important method is taking quotients for equivalences that are compatible with certain operations.

Definition 6.73 (Congruence) If f be an n-ary operation on A, and R an equivalence on A, then R is a **congruence** for f (or: R is **compatible with** f) if for all

$$x_1, \ldots, x_n, y_1, \ldots, y_n \in A : \quad x_1 R y_1, \ldots, x_n R y_n$$

imply that

$$f(x_1, \ldots, x_n) R f(y_1, \ldots, y_n).$$

If R is a congruence for f, then the operation induced by f on A/R is the operation $f_R : (A/R)^n \to A/R$ given by

$$f_R(|a_1|_R, \ldots, |a_n|_R) := |f(a_1, \ldots, a_n)|_R.$$

If (A, f) is a set with an operation f on it and R is a congruence for f, then $(A/R, f_R)$ is the *quotient structure* defined by R.

Example 6.74 Consider the modulo n relation on $\mathbb{Z}$. Suppose $m \equiv_n m'$ and $k \equiv_n k'$. Then (Proposition 5.67) there are $a, b \in \mathbb{Z}$ with $m' = m + an$ and $k' = k + bn$. Thus $m' + k' = m + k + (a + b)n$, i.e., $m + k \equiv_n m' + k'$. This shows that $\equiv_n$ is a congruence for addition. Similarly, it can be shown that $\equiv_n$ is a congruence for subtraction. It follows that we can define $[m]_n + [k]_n := [m + k]_n$ and $[m]_n - [k]_n := [m - k]_n$.

Exercise 6.75 Show that $\equiv_n$ on $\mathbb{Z}$ is a congruence for multiplication, for any $n \in \mathbb{Z}$ with $n \neq 0$.

Example 6.76 Is $\equiv_n$ on $\mathbb{Z}$ ($n \neq 0$) a congruence for exponentiation, i.e., is it possible to define exponentiation of classes in $\mathbb{Z}_n$ by means of:

$$([k]_n)^{([m]_n)} := [k^m]_n, \text{ for } k \in \mathbb{Z}, m \in \mathbb{N}?$$

No, for consider the following example: $([2]_3)^{([1]_3)} = [2^1]_3 = [2]_3$. Since $1 \equiv_3 4$ we also have: $([2]_3)^{([1]_3)} = ([2]_3)^{([4]_3)} = [2^4]_3 = [16]_3 = [1]_3 \neq [2]_3$. What this shows is that the definition is not independent of the class representatives. Therefore, $\equiv_n$ is *not* a congruence for exponentiation.

Example 6.77 The definition of the integers from the natural numbers, in Section 7.2 below, uses the fact that the relation R on $\mathbb{N}^2$ given by

$$(m_1, m_2) R (n_1, n_2) :\equiv m_1 + n_2 = m_2 + n_1$$

is an equivalence relation, that the relation $<_R$ on $\mathbb{N}^2/R$ given by

$$|(m_1, m_2)|_R <_R |(n_1, n_2)|_R :\equiv m_1 + n_2 < m_2 + n_1$$

is properly defined, and moreover, that R is a congruence for addition and multiplication on $\mathbb{N}^2$.

To check that R is a congruence for addition on $\mathbb{N}^2$, where addition on $\mathbb{N}^2$ is given by

$$(m_1, m_2) + (n_1, n_2) := (m_1 + n_1, m_2 + n_2),$$

we have to show that $(m_1, m_2)R(p_1, p_2)$ and $(n_1, n_2)R(q_1, q_2)$ together imply that $(m_1, m_2) + (n_1, n_2)R(p_1, p_2) + (q_1, q_2)$.

Assume $(m_1, m_2)R(p_1, p_2)$ and $(n_1, n_2)R(q_1, q_2)$. Applying the definition of R, this gives

$$m_1 + p_2 = p_1 + m_2 \text{ and } n_1 + q_2 = q_1 + n_2,$$

whence

$$m_1 + n_1 + p_2 + q_2 = n_1 + p_1 + m_2 + n_2. \quad (*)$$

We now have:

$$(m_1, m_2) + (n_1, n_2) = (m_1 + n_1, m_2 + n_2),$$
$$(p_1, p_2) + (q_1, q_2) = (p_1 + q_1, p_2 + q_2),$$

and by the definition of R we get from (*) that

$$(m_1, m_2) + (n_1, n_2)R(p_1, p_2) + (q_1, q_2).$$

This proves that R is a congruence for addition.

Exercise 6.78* Show that the relation R on $\mathbb{N}^2$ from Example 6.77 is a congruence for the multiplication operation on $\mathbb{N}^2$ given by:

$$(m_1, m_2) \cdot (n_1, n_2) := (m_1 n_1 + m_2 n_2, m_1 n_2 + n_1 m_2).$$

6.9 Further Reading

More on functions in the context of set theory in [DvDdS78]. A logical theory of functions is developed in the lambda calculus. See [Bar84]. A more compact presentation of lambda calculus, with applications in computer science, can be found in [Han04].

Chapter 7

Induction and Recursion

Preview

A very important proof method that is not covered by the recipes from Chapter 3 is the method of proof by Mathematical Induction. Roughly speaking, mathematical induction is a method to prove things about objects that can be built from a finite number of ingredients in a finite number of steps. Such objects can be thought of as construed by means of recursive definitions. Thus, as we will see in this chapter, recursive definitions and inductive proofs are two sides of one coin.

```
module IAR

where

import List
import STAL (display)
```

7.1 Mathematical Induction

Mathematical induction is a proof method that can be used to establish the truth of a statement for an infinite sequence of cases 0, 1, 2, ... Let $P(n)$ be a property of natural numbers. To prove a goal of the form $\forall n \in \mathbb{N} : P(n)$ one can proceed as follows:

1. *Basis.* Prove that 0 has the property P.

2. *Induction step.* Assume the *induction hypothesis* that n has property P. Prove on the basis of this that $n+1$ has property P.

That's all. The goal $\forall n \in \mathbb{N} : P(n)$ follows from this by the principle of mathematical induction.

By the principle of mathematical induction we mean the following fact:

Fact 7.1 For every set $X \subseteq \mathbb{N}$, we have that:
if $0 \in X$ and $\forall n \in \mathbb{N}(n \in X \Rightarrow n+1 \in X)$, then $X = \mathbb{N}$.

This fact is obviously true.

The best way to further explain mathematical induction is by way of examples.

Example 7.2 Sum of the Angles of a Convex Polygon.

Suppose we want to prove that the sum of the angles of a convex polygon of $n+3$ sides is $(n+1)\pi$ radians.

We can show this by mathematical induction with respect to n, as follows:

Basis For $n = 0$, the statement runs: the sum of the angles of a convex polygon of 3 sides, i.e., of a triangle, is π radians. We know from elementary geometry that this is true.

Induction step Assume that the sum of the angles of a convex polygon of $n+3$ sides is $(n+1)\pi$ radians. Take a convex polygon P of $n+4$ sides. Then, since P is convex, we can decompose P into a triangle T and a convex polygon P' of $n+3$ sides (just connect edges 1 and 3 of P). The sum of the angles of P equals the sum of the angles of T, i.e. π radians, plus the sum of the angles of P', i.e., by the induction hypothesis, $(n+1)\pi$ radians. This shows that the sum of the angles of P is $(n+2)\pi$ radians.

From 1. and 2. and the principle of mathematical induction the statement follows.

7.1. MATHEMATICAL INDUCTION

Notation The next examples all involve sums of the general form $a_1 + a_2 + \cdots + a_n$, written in summation notation (cf. Example 2.29 above) as $\Sigma_{k=1}^{n} a_k$. Note that sum is the computational counterpart to $\sum$. We agree that the "empty sum" $\Sigma_{k=1}^{0} a_k$ yields 0. The same convention is followed in computations with sum, for sum [] has value 0.

Example 7.3 The Sum of the First n Odd Numbers.

The sum of the first n odd numbers equals n^2. More formally:

$$\sum_{k=1}^{n}(2k-1) = n^2.$$

Proof by induction:

Basis For $n = 1$, we have $\sum_{k=1}^{1}(2k-1) = 1 = 1^2$.

Induction step Assume $\sum_{k=1}^{n}(2k-1) = n^2$. We have to show $\sum_{k=1}^{n+1}(2k-1) = (n+1)^2$. Indeed, $\sum_{k=1}^{n+1}(2k-1) = \sum_{k=1}^{n}(2k-1) + 2(n+1) - 1$. Using the induction hypothesis, this gives:

$$\sum_{k=1}^{n+1}(2k-1) = n^2 + 2n + 1 = (n+1)^2.$$

The closed formula gives us an improved computation procedure for summing the odd numbers: sumOdds performs better on large input than sumOdds'.

```
sumOdds' :: Integer -> Integer
sumOdds' n = sum [ 2*k - 1 | k <- [1..n] ]

sumOdds :: Integer -> Integer
sumOdds n = n^2
```

Note that the method of proof by mathematical induction may obscure the process of finding the relation in the first place. To see why the sum of the first n odd numbers equals n^2, it is instructive to look at the following picture of the sum $1 + 3 + 5 + 7 + 9$.

Example 7.4 The Sum of the First n Even Numbers.

What about the sum of the first n even natural numbers? Again, a picture suggests the answer. Look at the following representation of $2 + 4 + 6 + 8 + 10$:

Again, a picture is not (quite) a proof. Here is proof by mathematical induction of the fact that $\sum_{k=1}^{n} 2k = n(n+1)$.

Basis Putting $k = 1$ gives $2 = 1 \cdot 2$, which is correct.

Induction step Assume $\sum_{k=1}^{n} 2k = n(n+1)$. Then $\sum_{k=1}^{n+1} 2k = \sum_{k=1}^{n} 2k + 2(n+1)$. Using the induction hypothesis we see that this is equal to $n(n+1) + 2(n+1) = n^2 + 3n + 2 = (n+1)(n+2)$, and we are done.

Notice that we left the term for $k = 0$ out of the sum. We might as well have included it, for it holds that $\sum_{k=0}^{n} 2k = n(n+1)$. By the convention about empty sums, the two versions are equivalent.

From what we found for the sum of the first n even natural numbers the formula for the sum of the first n positive natural numbers follows immediately:

$$\sum_{k=1}^{n} k = \frac{n(n+1)}{2}.$$

Again, we get improved computation procedures from the closed forms that we found:

```
sumEvens' :: Integer -> Integer
sumEvens' n = sum [ 2*k | k <- [1..n] ]

sumEvens :: Integer -> Integer
sumEvens n = n * (n+1)

sumInts :: Integer -> Integer
sumInts n = (n * (n+1)) `div` 2
```

7.1. MATHEMATICAL INDUCTION

Example 7.5 Summing Squares.

Consider the problem of finding a closed formula for the sum of the first n squares (a closed formula, as opposed to a *recurrence* $f(0) = 0, f(n) = f(n-1) + n^2, n > 0$). By direct trial one might find the following:

$$1^2 + 2^2 = 5 = \frac{2 \cdot 3 \cdot 5}{6}.$$

$$1^2 + 2^2 + 3^2 = 5 + 9 = 14 = \frac{3 \cdot 4 \cdot 7}{6}.$$

$$1^2 + 2^2 + 3^2 + 4^2 = 14 + 16 = 30 = \frac{4 \cdot 5 \cdot 9}{6}.$$

$$1^2 + 2^2 + 3^2 + 4^2 + 5^2 = 30 + 25 = 55 = \frac{5 \cdot 6 \cdot 11}{6}.$$

This suggests a general rule:

$$1^2 + \ldots + n^2 = \frac{n(n+1)(2n+1)}{6}.$$

But the trial procedure that we used to *guess* the rule is different from the procedure that is needed to *prove* it.

Note that the fact that one can use mathematical induction to prove a rule gives no indication about how the rule was found in the first place. We will return to the issue of guessing closed forms for polynomial sequences in Sections 9.1 and 9.2.

In Haskell, you can compute sums of squares in a naive way or in a sophisticated way, as follows:

```
sumSquares' :: Integer -> Integer
sumSquares' n = sum [ k^2 | k <- [1..n] ]

sumSquares :: Integer -> Integer
sumSquares n = (n*(n+1)*(2*n+1)) `div` 6
```

Again, the insight that the two computation procedures will always give the same result can be proved by mathematical induction:

Exercise 7.6 Summing Squares. Prove by mathematical induction:

$$\sum_{k=1}^{n} k^2 = \frac{n(n+1)(2n+1)}{6}.$$

Example 7.7 Summing Cubes.

Let us move on to the problem of summing cubes. By direct trial one finds:
$$1^3 + 2^3 = 9 = (1+2)^2.$$
$$1^3 + 2^3 + 3^3 = 9 + 27 = 36 = (1+2+3)^2.$$
$$1^3 + 2^3 + 3^3 + 4^3 = 36 + 64 = 100 = (1+2+3+4)^2.$$
$$1^3 + 2^3 + 3^3 + 4^3 + 5^3 = 100 + 125 = 225 = (1+2+3+4+5)^2.$$

This suggests a general rule:
$$1^3 + \cdots + n^3 = (1 + \cdots + n)^2.$$

We saw in Example 7.4 that
$$\sum_{k=1}^{n} k = \frac{n(n+1)}{2},$$
so the general rule reduces to:
$$1^3 + \cdots + n^3 = \left(\frac{n(n+1)}{2}\right)^2.$$

So much about finding a rule for the sum of the first n cubes. In Sections 9.1 and 9.2 we will give an algorithm for generating closed forms for polynomial sequences.

What we found is that sumCubes defines the same function as the naive procedure sumCubes' for summing cubes:

```
sumCubes' :: Integer -> Integer
sumCubes' n = sum  [ k^3 |  k <- [1..n] ]

sumCubes :: Integer -> Integer
sumCubes n = (n*(n+1) 'div' 2)^2
```

Again, the relation we found suggests a more sophisticated computation procedure, and proving the general relationship between the two procedures is another exercise in mathematical induction.

Exercise 7.8 Summing Cubes.
Prove by mathematical induction:

$$\sum_{k=1}^{n} k^3 = \left(\frac{n(n+1)}{2}\right)^2.$$

Exercise 7.9 Prove that for all $n \in \mathbb{N}$: $3^{2n+3} + 2^n$ is divisible by 7.

Remark. If one compares the proof strategy needed to establish a principle of the form $\forall n \in \mathbb{N} : P(n)$ with that for an ordinary universal statement $\forall x \in A : P(x)$, where A is some domain of discourse, then the difference is that in the former case we can make use of what we know about the structure of $\mathbb{N}$. In case we know nothing about A, and we have to prove P for an arbitrary element from A, we have to take our cue from P. In case we have to prove something about an arbitrary element n from $\mathbb{N}$ we know a lot more: we know that either $n = 0$ or n can be reached from 0 in a finite number of steps. The key property of $\mathbb{N}$ that makes mathematical induction work is the fact that the relation $<$ on $\mathbb{N}$ is well-founded: any sequence $m_0 > m_1 > m_2 > \cdots$ terminates. This guarantees the existence of a starting point for the induction process.

For any A that is well-founded by a relation $\prec$ the following principle holds. Let $X \subseteq A$. If

$$\forall a \in A (\forall b \prec a (b \in X) \Rightarrow a \in X),$$

then $X = A$. In Section 11.1 we will say more about the use of well-foundedness as an induction principle. ∎

7.2 Recursion over the Natural Numbers

Why does induction over the natural numbers work? Because we can think of any natural number n as the the result of starting from 0 and applying the successor operation $+1$ a finite number of times. Let us use this fact about the natural numbers to give our own Haskell implementation, as follows:

```
data Natural = Z | S Natural
     deriving (Eq, Show)
```

Here Z is our representation of 0, while S n is a representation of $n+1$. The number 4 looks in our representation like S (S (S (S Z))).

The symbol | is used to specify alternatives for the data construction.

With deriving (Eq, Show) one declares Natural as a type in the classes Eq and Show. This ensures that objects of the type can be compared for equality and displayed on the screen without further ado.

We can define the operation of addition on the natural numbers *recursively* in terms of the successor operation +1 and addition for smaller numbers:

$$m + 0 := m$$
$$m + (n+1) := (m+n) + 1$$

This definition of the operation of addition is called *recursive* because the operation + that is being defined is used in the defining clause, but for a smaller second argument. Recursive definitions always have a *base case* (in the example: the first line of the definition, where the second argument of the addition operation equals 0) and a *recursive case* (in the example: the second line of the definition, where the second argument of the addition operation is greater than 0, and the operation that is being defined appears in the right hand side of the definition).

In proving things about recursively defined objects the idea is to use mathematical induction with the basic step justified by the base case of the recursive definition, the induction step justified by the recursive case of the recursive definition. This we will now demonstrate for properties of the recursively defined operation of addition for natural numbers.

Here is the Haskell version of the definition of +, in prefix notation:

```
plus m Z = m
plus m (S n) = S (plus m n)
```

If you prefer infix notation, just write m 'plus' n instead of plus m n. The back quotes around plus transform the two placed prefix operator into an infix operator. This gives the following equivalent version of the definition:

```
m 'plus' Z = m
m 'plus' (S n) = S (m 'plus' n)
```

7.2. RECURSION OVER THE NATURAL NUMBERS

Now, with diligence, we can prove the following list of fundamental laws of addition from the definition.

$$
\begin{aligned}
m + 0 &= m & \text{(0 is identity element for +)} \\
m + n &= n + m & \text{(commutativity of +)} \\
m + (n + k) &= (m + n) + k & \text{(associativity of +)}
\end{aligned}
$$

The first fact follows immediately from the definition of +.

In proving things about a recursively defined operator $\oplus$ it is convenient to be able to refer to clauses in the recursive definition, as follows: $\oplus.1$ refers to the first clause in the definition of $\oplus$, $\oplus.2$ to the second clause, and so on.

Here is a proof by mathematical induction of the associativity of +:

Proposition 7.10 $\forall m, n, k \in \mathbb{N}: (m + n) + k = m + (n + k)$.

Proof. Induction on k.

Basis $(m + n) + 0 \stackrel{+.1}{=} m + n \stackrel{+.1}{=} m + (n + 0)$.

Induction step Assume $(m+n)+k = m+(n+k)$. We show $(m+n)+(k+1) = m + (n + (k + 1))$:

$$
\begin{aligned}
(m + n) + (k + 1) &\stackrel{+.2}{=} ((m + n) + k) + 1 \\
&\stackrel{\text{i.h.}}{=} (m + (n + k)) + 1 \\
&\stackrel{+.2}{=} m + ((n + k) + 1) \\
&\stackrel{+.2}{=} m + (n + (k + 1)).
\end{aligned}
$$

∎

The inductive proof of commutativity of + uses the associativity of + that we just established.

Proposition 7.11 $\forall m, n \in \mathbb{N}: m + n = n + m$.

Proof. Induction on n.

Basis Induction on m.

    Basis $0 + 0 = 0 + 0$.

Induction Step Assume $m + 0 = 0 + m$. We show $(m + 1) + 0 = 0 + (m + 1)$:

$$
\begin{aligned}
(m+1) + 0 &\stackrel{+.1}{=} m + 1 \\
&\stackrel{+.1}{=} (m+0) + 1 \\
&\stackrel{ih}{=} (0+m) + 1 \\
&\stackrel{\text{prop } 7.10}{=} 0 + (m+1).
\end{aligned}
$$

Induction step Assume $m+n = n+m$. We show $m+(n+1) = (n+1)+m$:

$$
\begin{aligned}
m + (n+1) &\stackrel{+.2}{=} (m+n) + 1 \\
&\stackrel{ih}{=} (n+m) + 1 \\
&\stackrel{+.2}{=} n + (m+1) \\
&\stackrel{ih}{=} n + (1+m) \\
&\stackrel{\text{prop } 7.10}{=} (n+1) + m.
\end{aligned}
$$

■

Once we have addition, we can define multiplication in terms of it, again following a recursive definition:

$$
\begin{aligned}
m \cdot 0 &:= 0 \\
m \cdot (n+1) &:= (m \cdot n) + m
\end{aligned}
$$

We call $\cdot$ the multiplication operator. It is common to use mn as shorthand for $m \cdot n$, or, in other words, one usually does not write the multiplication operator.

Here is a Haskell implementation for our chosen representation (this time we give just the infix version):

```
m 'mult' Z = Z
m 'mult' (S n) = (m 'mult' n) 'plus' m
```

Let us try this out:

```
IAR> (S (S Z)) 'mult' (S (S (S Z)))
S (S (S (S (S (S Z)))))
```

7.2. RECURSION OVER THE NATURAL NUMBERS

The following laws hold for $\cdot$, and for the interaction of $\cdot$ and $+$:

$$\begin{array}{rcll} m \cdot 1 & = & m & \text{(1 is identity element for } \cdot) \\ m \cdot (n+k) & = & m \cdot n + m \cdot k & \text{(distribution of } \cdot \text{ over } +) \\ m \cdot (n \cdot k) & = & (m \cdot n) \cdot k & \text{(associativity of } \cdot) \\ m \cdot n & = & n \cdot m & \text{(commutativity of } \cdot) \end{array}$$

Exercise 7.12 Prove these laws from the recursive definitions of $+$ and $\cdot$, plus the laws that were established about $+$.

If we now wish to implement an operation `expn` for exponentiation on naturals, the only thing we have to do is find a recursive definition of exponentiation, and implement that. Here is the definition:

$$\begin{array}{rcl} m^0 & := & 1 \\ m^{n+1} & := & (m^n) \cdot m \end{array}$$

This leads immediately to the following implementation:

```
expn m Z = (S Z)
expn m (S n) = (expn m n) 'mult' m
```

This gives:

```
IAR> expn (S (S Z)) (S (S (S Z)))
S (S (S (S (S (S (S (S Z))))))))
```

Exercise 7.13 Prove by mathematical induction that $k^{m+n} = k^m \cdot k^n$.

We can define the relation $\leqslant$ on $\mathbb{N}$ as follows:

$$m \leqslant n \; :\equiv \; \text{there is a } k \in \mathbb{N} : m + k = n$$

Instead of $m \leqslant n$ we also write $n \geqslant m$, with the same meaning. We use $m < n$ for $m \leqslant n$ and $m \neq n$. Instead of $m < n$ we also write $n > m$.

This shows that we can define $<$ or $\leqslant$ in terms of addition. On further reflection, successor ($+1$) is enough to define $\leqslant$, witness the following recursive definition:

$$\begin{array}{l} 0 \leqslant m, \\ m+1 \leqslant n+1 \quad \text{if} \quad m \leqslant n. \end{array}$$

This is translated into Haskell as follows:

```
leq Z _     = True
leq (S _) Z = False
leq (S m) (S n) = leq m n
```

Note the use of _ for an anonymous variable: `leq Z _ = True` means that `leq`, applied to the zero element `Z` and to *anything else*, gives the value `True`. The expression `(S _)` specifies a *pattern*: it matches any natural that is a successor natural.

To fully grasp the Haskell definition of `leq` one should recall that the three equations that make up the definition are read as a list of three mutually exclusive cases. The Haskell system tries the equations one by one, from top to bottom, until it finds one that applies. Thus, `leq Z _` applies to any pair of naturals with the first member equal to `Z`. `leq (S _) Z` applies to any pair of naturals with the first one starting with an `S`, and the second one equal to `Z`. The third equation, finally, applies to those pairs of naturals where both the first and the second member of the pair start with an `S`.

We can now define `geq`, `gt`, `lt`, in terms of `leq` and negation:

```
geq m n = leq n m
gt  m n = not (leq m n)
lt  m n = gt n m
```

Exercise 7.14 Implement an operation for cut-off subtraction `subtr` on naturals: the call `subtr (S (S (S Z))) (S (S (S (S Z))))` should yield Z.

Exercise 7.15 Implement operations `quotient` and `remainder` on naturals. Dividing a by b yields quotient q and remainder r with $0 \leqslant r < b$, according to the formula $a = q \cdot b + r$. (Hint: you will need the procedure `subtr` from Exercise 7.14.)

7.3 The Nature of Recursive Definitions

Not any set of equations over natural numbers can serve as a definition of an operation on natural numbers. Consider

$$f(0) := 1$$
$$f(n+1) := f(n+2).$$

This does not define unique values for $f(1)$, $f(2)$, $f(3)$, ..., for the equations only require that all these values should be the same, not what the value should be.

The following format does guarantee a proper definition:

$$f(0) := c$$
$$f(n+1) := h(f(n)).$$

Here c is a description of a value (say of type A), and h is a function of type $A \to A$. A definition in this format is said to be a definition by *structural recursion* over the natural numbers. The function f defined by this will be of type $\mathbb{N} \to A$.

This format is a particular instance of a slightly more general one called *primitive recursion* over the natural numbers. Primitive recursion allows c to depend on a number of parameters, so the function f will also depend on these parameters. But we will leave those parameters aside for now.

Definition by structural recursion of f from c and h works like this: take a natural number n, view it as

$$\underbrace{1 + \cdots + 1}_{n \text{ times}} + 0,$$

replace 0 by c, replace each successor step $1+$ by h, and evaluate the result:

$$\underbrace{h(\ \cdots\ (h(c))\cdot)}_{n \text{ times}}.$$

This general procedure is easily implemented in an operation `foldn`, defined as follows:

```
foldn :: (a -> a) -> a -> Natural -> a
foldn h c Z = c
foldn h c (S n) = h (foldn h c n)
```

Here is a first example application; please make sure you understand how and why this works.

```
exclaim :: Natural -> String
exclaim = foldn ('!':) []
```

Now a function 'adding m' can be defined by taking m for c, and successor for h. For this we should be able to refer to the successor function as an object in its own right. Well, it is easily defined by lambda abstraction as (\ n -> S n). This is used in our alternative definition of plus, in terms of foldn. Note that there is no need to mention the two arguments of plus in the definition.

```
plus :: Natural -> Natural -> Natural
plus = foldn (\ n -> S n)
```

Similarly, we can define an alternative for mult in terms of foldn. The recipe for the definition of mult m ('multiply by m') is to take Z for c and plus m for h:

```
mult :: Natural -> Natural -> Natural
mult m = foldn (plus m) Z
```

Finally, for exponentiation expn m ('raise m to power ...') we take (S Z) for c and mult m for h:

```
expn :: Natural -> Natural -> Natural
expn m = foldn (mult m) (S Z)
```

7.3. THE NATURE OF RECURSIVE DEFINITIONS

Exercise 7.16 Implement the operation of cut-off subtraction (`subtr m` for 'subtract from m'; Exercise 7.14) in terms of `foldn` and a function for predecessor, on the basis of the following definition:

$$x \dotdiv 0 := x$$
$$x \dotdiv (n+1) := p(x \dotdiv n),$$

where p is the function for predecessor given by $p(0) := 0, p(n+1) := n$. Call the predecessor function `pre`.

Exercise 7.17 The Fibonacci numbers are given by the following recursion:

$$F_0 = 0, \quad F_1 = 1, \quad F_{n+2} = F_{n+1} + F_n \text{ for } n \geqslant 0.$$

This gives:

$$0, 1, 1, 2, 3, 5, 8, 13, 21, 34, 55, 89, 144, 233, 377, 610, 987, 1597, 2584, 4181, \ldots$$

Prove with induction that for all $n > 1$:

$$F_{n+1}F_{n-1} - F_n^2 = (-1)^n.$$

Exercise 7.18 A bitlist is a list of zeros and ones. Consider the following code `bittest` for selecting the bitlists without consecutive zeros.

```
bittest :: [Int] -> Bool
bittest []       = True
bittest [0]      = True
bittest (1:xs)   = bittest xs
bittest (0:1:xs) = bittest xs
bittest _        = False
```

1. How many bitlists of length 0 satisfy `bittest`? How many bitlists of length 1 satisfy `bittest`? How many bitlists of length 2 satisfy `bittest`? How many bitlists of length 3 satisfy `bittest`?

2. Let a_n be the number of bitlists of length n without consecutive zeros. Give an induction proof to show that for every $n \geqslant 0$ it holds that $a_n = F_{n+2}$, where F_n is the n-th Fibonacci number. Take care: you will need two base cases ($n = 0, n = 1$), and an induction hypothesis of the form: "assume that the formula holds for n and for $n + 1$." A further hint: the code for `bittest` points the way to the solution.

Exercise 7.19* Consider the following two definitions of the Fibonacci numbers (Exercise 7.17):

```
fib 0 = 0
fib 1 = 1
fib n = fib (n-1) + fib (n-2)

fib' n = fib2 0 1 n where
   fib2 a b 0 = a
   fib2 a b n = fib2 b (a+b) (n-1)
```

Use an induction argument to establish that `fib` and `fib'` define the same function. Hint: establish the more general claim that for all `i,n` it holds that

   `fib2 (fib i) (fib (i+1)) n = fib (i+n)`

by induction on n.

Exercise 7.20 The Catalan numbers are given by the following recursion:

$$C_0 = 1, \quad C_{n+1} = C_0 C_n + C_1 C_{n-1} + \cdots + C_{n-1} C_1 + C_n C_0.$$

This gives:

$[1, 1, 2, 5, 14, 42, 132, 429, 1430, 4862, 16796, 58786, 208012, 742900, 2674440, \ldots$

Use this recursion to give a Haskell implementation. Can you see why this is not a very efficient way to compute the Catalan numbers?

Exercise 7.21 Let $x_0, \ldots, x_n$ be a sequence of $n+1$ variables. Suppose their product is to be computed by doing n multiplications. The number of ways to do the multiplications corresponds to the number of bracketings for the sequence. For instance, if $n = 3$ there are four variables x_0, x_1, x_2, x_3, and five possible bracketings:

$(x_1 x_2)(x_3 x_4), \quad ((x_1 x_2) x_3) x_4, \quad (x_1 (x_2 x_3)) x_4, \quad x_1 ((x_2 x_3) x_4), \quad x_1 (x_2 (x_3 x_4)).$

Show that the number of bracketings for $n+1$ variables is given by the Catalan number C_n. (Hint: you will need strong induction, so-called because of its strengthened induction hypothesis. Your induction hypothesis should run: "For any i with $0 \leqslant i \leqslant n$, for any sequence of $i+1$ variables $x_0 \cdots x_i$ it holds that C_i gives the number of bracketings for that sequence."

7.4. INDUCTION AND RECURSION OVER TREES

Example 7.22 Balanced sequences of parentheses of length $2n$ are defined recursively as follows: the empty sequence is balanced; if sequence w is balanced then (w) is balanced; if sequences w and v are balanced then wv is balanced. Thus, $())(()$ is not balanced. The balanced sequences involving 3 left and 3 right parentheses are:

$$()()(),\ (())(),\ ()(()),\ (()()),\ ((())).$$

There is a one-to-one mapping between bracketings for sequences of $n+1$ variables and balanced sequences of parentheses with $2n$ parentheses. Let a bracketing for $x_0 \cdots x_n$ be given. This can be changed into a balanced sequence of parentheses as follows: We illustrate with the example $x_0((x_2x_3)x_4)$.

1. Put one extra pair of parentheses around the bracketing: $(x_0((x_2x_3)x_4))$.

2. Insert multiplication operators at the appropriate places: $(x_0 \cdot ((x_2 \cdot x_3) \cdot x_4))$.

3. Erase the variables and the left-brackets: $\cdots) \cdot))$.

4. Replace the $\cdot$'s with left-brackets: $(()())$.

This mapping gives a one-to-one correspondence between variable bracketings and balanced parentheses strings, so we get from the previous exercise that there are C_n different balanced parentheses strings of length $2n$.

7.4 Induction and Recursion over Trees

Here is a recursive definition of binary trees:

- A single leaf node • is a binary tree.

- If t_1 and t_2 are binary trees, then the result of joining t_1 and t_2 under a single node (called the *root node*) is a binary tree. A notation for this is: $(\bullet\ t_1\ t_2)$

- Nothing else is a binary tree.

The depth of a binary tree is given by:

- The depth of • is 0.

- The depth of $(\bullet\ t_1\ t_2)$ is 1+ the maximum of the depths of t_1 and t_2.

A binary tree is balanced if it either is a single leaf node •, or it has the form (• t_1 t_2), with both t_1 and t_2 balanced, and having the same depth.

We see the following: A balanced binary tree of depth 0 is just a single leaf node •, so its number of nodes is 1. A balanced binary tree of depth 1 has one internal node and two leaves, so it has 3 nodes. A balanced binary tree of depth 2 has 3 internal nodes and 4 leaf nodes, so it has 7 nodes. A binary tree of depth 3 has 7 internal nodes plus 8 leaf nodes, so it has 15 nodes.

Recursion and induction over binary trees are based on two cases $t = \bullet$ and $t = (\bullet\ t_1\ t_2)$.

Example 7.23 Suppose we want to find a formula for the number of nodes in a balanced binary tree of depth d.

Suppose we want to show in general that the number of nodes of a binary tree of depth n is $2^{n+1} - 1$. Then a proof by mathematical induction is in order. We proceed as follows:

Basis If $n = 0$, then $2^{n+1} - 1 = 2^1 - 1 = 1$. This is indeed the number of nodes of a binary tree of depth 0.

Induction step Assume the number of nodes of a binary tree of depth n is $2^{n+1} - 1$. We have to show that the number of nodes of a binary tree of depth $n + 1$ equals $2^{n+2} - 1$.

A binary tree of depth $n + 1$ can be viewed as a set of *internal nodes* constituting a binary tree of depth n, plus a set of leaf nodes, consisting of two new leaf nodes for every old leaf node from the tree of depth n. By induction hypothesis, we know that a tree of depth n has $2^{n+1} - 1$ nodes, so a tree of depth $n + 1$ has $2^{n+1} - 1$ internal nodes. It is easy to see that a tree of depth $n + 1$ has 2^{n+1} leaf nodes. The total number of nodes of a tree of depth $n + 2$ is therefore $2^{n+1} - 1 + 2^{n+1} = 2 \cdot 2^{n+1} - 1 = 2^{n+2} - 1$, and we have proved our induction step.

To illustrate trees and tree handling a bit further, here is a Haskell definition of binary trees, with a procedure for making balanced trees of an arbitrary depth n, and a procedure for counting their nodes in a naive way.

7.4. INDUCTION AND RECURSION OVER TREES

We use L for a single leaf •. The `data` declaration specifies that a `BinTree` either is an object L (a single leaf), or an object constructed by applying the constructor N to two `BinTree` objects (the result of constructing a new binary tree (• t_1 t_2) from two binary trees t_1 and t_2).

The addition `deriving Show` ensures that data of this type can be displayed on the screen.

```
data BinTree = L | N BinTree BinTree deriving Show

makeBinTree :: Integer -> BinTree
makeBinTree 0 = L
makeBinTree (n + 1) = N (makeBinTree n) (makeBinTree n)

count :: BinTree -> Integer
count L = 1
count (N t1 t2) = 1 + count t1 + count t2
```

With this code you can produce binary trees as follows:

```
IAR> makeBinTree 6
N (N (N (N (N (N L L) (N L L)) (N (N L L) (N L L))) (N (N (N L L) (N
L L)) (N (N L L) (N L L)))) (N (N (N (N L L) (N L L)) (N (N L L) (N L
L))) (N (N (N L L) (N L L)) (N (N L L) (N L L))))) (N (N (N (N (N L L)
(N L L)) (N (N L L) (N L L))) (N (N (N L L) (N L L)) (N (N L L) (N L
L)))) (N (N (N (N L L) (N L L)) (N (N L L) (N L L))) (N (N (N L L) (N
L L)) (N (N L L) (N L L)))))
IAR>
```

If you want to check that the depth of the result of `maketree 6` is indeed 6, or that `maketree 6` is indeed balanced, here are some procedures. Note that the procedures follow the definitions of *depth* and *balanced* to the letter:

```
depth :: BinTree -> Integer
depth L = 0
depth (N t1 t2) = (max (depth t1) (depth t2)) + 1
```

```
balanced :: BinTree -> Bool
balanced L = True
balanced (N t1 t2) = (balanced t1)
                  && (balanced t2)
                  && depth t1 == depth t2
```

The programs allow us to check the relation between count (makeBinTree n) and 2^(n+1) - 1 for individual values of n:

```
IAR>    count (makeBinTree 6) ==  2^7 - 1
True
```

What the proof by mathematical induction provides is an insight that the relation holds in general. Mathematical induction does not give as clue as to how to *find* a formula for the number of nodes in a tree. It only serves as a method of proof once such a formula is found.

So how does one *find* a formula for the number of nodes in a binary tree in the first place? By noticing how such trees grow. A binary tree of depth 0 has 1 node, and this node is a leaf node. This leaf grows two new nodes, so a binary tree of depth 1 has $1 + 2 = 3$ nodes. In the next step the 2 leaf nodes grow two new nodes each, so we get $2^2 = 4$ new leaf nodes, and the number of nodes of a binary tree of depth 2 equals $1 + 2 + 4 = 7$. In general, a tree of depth $n-1$ is transformed into one of depth n by growing 2^n new leaves, and the total number of leaves of the new tree is given by $2^0 + 2^1 + \cdots + 2^n$. In other words, the number of nodes of a balanced binary tree of depth n is given by : $\sum_{k=0}^{n} 2^k$. To get a value for this, here is a simple trick:

$$\sum_{k=0}^{n} 2^k = 2 \cdot \sum_{k=0}^{n} 2^k - \sum_{k=0}^{n} 2^k = (2 \cdot 2^n + 2^n \cdots + 2) - (2^n + \cdots + 1) =$$

$$= 2 \cdot 2^n - 1 = 2^{n+1} - 1.$$

Example 7.24 Counting the Nodes of a Balanced Ternary Tree.

Now suppose we want to find a formula for the number of nodes in a balanced *ternary* tree of depth n. The number of leaves of a balanced ternary tree of depth n is 3^n, so the total number of nodes of a balanced ternary tree of depth n is given by $\sum_{k=0}^{n} 3^k$.

We prove by induction that $\sum_{k=0}^{n} 3^k = \frac{3^{n+1}-1}{2}$.

7.4. INDUCTION AND RECURSION OVER TREES

Basis A ternary tree of depth $n = 0$ consists of just a single node. Indeed,

$$\sum_{k=0}^{0} 3^k = 1 = \frac{3^1 - 1}{2}.$$

Induction step Assume that the number of nodes of a ternary tree of depth n is $\frac{3^{n+1}-1}{2}$. The number of leaf nodes is 3^n, and each of these leaves grows 3 new leaves to produce the ternary tree of depth $n+1$. Thus, the number of leaves of the tree of depth $n+1$ is given by

$$\sum_{k=0}^{n+1} 3^k = \sum_{k=0}^{n} 3^k + 3^{n+1}.$$

Using the induction hypothesis, we see that this is equal to

$$\frac{3^{n+1} - 1}{2} + 3^{n+1} = \frac{3^{n+1} - 1}{2} + \frac{2 \cdot 3^{n+1}}{2} = \frac{3^{n+2} - 1}{2}.$$

But how did we get at $\sum_{k=0}^{n} 3^k = \frac{3^{n+1}-1}{2}$ in the first place? In the same way as in the case of the binary trees:

$$2 \cdot \sum_{k=0}^{n} 3^k = 3 \cdot \sum_{k=0}^{n} 3^k - \sum_{k=0}^{n} 3^k = (3 \cdot 3^n + 3^n \cdots + 3) - (3^n + \cdots + 1) =$$

$$= 3 \cdot 3^n - 1 = 3^{n+1} - 1.$$

Therefore
$$\sum_{k=0}^{n} 3^k = \frac{3^{n+1} - 1}{2}.$$

Exercise 7.25 Write a Haskell definition of ternary trees, plus procedures for generating balanced ternary trees and counting their node numbers.

Example 7.26 Counting the Nodes of a Balanced m-ary Tree. The number of nodes of a balanced m-ary tree of depth n (with $m > 1$) is given by $\sum_{k=0}^{n} m^k = \frac{m^{n+1}-1}{m-1}$. Here is a proof by mathematical induction.

Basis An m-ary tree of depth 0 consists of just a single node. In fact, $\sum_{k=0}^{0} m^k = 1 = \frac{m-1}{m-1}$.

Induction step Assume that the number of nodes of an m-ary tree of depth n is $\frac{m^{n+1}-1}{m-1}$. The number of leaf nodes is m^n, and each of these leaves grows m new leaves to produce the ternary tree of depth $n+1$. Thus, the number of leaves of the tree of depth $n+1$ is given by

$$\sum_{k=0}^{n+1} m^k = \sum_{k=0}^{n} m^k + m^{n+1}.$$

Using the induction hypothesis, we see that this is equal to

$$\frac{m^{n+1}-1}{m-1} + m^{n+1} = \frac{m^{n+1}-1}{m-1} + \frac{m^{n+2}-m^{n+1}}{m-1} = \frac{m^{n+2}-1}{m-1}.$$

Note that the proofs by mathematical induction do not tell you how to *find* the formulas for which the induction proof works in the first place. This is an illustration of the fact that mathematical induction is a method of verification, not a method of invention. Indeed, mathematical induction is no replacement for the use of creative intuition in the process of finding meaningful relationships in mathematics.

Exercise 7.27 Geometric Progressions.
Prove by mathematical induction (assuming $q \neq 1, q \neq 0$):

$$\sum_{k=0}^{n} q^k = \frac{q^{n+1}-1}{q-1}.$$

Note that this exercise is a further generalization of Example 7.26.

To get some further experience with tree processing, consider the following definition of binary trees with integer numbers at the internal nodes:

```
data Tree = Lf | Nd Int Tree Tree deriving Show
```

We say that such a tree is *ordered* if it holds for each node N of the tree that the integer numbers in the left subtree below N are all smaller than the number at node N, and the number in the right subtree below N are all greater than the number at N.

Exercise 7.28 Write a function that inserts a number n in an ordered tree in such a way that the tree remains ordered.

7.4. INDUCTION AND RECURSION OVER TREES

Exercise 7.29 Write a function `list2tree` that converts a list of integers to an ordered tree, with the integers at the tree nodes. The type is `[Int] -> Tree`. Also, write a function `tree2list` for conversion in the other direction.

Exercise 7.30 Write a function that checks whether a given integer i occurs in an ordered tree.

Exercise 7.31 Write a function that merges two ordered trees into a new ordered tree containing all the numbers of the input trees.

Exercise 7.32 Write a function that counts the number of steps that are needed to reach a number i in an ordered tree. The function should give 0 if i is at the top node, and -1 if i does not occur in the tree at all.

A general data type for binary trees with information at the internal nodes is given by:

```
data Tr a = Nil | T a (Tr a) (Tr a) deriving (Eq,Show)
```

Exercise 7.33 Write a function `mapT :: (a -> b) -> Tr a -> Tr b` that does for binary trees what `map` does for lists.

Exercise 7.34 Write a function

`foldT :: (a -> b -> b -> b) -> b -> (Tr a) -> b`

that does for binary trees what `foldn` does for naturals.

Exercise 7.35 Conversion of a tree into a list can be done in various ways, depending on when the node is visited:

Preorder traversal of a tree is the result of first visiting the node, next visiting the left subtree, and finally visiting the right subtree.

Inorder traversal of a tree is the result of first visiting the left subtree, next visiting the node, and finally visiting the right subtree.

Postorder traversal of a tree is the result of first visiting the left subtree, next visiting the right subtree, and finally visiting the node.

Define these three conversion functions from trees to lists in terms of the `foldT` function from Exercise 7.34.

Exercise 7.36 An ordered tree is a tree with information at the nodes given in such manner that the item at a node must be bigger than all items in the left subtree and smaller than all items in the right subtree. A tree is ordered iff the list resulting from its inorder traversal is ordered and contains no duplicates. Give an implementation of this check.

Exercise 7.37 An ordered tree (Exercise 7.36) can be used as a dictionary by putting items of of type (String,String) at the internal nodes, and defining the ordering as: $(v,w) \leqslant (v',w')$ iff $v \leqslant v'$. Dictionaries get the following type:

```
type Dict = Tr (String,String)
```

Give code for looking up a word definition in a dictionary. The type declaration is:

```
lookupD :: String -> Dict -> [String]
```

If (v,w) occurs in the dictionary, the call `lookupD v d` should yield [w], otherwise []. Use the order on the dictionary tree.

Exercise 7.38 For efficient search in an ordered tree (Exercises 7.36 and 7.37) it is crucial that the tree is balanced: the left and right subtree should have (nearly) the same depth and should themselves be balanced.

The following auxiliary function splits non-empty lists into parts of (roughly) equal lengths.

```
split :: [a] -> ([a],a,[a])
split xs = (ys1,y,ys2)
   where
   ys1       = take n xs
   (y:ys2)   = drop n xs
   n         = length xs 'div' 2
```

7.4. INDUCTION AND RECURSION OVER TREES

Use this to implement a function buildTree :: [a] -> Tr a for transforming an ordered list into an ordered and balanced binary tree.

Here is a data type LeafTree for binary leaf trees (binary trees with information at the leaf nodes):

```
data LeafTree a = Leaf a
                | Node (LeafTree a) (LeafTree a) deriving Show
```

Here is an example leaf tree:

```
ltree :: LeafTree String
ltree = Node
          (Leaf "I")
          (Node
             (Leaf "love")
             (Leaf "you"))
```

Exercise 7.39 Repeat Exercise 7.33 for leaf trees. Call the new map function mapLT.

Exercise 7.40 Give code for mirroring a leaf tree on its vertical axis. Call the function reflect. In the mirroring process, the left- and right branches are swapped, and the same swap takes place recursively within the branches. The reflection of

Node (Leaf 1) (Node (Leaf 2) (Leaf 3))

is

Node (Node (Leaf 3) (Leaf 2)) (Leaf 1).

Exercise 7.41 Let reflect be the function from Exercise 7.40. Prove with induction on tree structure that reflect (reflect t) == t holds for every leaf tree t.

A data type for trees with arbitrary numbers of branches (rose trees), with information of type a at the buds, is given by:

```
data Rose a = Bud a | Br [Rose a] deriving (Eq,Show)
```

Here is an example rose:

```
rose = Br [Bud 1, Br [Bud 2, Bud 3, Br [Bud 4, Bud 5, Bud 6]]]
```

Exercise 7.42 Write a function mapR :: (a -> b) -> Rose a -> Rose b that does for rose trees what map does for lists. For the example rose, we should get:

```
IAR> rose
Br [Bud 1, Br [Bud 2, Bud 3, Br [Bud 4, Bud 5, Bud 6]]]
IAR> mapR succ rose
Br [Bud 2, Br [Bud 3, Bud 4, Br [Bud 5, Bud 6, Bud 7]]]
```

7.5 Induction and Recursion over Lists

Induction and recursion over natural numbers are based on the two cases $n = 0$ and $n = k + 1$. Induction and recursion over lists are based on the two cases $l = []$ and $l = $ x:xs, where x is an item and xs is the tail of the list. An example is the definition of a function len that gives the length of a list. In fact, Haskell has a predefined function length for this purpose; our definition of len is just for purposes of illustration.

```
len []     = 0
len (x:xs) = 1 + len xs
```

Similarly, Haskell has a predefined operation ++ for concatenation of lists. For purposes of illustration we repeat our version from Section 4.8.

7.5. INDUCTION AND RECURSION OVER LISTS

```
cat :: [a] -> [a] -> [a]
cat []     ys = ys
cat (x:xs) ys = x : (cat xs ys)
```

As an example of an inductive proof over lists, we show that concatenation of lists is associative.

Proposition 7.43 For all lists xs, ys and zs over the same type a:

cat (cat xs ys) zs = cat xs (cat ys zs).

Proof. Induction on xs.

Basis

$$\text{cat (cat [] ys) zs} \stackrel{\text{cat.1}}{=} \text{cat ys zs}$$
$$\stackrel{\text{cat.1}}{=} \text{cat [] (cat ys zs)}.$$

Induction step

$$\text{cat x:xs (cat ys zs)} \stackrel{\text{cat.2}}{=} \text{x:(cat xs (cat ys zs))}$$
$$\stackrel{\text{i.h.}}{=} \text{x:(cat (cat xs ys) zs)}$$
$$\stackrel{\text{cat.2}}{=} \text{cat x:(cat xs ys) zs}$$
$$\stackrel{\text{cat.2}}{=} \text{cat (cat x:xs ys) zs}.$$

∎

Exercise 7.44 Prove by induction that cat xs [] = cat [] xs.

Exercise 7.45 Prove by induction:

len (cat xs ys) = (len xs) + (len ys).

A general scheme for structural recursion over lists (without extra parameters) is given by:

$$f\ [] := z$$
$$f(x : \text{xs}) := h\ x\ (f\ \text{xs})$$

For example, the function s that computes the sum of the elements in a list of numbers is defined by:

$$s\ [] := 0$$
$$s(n : \text{xs}) := n + s\ \text{xs}$$

Here 0 is taken for z, and $+$ for h.

As in the case for natural numbers, it is useful to implement an operation for this general procedure. In fact, this is predefined in Haskell, as follows (from the Haskell file *Prelude.hs*):

```
foldr            :: (a -> b -> b) -> b -> [a] -> b
foldr f z []     = z
foldr f z (x:xs) = f x (foldr f z xs)
```

In general, z is the identity element of the operation f, i.e., the value you would start out with in the base case of a recursive definition of the operation. The identity element for addition is 0, for multiplication it is 1 (see Section 7.3).

The base clause of the definition of foldr says that if you want to fold an empty list for operation f, the result is the identity element for f. The recursive clause says that to fold a non-empty list for f, you perform f to its first element and to the result of folding the remainder of the list for f.

The following informal version of the definition of foldr may further clarify its meaning:

$$\text{foldr}\ (\oplus)\ z\ [x_1, x_2, \ldots, x_n] := x_1 \oplus (x_2 \oplus (\cdots (x_n \oplus z) \cdots)).$$

The function add :: [Natural] -> Natural can now be defined as:

```
add = foldr plus Z
```

The function mlt :: [Natural] -> Natural is given by:

7.5. INDUCTION AND RECURSION OVER LISTS

```
mlt = foldr mult (S Z)
```

Here is an alternative definition of list length, with values of type Natural. Note that the function (\ _ n -> S n) ignores its first argument (you don't have to *look* at the items in a list in order to count them) and returns the successor of its second argument (for this second argument represents the number of items that were counted so far).

```
ln :: [a] -> Natural
ln = foldr (\ _ n -> S n) Z
```

It is also possible to use `foldr` on the standard type for integers. Computing the result of adding or multiplying all elements of a list of integers can be done as follows:

```
Prelude> foldr (+) 0 [1..10]
55
Prelude> foldr (*) 1 [1..10]
3628800
```

Exercise 7.46 Use `foldr` to give a new implementation of generalized union and `foldr1` to give a new implementation of generalized intersection for lists. (Look up the code for `foldr1` in the Haskell prelude. Compare with Exercise 4.53.)

Consider the following definitions of generalized conjunction and disjunction:

```
or :: [Bool] -> Bool
or [] = False
or (x:xs) = x || or xs

and :: [Bool] -> Bool
and [] = True
and (x:xs) = x && and xs
```

The function or takes a list of truth values and returns True if at least one member of the list equals True, while and takes a list of truth values and returns True if all members of the list equal True. (We have encountered and before, in Section 2.2.)

In fact, Haskell has predefined these operations, in terms of foldr. To see how we can use foldr to implement generalized conjunction and disjunction, we only need to know what the appropriate identity elements of the operations are. Should the conjunction of all elements of [] count as true or false? As true, for it is indeed (trivially) the case that all elements of [] are true. So the identity element for conjunction is True. Should the disjunction of all elements of [] count as true or false? As false, for it is false that [] contains an element which is true. Therefore, the identity element for disjunction is False. This explains the following Haskell definition in *Prelude.hs*:

```
and, or          :: [Bool] -> Bool
and              = foldr (&&) True
or               = foldr (||) False
```

Exercise 7.47 Define a function srt that sorts a list of items in class Ord a by folding the list with a function insrt.

The operation foldr folds 'from the right'. Folding 'from the left' can be done with its cousin foldl, predefined in *Prelude.hs* as follows:

```
foldl            :: (a -> b -> a) -> a -> [b] -> a
foldl f z []     = z
foldl f z (x:xs) = foldl f (f z x) xs
```

An informal version may further clarify this:

$$\text{foldl } (\oplus) \ z \ [x_1, x_2, \ldots, x_n] := (\cdots((z \oplus x_1) \oplus x_2) \oplus \cdots) \oplus x_n.$$

This can be used to flesh out the following recursion scheme:

$$f \ z \ [] \ := \ z$$
$$f \ z \ (x : \text{xs}) \ := \ f \ (h \ z \ x) \ \text{xs}$$

7.5. INDUCTION AND RECURSION OVER LISTS

This boils down to recursion over lists with an extra parameter, and in fact foldl can be used to speed up list processing. The case of reversing a list is an example:

$$r \; zs \; [] \; := \; zs$$
$$r \; zs \; (x : xs) \; := \; r \; (\text{prefix } zs \; x) \; xs,$$

where *prefix* is given by prefix ys y = y:ys. Here is a definition in terms of foldl:

```
rev = foldl (\ xs x -> x:xs) []
```

The list [1,2,3] is reversed as follows:

```
rev [1,2,3] = foldl (\ xs x -> x:xs) [] [1,2,3]
            = foldl (\ xs x -> x:xs) ((\ xs x -> x:xs) [] 1) [2,3]
            = foldl (\ xs x -> x:xs) [1] [2,3]
            = foldl (\ xs x -> x:xs) ((\ xs x -> x:xs) [1] 2) [3]
            = foldl (\ xs x -> x:xs) [2,1] [3]
            = foldl (\ xs x -> x:xs) ((\ xs x -> x:xs) [2,1] 3) []
            = foldl (\ xs x -> x:xs) [3,2,1] []
            = [3,2,1]
```

Note that (\ xs x -> x:xs) has type [a] -> a -> [a]. An alternative definition of rev, in terms of foldr, would need a swap function
\ x xs -> xs ++ [x] :: a -> [a] -> [a]
and would be much less efficient:

```
rev' = foldr (\ x xs -> xs ++ [x]) []
```

The inefficiency resides in the fact that ++ itself is defined recursively, as follows:

```
[]     ++ ys = ys
(x:xs) ++ ys = x : (xs ++ ys)
```

To see why rev' is less efficient than rev, look at the following, where we write postfix for (\ x xs -> xs ++[x]).

```
rev' [1,2,3] = foldr postfix [] [1,2,3]
             = postfix 1 (foldr postfix [] [2,3])
             = (foldr postfix [] [2,3]) ++ [1]
             = (postfix 2 (foldr postfix [] [3])) ++ [1]
             = (foldr postfix [] [3]) ++ [2] ++ [1]
             = (postfix 3 (foldr postfix [] [])) ++ [2] ++ [1]
             = (foldr postfix [] []) ++ [3] ++ [2] ++ [1]
             = [] ++ [3] ++ [2] ++ [1]
             = ([3] ++ [2]) ++ [1]
             = 3:([] ++ [2]) ++ [1]
             = [3,2] ++ [1]
             = 3:([2] ++ [1])
             = 3:2:([] ++ [1])
             = [3,2,1]
```

If we compare the two list recursion schemes that are covered by `foldr` and `foldl`, then we see that the two folding operations h and h' are close cousins:

$$f\;[] := z \qquad\qquad f'\;[] := z$$
$$f(x:\text{xs}) := h\;x\;(f\;\text{xs}) \qquad f'(x:\text{xs}) := h'\;(f'\;\text{xs})\;x$$

An obvious question to ask now is the following: what requirements should h and h' satisfy in order to guarantee that f and f' define the same function? Exercise 7.48 and Example 7.49 invite you to further investigate this issue of the relation between the schemes.

Exercise 7.48 Let h :: a -> b -> b and h' :: b -> a -> b, and z :: b. Assume that for all x :: a it holds that h x (h' y z) = h' z (h x y). Show that for every x,y :: a and every finite xs :: [a] the following holds:

```
h x (foldl h' y xs) = foldl h' (h x y) xs
```

Use induction on xs.

Example 7.49 Let h :: a -> b -> b and h' :: b -> a -> b, and z :: b. Assume we have for all x,y :: a and all xs :: [a] the following:

```
h x z = h' z x
h x (h' y z) = h' z (h x y)
```

We show that we have for all finite xs :: [a] that

```
foldr h z xs = foldl h' z xs
```

7.5. INDUCTION AND RECURSION OVER LISTS

We use induction on the structure of xs.

Basis Immediate from the definitions of `foldr` and `foldl` we have that:

```
foldr h z [] = z = foldl h' z []
```

Induction step Assume the induction hypothesis `foldr h z xs = foldl h' z xs`. We have to show that `foldr h z x:xs = foldl h' z x:xs`. Here is the reasoning:

$$
\begin{aligned}
\text{foldr } h \ z \ \text{x:xs} &\stackrel{\text{foldr}}{=} h \ x \ (\text{foldr } h \ z \ \text{xs}) \\
&\stackrel{\text{IH}}{=} h \ x \ (\text{foldl } h \ z \ \text{xs}) \\
&\stackrel{7.48}{=} \text{foldl } h' \ (h \ x \ z) \ \text{xs} \\
&\stackrel{\text{given}}{=} \text{foldl } h' \ (h' \ z \ x) \ \text{xs} \\
&\stackrel{\text{foldl}}{=} \text{foldl } h' \ z \ \text{x:xs}
\end{aligned}
$$

For an application of Example 7.49, note that the functions `postfix` and `prefix` are related by:

```
postfix x [] = [] ++ [x] = [x] == prefix [] x

postfix x (prefix xs y) = (prefix xs y) ++ [x]
                       = y:(xs ++ [x])
                       = y:(postfix x xs)
                       = prefix (postfix x xs) y
```

It follows from this and the result of the example that `rev` and `rev'` indeed compute the same function.

Exercise 7.50 Consider the following version of `rev`.

```
rev1 :: [a] -> [a]
rev1 xs  = rev2 xs []
  where
  rev2 []     ys = ys
  rev2 (x:xs) ys = rev2 xs (x:ys)
```

Which version is more efficient, the original `rev`, the version `rev'`, or this version? Why?

Exercise 7.51 Define an alternative version `ln'` of `ln` using `foldl` instead of `foldr`.

In Section 1.8 you got acquainted with the `map` and `filter` functions. The two operations `map` and `filter` can be combined:

```
Prelude> filter (>4) (map (+1) [1..10])
[5,6,7,8,9,10,11]
Prelude> map (+1) (filter (>4) [1..10])
[6,7,8,9,10,11]
```

These outcomes are different. This is because the test (>4) yields a different result after all numbers are increased by 1. When we make sure that the test used in the filter takes this change into account, we get the same answers:

```
Prelude> filter (>4) (map (+1) [1..10])
[5,6,7,8,9,10,11]
Prelude> map (+1) (filter ((>4).(+1)) [1..10])
[5,6,7,8,9,10,11]
```

Here (f . g) denotes the result of first applying g and next f. Note: ((>4).(+1)) defines the same property as (>3). Exercise 7.52 gives you an opportunity to show that these identical answers are no accident.

Exercise 7.52 Let `xs :: [a]`, let `f :: a -> b`, and let `p :: b -> Bool` be a total predicate. Show that the following holds:

$$\text{filter } p \text{ (map } f \text{ xs)} = \text{map } f \text{ (filter } (p \cdot f) \text{ xs)}.$$

Note: a predicate `p :: b -> Bool` is total if for every object `x :: b`, the application `p x` gives either true or false. In particular, for no `x :: b` does `p x` raise an error.

7.6 Some Variations on the Tower of Hanoi

The Tower of Hanoi is a tower of 8 disks of different sizes, stacked in order of decreasing size on a peg. Next to the tower, there are two more pegs. The task is to transfer the whole stack of disks to one of the other pegs (using the third peg as an auxiliary) while keeping to the following rules: (i) move only one disk at a time, (ii) never place a larger disk on top of a smaller one.

7.6. SOME VARIATIONS ON THE TOWER OF HANOI

Figure 7.1: The Tower of Hanoi.

Exercise 7.53 In this exercise, you are required to invent your own solution, and next prove it by mathematical induction. Make sure that the reasoning also establishes that the formula you find for the number of moves is the *best* one can do.

1. How many moves does it take to completely transfer a tower consisting of n disks?
2. Prove by mathematical induction that your answer to the previous question is correct.
3. How many moves does it take to completely transfer the tower of Hanoi?

Exercise 7.54 Can you also find a formula for the number of moves of the disk of size k during the transfer of a tower with disks of sizes $1, \ldots, n$, and $1 \leqslant k \leqslant n$? Again, you should prove by mathematical induction that your formula is correct.

According to legend, there exists a much larger tower than the tower of Hanoi, the tower of Brahma, with 64 disks. Monks are engaged in transferring the disks of the Tower of Brahma, from the beginning of the universe to the present day. As soon as they will have completed their task the tower will crumble and the world will end.

Exercise 7.55 How long will the universe last, given that the monks move one disk per day?

For an implementation of the disk transfer procedure, an obvious way to represent the starting configuration of the tower of Hanoi is (`[1,2,3,4,5,6,7,8],[],[]`). For clarity, we give the three pegs names A, B and C. and we declare a type `Tower`:

```
IAR> (display 88 . show . take 200 . hanoi) 8
[([1,2,3,4,5,6,7,8],[],[]),([2,3,4,5,6,7,8],[1],[]),([3,4,5,6,7,8],[1],[2]),([3,4,5,6,7,
8],[],[1,2]),([4,5,6,7,8],[3],[1,2]),([1,4,5,6,7,8],[3],[2]),([1,4,5,6,7,8],[2,3],[]),([
4,5,6,7,8],[1,2,3],[]),([5,6,7,8],[1,2,3],[4]),([5,6,7,8],[2,3],[1,4]),([2,5,6,7,8],[3],
[1,4]),([1,2,5,6,7,8],[3],[4]),([1,2,5,6,7,8],[],[3,4]),([2,5,6,7,8],[1],[3,4]),([5,6,7,
8],[1],[2,3,4]),([5,6,7,8],[],[1,2,3,4]),([6,7,8],[5],[1,2,3,4]),([1,6,7,8],[5],[2,3,4])
,([1,6,7,8],[2,5],[3,4]),([6,7,8],[1,2,5],[3,4]),([3,6,7,8],[1,2,5],[4]),([3,6,7,8],[2,5
],[1,4]),([2,3,6,7,8],[5],[1,4]),([1,2,3,6,7,8],[5],[4]),([1,2,3,6,7,8],[4,5],[]),([2,3,
6,7,8],[1,4,5],[]),([3,6,7,8],[1,4,5],[2]),([3,6,7,8],[4,5],[1,2]),([6,7,8],[3,4,5],[1,2
]),([1,6,7,8],[3,4,5],[2]),([1,6,7,8],[2,3,4,5],[]),([6,7,8],[1,2,3,4,5],[]),([7,8],[1,2
,3,4,5],[6]),([7,8],[2,3,4,5],[1,6]),([2,7,8],[3,4,5],[1,6]),([1,2,7,8],[3,4,5],[6]),([1
,2,7,8],[4,5],[3,6]),([2,7,8],[1,4,5],[3,6]),([7,8],[1,4,5],[2,3,6]),([7,8],[4,5],[1,2,3
,6]),([4,7,8],[5],[1,2,3,6]),([1,4,7,8],[5],[2,3,6]),([1,4,7,8],[2,5],[3,6]),([4,7,8],[1
,2,5],[3,6]),([3,4,7,8],[1,2,5],[6]),([3,4,7,8],[2,5],[1,6]),([2,3,4,7,8],[5],[1,6]),([1
,2,3,4,7,8],[5],[6]),([1,2,3,4,7,8],[],[5,6]),([2,3,4,7,8],[1],[5,6]),([3,4,7,8],[1],[2,
5,6]),([3,4,7,8],[],[1,2,5,6]),([4,7,8],[3],[1,2,5,6]),([1,4,7,8],[3],[2,5,6]),([1,4,7,8
],[2,3],[5,6]),([4,7,8],[1,2,3],[5,6]),([7,8],[1,2,3],[4,5,6]),([7,8],[2,3],[1,4,5,6]),(
[2,7,8],[3],[1,4,5,6]),([1,2,7,8],[3],[4,5,6]),([1,2,7,8],[],[3,4,5,6]),([2,7,8],[1],[3,
4,5,6]),([7,8],[1],[2,3,4,5,6]),([7,8],[],[1,2,3,4,5,6]),([8],[7],[1,2,3,4,5,6]),([1,8],
[7],[2,3,4,5,6]),([1,8],[2,7],[3,4,5,6]),([8],[1,2,7],[3,4,5,6]),([3,8],[1,2,7],[4,5,6])
,([3,8],[2,7],[1,4,5,6]),([2,3,8],[7],[1,4,5,6]),([1,2,3,8],[7],[4,5,6]),([1,2,3,8],[4,7
],[5,6]),([2,3,8],[1,4,7],[5,6]),([3,8],[1,4,7],[2,5,6]),([3,8],[4,7],[1,2,5,6]),([8],[3
,4,7],[1,2,5,6]),([1,8],[3,4,7],[2,5,6]),([1,8],[2,3,4,7],[5,6]),([8],[1,2,3,4,7],[5,6])
,([5,8],[1,2,3,4,7],[6]),([5,8],[2,3,4,7],[1,6]),([2,5,8],[3,4,7],[1,6]),([1,2,5,8],[3,4
,7],[6]),([1,2,5,8],[4,7],[3,6]),([2,5,8],[1,4,7],[3,6]),([5,8],[1,4,7],[2,3,6]),([5,8],
[4,7],[1,2,3,6]),([4,5,8],[7],[1,2,3,6]),([1,4,5,8],[7],[2,3,6]),([1,4,5,8],[2,7],[3,6])
,([4,5,8],[1,2,7],[3,6]),([3,4,5,8],[1,2,7],[6]),([3,4,5,8],[2,7],[1,6]),([2,3,4,5,8],[7
],[1,6]),([1,2,3,4,5,8],[7],[6]),([1,2,3,4,5,8],[6,7],[]),([2,3,4,5,8],[1,6,7],[]),([3,4
,5,8],[1,6,7],[2]),([3,4,5,8],[6,7],[1,2]),([4,5,8],[3,6,7],[1,2]),([1,4,5,8],[3,6,7],[2
]),([1,4,5,8],[2,3,6,7],[]),([4,5,8],[1,2,3,6,7],[]),([5,8],[1,2,3,6,7],[4]),([5,8],[2,3
,6,7],[1,4]),([2,5,8],[3,6,7],[1,4]),([1,2,5,8],[3,6,7],[4]),([1,2,5,8],[6,7],[3,4]),([2
,5,8],[1,6,7],[3,4]),([5,8],[1,6,7],[2,3,4]),([5,8],[6,7],[1,2,3,4]),([8],[5,6,7],[1,2,3
,4]),([1,8],[5,6,7],[2,3,4]),([1,8],[2,5,6,7],[3,4]),([8],[1,2,5,6,7],[3,4]),([3,8],[1,2
,5,6,7],[4]),([3,8],[2,5,6,7],[1,4]),([2,3,8],[5,6,7],[1,4]),([1,2,3,8],[5,6,7],[4]),([1
,2,3,8],[4,5,6,7],[]),([2,3,8],[1,4,5,6,7],[]),([3,8],[1,4,5,6,7],[2]),([3,8],[4,5,6,7],
[1,2]),([8],[3,4,5,6,7],[1,2]),([1,8],[3,4,5,6,7],[2]),([1,8],[2,3,4,5,6,7],[]),([8],[1,
2,3,4,5,6,7],[]),([],[1,2,3,4,5,6,7],[8]),([],[2,3,4,5,6,7],[1,8]),([2],[3,4,5,6,7],[1,8
]),([1,2],[3,4,5,6,7],[8]),([1,2],[4,5,6,7],[3,8]),([2],[1,4,5,6,7],[3,8]),([],[1,4,5,6,
7],[2,3,8]),([],[4,5,6,7],[1,2,3,8]),([4],[5,6,7],[1,2,3,8]),([1,4],[5,6,7],[2,3,8]),([1
,4],[2,5,6,7],[3,8]),([4],[1,2,5,6,7],[3,8]),([3,4],[1,2,5,6,7],[8]),([3,4],[2,5,6,7],[1
,8]),([2,3,4],[5,6,7],[1,8]),([1,2,3,4],[5,6,7],[8]),([1,2,3,4],[6,7],[5,8]),([3,4],[1,6
,7],[5,8]),([3,4],[1,6,7],[2,5,8]),([3,4],[6,7],[1,2,5,8]),([4],[3,6,7],[1,2,5,8]),([1
,4],[3,6,7],[2,5,8]),([1,4],[2,3,6,7],[5,8]),([4],[1,2,3,6,7],[5,8]),([],[1,2,3,6,7],[4,
5,8]),([],[2,3,6,7],[1,4,5,8]),([2],[3,6,7],[1,4,5,8]),([1,2],[3,6,7],[4,5,8]),([1,2],[6
,7],[3,4,5,8]),([2],[1,6,7],[3,4,5,8]),([],[1,6,7],[2,3,4,5,8]),([],[6,7],[1,2,3,4,5,8])
,([6],[7],[1,2,3,4,5,8]),([1,6],[7],[2,3,4,5,8]),([1,6],[2,7],[3,4,5,8]),([6],[1,2,7],[3
,4,5,8]),([3,6],[1,2,7],[4,5,8]),([3,6],[2,7],[1,4,5,8]),([2,3,6],[7],[1,4,5,8]),([1,2,3
,6],[7],[4,5,8]),([1,2,3,6],[4,7],[5,8]),([2,3,6],[1,4,7],[5,8]),([3,6],[1,4,7],[2,5,8])
,([3,6],[4,7],[1,2,5,8]),([6],[3,4,7],[1,2,5,8]),([1,6],[3,4,7],[2,5,8]),([1,6],[2,3,4,7
],[5,8]),([6],[1,2,3,4,7],[5,8]),([5,6],[1,2,3,4,7],[8]),([5,6],[2,3,4,7],[1,8]),([2,5,6
],[3,4,7],[1,8]),([1,2,5,6],[3,4,7],[8]),([1,2,5,6],[4,7],[3,8]),([2,5,6],[1,4,7],[3,8])
,([5,6],[1,4,7],[2,3,8]),([5,6],[4,7],[1,2,3,8]),([4,5,6],[7],[1,2,3,8]),([1,4,5,6],[7],
[2,3,8]),([1,4,5,6],[2,7],[3,8]),([4,5,6],[1,2,7],[3,8]),([3,4,5,6],[1,2,7],[8]),([3,4,5
,6],[2,7],[1,8]),([2,3,4,5,6],[7],[1,8]),([1,2,3,4,5,6],[7],[8]),([1,2,3,4,5,6],[],[7,8]
),([2,3,4,5,6],[1],[7,8]),([3,4,5,6],[1],[2,7,8]),([3,4,5,6],[],[1,2,7,8]),([4,5,6],[3],
[1,2,7,8]),([1,4,5,6],[3],[2,7,8]),([1,4,5,6],[2,3],[7,8]),([4,5,6],[1,2,3],[7,8])]
```

Figure 7.2: Tower of Hanoi (first 200 configurations)

7.6. SOME VARIATIONS ON THE TOWER OF HANOI

```
data Peg   = A | B | C
type Tower = ([Int], [Int], [Int])
```

There are six possible single moves from one peg to another:

```
move :: Peg -> Peg -> Tower -> Tower
move A B (x:xs,ys,zs) = (xs,x:ys,zs)
move B A (xs,y:ys,zs) = (y:xs,ys,zs)
move A C (x:xs,ys,zs) = (xs,ys,x:zs)
move C A (xs,ys,z:zs) = (z:xs,ys,zs)
move B C (xs,y:ys,zs) = (xs,ys,y:zs)
move C B (xs,ys,z:zs) = (xs,z:ys,zs)
```

The procedure `transfer` takes three arguments for the pegs, an argument for the number of disks to move, and an argument for the tower configuration to move. The output is a list of tower configurations. The procedure `hanoi`, finally, takes a size argument and outputs the list of configurations to move a tower of that size.

```
transfer :: Peg -> Peg -> Peg -> Int -> Tower -> [Tower]
transfer _ _ _ 0 tower = [tower]
transfer p q r n tower = transfer p r q (n-1) tower
                         ++
                         transfer r q p (n-1) (move p q tower')
    where tower' = last (transfer p r q (n-1) tower)

hanoi :: Int -> [Tower]
hanoi n = transfer A C B n ([1..n],[],[])
```

The output for `hanoi 8` is given in Figure 7.2. Here is the output for `hanoi 4`:

```
IAR> hanoi 4
[([1,2,3,4],[],[]),([2,3,4],[1],[]),([3,4],[1],[2]),([3,4],[],[1,2]),
([4],[3],[1,2]),([1,4],[3],[2]),([1,4],[2,3],[]),([4],[1,2,3],[]),
([],[1,2,3],[4]),([],[2,3],[1,4]),([2],[3],[1,4]),([1,2],[3],[4]),
([1,2],[],[3,4]),([2],[1],[3,4]),([],[1],[2,3,4]),([],[],[1,2,3,4])]
```

If you key in `hanoi 64` and expect you can start meditating until the end of the world, you will discover that the program suffers from what functional programmers call a *space leak* or *black hole*: as the execution progresses, the list of tower configurations that is kept in memory grows longer and longer, and execution will abort with an 'out of memory' error. All computation ceases before the world ends.

Now consider the following. If there is a best way to transfer the tower of Hanoi (or the tower of Brahma), then in any given configuration it should be clear what the next move is. If they only have to look at the present configuration, the monks in charge of the tower of Brahma can go about their task in a fully enlightened manner, with complete focus on the here and now.

Observe that there are 3^n ways to stack n disks of decreasing sizes on 3 pegs in such a way that no disk is on top of a smaller disk. Since the number of moves to get from ([1..n],[],[]) to ([],[],[1..n]) is less than this, not all of these configurations are correct. How can we implement a check for correct configurations? Here is a checking procedure, with an argument for the size of the largest disk.

```
check :: Int -> Tower -> Bool
check 0 t = t == ([],[],[])
check n (xs,ys,zs)
   | xs /= [] && last xs == n = check (n-1) (init xs, zs, ys)
   | zs /= [] && last zs == n = check (n-1) (ys, xs, init zs)
   | otherwise                = False
```

Exercise 7.56 To see that the implementation of `check` is correct, we need an inductive proof. Give that proof.

Here is a function for finding the largest disk in a configuration, and a function `checkT` for checking a configuration of any size.

```
maxT :: Tower -> Int
maxT (xs, ys, zs) = foldr max 0 (xs ++ ys ++ zs)

checkT :: Tower -> Bool
checkT t = check (maxT t) t
```

7.6. SOME VARIATIONS ON THE TOWER OF HANOI

The following exercise gives a more direct criterion.

Exercise 7.57 Show that configuration (xs,ys,zs) with largest disk n is correct iff it holds that every disk m is either on top of a disk k with $k - m$ odd, or at the bottom of the source or destination peg, with $n - m$ even, or at the bottom of the auxiliary peg, with $n - k$ odd.

The previous exercise immediately gives a procedure for building correct configurations: put each disk according to the rule, starting with the largest disk. This will give exactly two possibilities for every disk k: the largest disk can either go at source or at destination; if $k + 1$ is already in place, k can either go on top of $k + 1$ or to the only other place with the same parity as $k + 1$.

For the implementation of a parity check, for the case where the largest disk has size n, we can take any configuration (xs,ys,zs), extend it to

(xs ++ [n+1], ys ++ [n], zs ++ [n+1])

and then define the parities by means of par (x:xs) = x mod 2. Here is the implementation:

```
parity :: Tower -> (Int,Int,Int)
parity (xs,ys,zs) = par (xs ++ [n+1], ys ++ [n],zs ++ [n+1])
  where
  n = maxT (xs, ys, zs)
  par (x:xs,y:ys,z:zs) = (mod x 2, mod y 2, mod z 2)
```

Exercise 7.58 Show that if (xs,ys,zs) is a correct configuration, then

$$\text{parity (xs,ys,zs)} \in \{(1,1,0),(1,0,1),(0,1,1)\}.$$

A little reflection shows that there are only two kinds of moves.

- Moves of the first kind move disk 1 (the smallest disk).
- Moves of the second kind move a disk other than 1.

Moves of the first kind are fully determined, as can be seen by the fact that disk 1 should always be moved to the place with parity 0.

Exercise 7.59 Give an argument for this.

Moves of the second kind are also fully determined, for if there is one empty peg, then there are disks $1, k$ on top of the other pegs, and we can only move k to the empty peg, otherwise there are disks $1 < k < m$ on top of the pegs, and we can only move k on top of m.

This gives a new algorithm for tower transfer, without space leak. We determine the target of a move of disk 1 by means of a parity check:

```
target :: Tower -> Peg
target t@(xs,ys,zs) | parity t == (0,1,1) = A
                    | parity t == (1,0,1) = B
                    | parity t == (1,1,0) = C
```

The moves of disk 1 are now given by:

```
move1 :: Tower -> Tower
move1 t@(1:_,ys,zs) = move A (target t) t
move1 t@(xs,1:_,zs) = move B (target t) t
move1 t@(xs,ys,1:_) = move C (target t) t
```

The moves of the middle disk are given by:

```
move2 :: Tower -> Tower
move2 t@(1:xs,[],zs) = move C B t
move2 t@(1:xs,ys,[]) = move B C t
move2 t@(1:xs,ys,zs) = if ys < zs then move B C t else move C B t
move2 t@([],1:ys,zs) = move C A t
move2 t@(xs,1:ys,[]) = move A C t
move2 t@(xs,1:ys,zs) = if xs < zs then move A C t else move C A t
move2 t@([],ys,1:zs) = move B A t
move2 t@(xs,[],1:zs) = move A B t
move2 t@(xs,ys,1:zs) = if xs < ys then move A B t else move B A t
```

The check for completion is:

7.6. SOME VARIATIONS ON THE TOWER OF HANOI 273

```
done :: Tower -> Bool
done ([],[], _) = True
done (xs,ys,zs) = False
```

Transfer of a tower takes place by alternation between the two kinds of moves, until the tower is completely transferred. Since the last move has to be a move1, we only have to check for complete transfer right after such moves, i.e., in the function `transfer2`:

```
transfer1, transfer2 :: Tower -> [Tower]
transfer1 t = t : transfer2 (move1 t)
transfer2 t = if done t then [t] else t : transfer1 (move2 t)
```

Here is our new Hanoi procedure:

```
hanoi' :: Int -> [Tower]
hanoi' n = transfer1 ([1..n],[],[])

zazen :: [Tower]
zazen = hanoi' 64
```

By now we know enough about correct tower configurations to be able to order them.

Exercise 7.60 Define and implement a total ordering on the list of all correct tower configurations.

Exercise 7.60 makes clear that it should be possible to define a bijection between the natural numbers and the correct tower configurations, in their natural order. For this, we first define a function for finding the k-th correct configuration in the list of transitions for ([1..n],[],[]).

```
hanoiCount :: Int -> Integer -> Tower
hanoiCount n k | k < 0          = error "argument negative"
               | k > 2^n - 1    = error "argument not in range"
               | k == 0         = ([1..n],[],[])
               | k == 2^n - 1   = ([],[],[1..n])
               | k < 2^(n-1)    = (xs ++ [n], zs, ys)
               | k >= 2^(n-1)   = (ys', xs', zs' ++ [n])
    where
    (xs,ys,zs)    = hanoiCount (n-1) k
    (xs',ys',zs') = hanoiCount (n-1) (k - 2^(n-1))
```

In terms of this we define the bijection. Note that for the definition we need the inverse of the function $\lambda n.2^n$, i.e., the function $\lambda n.\log_2 n$. The predefined Haskell function logBase gives logarithms, and logarithms to base 2 are given by logBase 2. Since the results are in class Floating (the class of floating point numbers), we need conversion to get back to class Integral (the class consisting of Int and Integer). For this we use truncate.

```
toTower :: Integer -> Tower
toTower n = hanoiCount k m
   where
   n' = fromInteger (n+1)
   k  = truncate (logBase 2 n')
   m  = truncate (n' - 2^k)
```

Exercise 7.61 The function hanoiCount gives us yet another approach to the tower transfer problem. Implement this as hanoi'' :: Int -> [Tower].

Exercise 7.62 Implement the function fromTower :: Tower -> Integer that is the inverse of toTower.

7.7 Other Data Structures

A standard way to prove properties of logical formulas is by induction on their syntactic structure. Consider e.g. the following Haskell data type for propositional formulas.

7.7. OTHER DATA STRUCTURES

```
data Form = P Int | Conj Form Form | Disj Form Form | Neg Form

instance Show Form where
  show (P i)         = 'P':show i
  show (Conj f1 f2)  = "(" ++ show f1 ++ " & " ++ show f2 ++ ")"
  show (Disj f1 f2)  = "(" ++ show f1 ++ " v " ++ show f2 ++ ")"
  show (Neg f)       = "~" ++ show f
```

The instance Show Form ensures that the data type is in the class Show, and the function show indicates how the formulas are displayed.

It is assumed that all proposition letters are from a list $P_0, P_1, \ldots$. Then $\neg(P_1 \vee \neg P_2)$ is represented as Neg (Disj (P 1) (Neg (P 2))), and shown on the screen as ~(P1 v ~P2), and so on.

We define the list of sub formulas of a formula as follows:

```
subforms :: Form -> [Form]
subforms (P n) = [(P n)]
subforms (Conj f1 f2) = (Conj f1 f2):(subforms f1 ++ subforms f2)
subforms (Disj f1 f2) = (Disj f1 f2):(subforms f1 ++ subforms f2)
subforms (Neg f) = (Neg f):(subforms f)
```

This gives, e.g.:

```
IAR> subforms (Neg (Disj (P 1) (Neg (P 2))))
[~(P1 v ~P2),(P1 v ~P2),P1,~P2,P2]
```

The following definitions count the number of connectives and the number of atomic formulas occurring in a given formula:

```
ccount :: Form -> Int
ccount (P n) = 0
ccount (Conj f1 f2) = 1 + (ccount f1) + (ccount f2)
ccount (Disj f1 f2) = 1 + (ccount f1) + (ccount f2)
ccount (Neg f) = 1 + (ccount f)
```

```
acount :: Form -> Int
acount (P n) = 1
acount (Conj f1 f2) = (acount f1) + (acount f2)
acount (Disj f1 f2) = (acount f1) + (acount f2)
acount (Neg f) = acount f
```

Now we can prove that the number of sub formulas of a formula equals the sum of its connectives and its atoms:

Proposition 7.63 For every member f of Form:

length (subforms f) = (ccount f) + (acount f).

Proof.

Basis If f is an atom, then subforms f = [f], so this list has length 1. Also, ccount f = 0 and acount f = 1.

Induction step If f is a conjunction or a disjunction, we have:

- length (subforms f) =
  1 + length (subforms f1) + length (subforms f2),
- ccount f = 1 + (ccount f1) + (ccount f2),
- acount f = (acount f1) + (acount f2),

where f1 and f2 are the two conjuncts or disjuncts. By induction hypothesis:

length (subforms f1) = (ccount f1) + (acount f1).
length (subforms f2) = (ccount f2) + (acount f2).

The required equality follows immediately from this.

If f is a negation, we have:

- length (subforms f) = 1 + length (subforms f1),
- ccount f = 1 + (ccount f1),
- acount f = (acount f1),

and again the required equality follows immediately from this and the induction hypothesis.

If one proves a property of formulas by induction on the structure of the formula, then the fact is used that every formula can be mapped to a natural number that indicates its constructive complexity: 0 for the atomic formulas, the maximum of rank(Φ) and rank(Ψ) plus 1 for a conjunction $\Phi \wedge \Psi$, and so on.

7.8 Further Reading

Induction and recursion are at the core of discrete mathematics. See [Bal91]. A splendid textbook that teaches concrete (and very useful) mathematical skills in this area is Graham, Knuth and Patashnik [GKP89]. Recursion is also crucial for algorithm design. If you are specifically interested in the design and analysis of algorithms you should definitely read Harel [Har87]. Algorithm design in Haskell is the topic of [RL99].

Chapter 8

Working with Numbers

Preview

When reasoning about mathematical objects we make certain assumptions about the existence of things to reason about. In the course of this chapter we will take a look at integers, rational numbers, real numbers, and complex numbers. We will recall what they all look like, and we will demonstrate that reasoning about mathematical objects can be put to the practical test by an *implementation* of the definitions that we reason about. The implementations turn the definitions into procedures for handling representations of the mathematical objects. This chapter will also present some illuminating examples of the art of mathematical reasoning.

```
module WWN

where

import List
import Nats
```

8.1 Natural Numbers

The natural numbers are the conceptual basis of more involved number systems. The implementation of `Natural` from the previous chapter has the advantage that the unary representation makes the two cases for inductive proofs and recursive definitions, Z and S n, very clear. We repeat the code of our implementation of Natural, wrapped up this time in the form of a module, and integrated into the Haskell type system: see Figure 8.1. Note that an implementation of quotRem was added: quotRem n m returns a pair (q,r) consisting of a quotient and a remainder of the process of dividing n by m, i.e., q and r satisfy $0 \leqslant r < m$ and $q \times m + r = n$.

The integration of Natural in the system of Haskell types is achieved by means of instance declarations. E.g., the declaration instance Ord Natural makes the type Natural an instance of the class Ord. From an inspection of *Prelude.hs* we get that (i) a type in this class is assumed to also be in the class Eq, and (ii) a minimal complete definition of a type in this class assumes definitions of <= or compare. Condition (i) is fulfilled for Natural by the deriving Eq statement, while (ii) is taken care of by the code for compare in the module. The general code for class Ord provides methods for the relations <=, <, >=, >, and for the comparison operators max and min, so putting Natural in class Ord provides us with all of these.

Similarly, the declaration instance Enum Natural ensures that Natural is in the class Enum. Definitions for the following functions for types in this class are provided in the module:

```
succ     :: Natural -> Natural
pred     :: Natural -> Natural
toEnum   :: Int -> Natural
fromEnum :: Natural -> Int
enumFrom :: Natural -> [Natural]
```

The general code for the class provides, e.g., enumFromTo, so we get:

```
Nats> enumFromTo Z (toEnum 5)
[Z,S Z,S (S Z),S (S (S Z)),S (S (S (S Z))),S (S (S (S (S Z))))]
```

Next, the declaration instance Num Natural ensures that Natural is in the class Num. This is where the functions for addition, multiplication, sign, absolute value belong, as well as a type conversion function fromInteger. Putting in the appropriate definitions allows the use of the standard names for these operators. In terms of these, *Prelude.hs* defines subtraction, the type conversion function fromInt, and negate. Since negate is defined in terms of subtraction, and subtraction for naturals is cut-off subtraction, we get:

```
Nats> negate (S (S (S (S Z))))
Z
```

8.1. NATURAL NUMBERS

```
module Nats where

data Natural = Z | S Natural deriving (Eq, Show)

instance Ord Natural where
  compare Z Z = EQ
  compare Z _ = LT
  compare _ Z = GT
  compare (S m) (S n) = compare m n

instance Enum Natural where
  succ = \ n -> S n
  pred Z = Z
  pred (S n) = n
  toEnum   = fromInt
  fromEnum = toInt
  enumFrom n = map toEnum [(fromEnum n)..]

instance Num Natural where
  (+) = foldn succ
  (*) = \m -> foldn (+m) Z
  (-) = foldn pred
  abs = id
  signum Z = Z
  signum n = (S Z)
  fromInteger n | n < 0     = error "no negative naturals"
                | n == 0    = Z
                | otherwise = S (fromInteger (n-1))

foldn :: (a -> a) -> a -> Natural -> a
foldn h c Z = c
foldn h c (S n) = h (foldn h c n)

instance Real Natural where toRational x = toInteger x % 1

instance Integral Natural where
   quotRem n d | d > n     = (Z,n)
               | otherwise = (S q, r)
           where (q,r) = quotRem (n-d) d
   toInteger = foldn succ 0
```

Figure 8.1: A Module for Natural Numbers.

The next instance declaration puts Natural in the class Real. The only code we have to provide is that for the type conversion function toRational. Putting Natural in this class is necessary for the next move: the instance declaration that puts Natural in the class Integral: types in this class are constrained by *Prelude.hs* to be in the classes Real and Enum. The benefits of making Natural an integral type are many: we now get implementations of even, odd, quot, rem, div, mod, toInt and many other functions and operations for free (see the contents of *Prelude.hs* for full details).

The implicit type conversion also allows us to introduce natural numbers in shorthand notation:

```
Nats> 12 :: Natural
S (S (S (S (S (S (S (S (S (S (S (S Z))))))))))))
```

The standard representation for natural numbers that we are accustomed to, decimal string notation, is built into the Haskell system for the display of integers. To display naturals in decimal string notation, all we have to do is use type conversion:

```
Nats> toInt (S (S (S Z)))
3
Nats> (S (S Z)) ^ (S (S (S Z)))
S (S (S (S (S (S (S (S Z))))))))
Nats> toInt (S (S Z)) ^ (S (S (S Z)))
8
```

For other representations, i.e., binary string representation of an integral number n, divide by 2 repeatedly, and collect the remainders in a list. The reversal of this list gives the binary representation $b_k \ldots b_0$ satisfying $n = b_k \cdot 2^k + b_{k-1} \cdot 2^{k-1} + \cdots + b_0$. The nice thing is that we can provide this code for the whole class of Integral types:

```
binary :: Integral a => a -> [Int]
binary x = reverse (bits x)
  where bits 0 = [0]
        bits 1 = [1]
        bits n = toInt (rem n 2) : bits (quot n 2)
```

To display this on the screen, we need intToDigit for converting integers into digital characters:

```
showDigits :: [Int] -> String
showDigits = map intToDigit

bin :: Integral a => a -> String
bin = showDigits . binary
```

Exercise 8.1 Give a function `hex` for displaying numbers in type class `Integral` in hexadecimal form, i.e., in base 16 representation. The extra digits a, b, c, d, e, f for 10, 11, 12, 13, 14, 15 that you need are provided by `intToDigit`. The call `hex 31` should yield `"1f"`. (Hint: it is worthwhile to provide a more general function `toBase` for converting to a list of digits in any base in $\{2, \ldots, 16\}$, and then define `hex` in terms of that.)

8.2 GCD and the Fundamental Theorem of Arithmetic

The fundamental theorem of arithmetic, stating that every $n \in \mathbb{N}$ with $n > 1$ has a unique prime factorization, was known in Antiquity. Let us reflect a bit to see why it is true. First note that the restriction on n is necessary, for $m \cdot 0 = 0$ for all $m \in \mathbb{N}$, so a factorization of 0 can never be unique, and since we have ruled out 1 as a prime, 1 does not have a prime factorization at all. The fundamental theorem of arithmetic constitutes the reason, by the way, for ruling out 1 as a prime number. We have $n = 1^m \cdot n$ for any $n, m \in \mathbb{N}$, so a factorization of n that admits 1^m as a factor can never be unique.

From the prime factorization algorithm (Section 1.7) we know that prime factorizations of every natural number > 1 exist. To show uniqueness, we still have to establish that no number has more than one prime factorization.

Euclid's GCD algorithm The greatest common divisor of two natural numbers a, b, notation $\text{GCD}(a, b)$, is the natural number d with the property that d divides both a and b, and for all natural numbers d' that divide both a and b it holds that d' divides d. For example, $\text{GCD}(30, 84) = 6$, for 6 divides 30 and 84, and every other common divisor of 30 and 84 divides 6 (these other common divisors being 1, 2 and 3).

Clearly, if such d exists, it is unique, for suppose that d' also qualifies. Then because d' divides a and b, it follows from the definition of d that

d' divides d. Similarly, because d divides a and b, it follows from the definition of d' that d divides d'. But if d divides d' and d' divides d then it follows that $d = d'$.

The greatest common divisor of a and b can be found from the prime factorizations of a and b as follows. Let $a = p_1^{\alpha_1} \cdots p_k^{\alpha_k}$ and $b = p_1^{\beta_1} \cdots p_k^{\beta_k}$ be prime factorizations of a and b, with $p_1, \ldots, p_k$ distinct primes, and $\alpha_1, \ldots, \alpha_k, \beta_1, \ldots, \beta_k$ natural numbers. Then the greatest common divisor of a and b equals the natural number $p_1^{\gamma_1} \cdots p_k^{\gamma_k}$, where each γ_i is the minimum of α_i and β_i.

For example, the greatest common divisor of $30 = 2^1 \cdot 3^1 \cdot 5^1 \cdot 7^0$ and $84 = 2^2 \cdot 3^1 \cdot 5^0 \cdot 7^1$ is given by $2^1 \cdot 3^1 \cdot 5^0 \cdot 7^0 = 6$.

But there is an easier way to find $\text{GCD}(a, b)$. Here is Euclid's famous algorithm (assume neither of a, b equals 0):

WHILE $a \neq b$ DO IF $a > b$ THEN $a := a - b$ ELSE $b := b - a$.

Let us run this for the example case $a = 30$, $b = 84$. We get the following (conventions about variable names as in Section 1.7).

$$
\begin{array}{lll}
 & a_0 = 30 & b_0 = 84 \\
a_0 < b_0 & a_1 = 30 & b_1 = 84 - 30 = 54 \\
a_1 < b_1 & a_2 = 30 & b_2 = 54 - 30 = 24 \\
a_2 > b_2 & a_3 = 30 - 24 = 6 & b_3 = 24 \\
a_3 < b_3 & a_4 = 6 & b_4 = 24 - 6 = 18 \\
a_4 < b_4 & a_5 = 6 & b_5 = 18 - 6 = 12 \\
a_5 < b_5 & a_6 = 6 & b_6 = 12 - 6 = 6 \\
a_6 = b_6 = 6 & &
\end{array}
$$

Now *why* does this work? The key observation is the following.

If d divides a and b and $a > b$ then d divides $a - b$ (for then there are natural numbers m, n with $m > n$ and $a = md$, $b = nd$, and therefore $a - b = md - nd = d(m - n)$), and similarly, if d divides a and b and $a < b$ then d divides $b - a$.

Conversely, if $a > b$ and d divides $a - b$ and b, then d divides a (for then there are natural numbers m, n with $a - b = md$ and $b = nd$, hence $a = md + nd = d(m + n)$), and similarly, if $a > b$ and d divides $b - a$ and a, then d divides a.

Thus, if $a > b$ then the set of common divisors of a and b equals the set of common divisors of $a - b$ and b, and if $b > a$ then the set of common divisors of a and b equals the set of common divisors of a and $b - a$.

Since the sets of all common divisors are equal, the greatest common divisors must be equal as well. Therefore we have: if $a > b$ then $\text{GCD}(a, b) = \text{GCD}(a - b, b)$, and if $b > a$ then $\text{GCD}(a, b) = \text{GCD}(a, b - a)$.

8.2. GCD AND THE FUNDAMENTAL THEOREM OF ARITHMETIC

Using this we see that every iteration through the loop preserves the greatest common divisor in the following sense: $\text{GCD}(a_i, b_i) = \text{GCD}(a_{i+1}, b_{i+1})$. Since we know that the algorithm terminates we have: there is some k with $a_k = b_k$. Therefore $a_k = \text{GCD}(a_k, b_k) = \text{GCD}(a, b)$.

Haskell contains a standard function gcd for the greatest common divisor of a pair of objects of type Integral:

```
gcd             :: Integral a => a -> a -> a
gcd 0 0         = error "Prelude.gcd: gcd 0 0 is undefined"
gcd x y         = gcd' (abs x) (abs y)
                  where gcd' x 0 = x
                        gcd' x y = gcd' y (x `rem` y)
```

Exercise 8.2 If you compare the code for gcd to Euclid's algorithm, you see that Euclid uses repeated subtraction where the Haskell code uses rem. Explain as precisely as you can how the Haskell version of the GCD algorithm is related to Euclid's method for finding the GCD.

We can use the GCD to define an interesting relation. Two natural numbers n and m are said to be *co-prime* or *relatively prime* if $\text{GCD}(m, n) = 1$. Here is an implementation:

```
coprime :: Integer -> Integer -> Bool
coprime m n = (gcd m n) == 1
```

Exercise 8.3 Consider the following experiment:

```
WWN> coprime 12 25
True
WWN> 12 + 25
37
WWN> coprime 25 37
True
WWN> 25 + 37
62
WWN> coprime 37 62
True
```

```
WWN> 37 + 62
99
WWN> coprime 62 99
True
WWN> 62 + 99
161
WWN> coprime 99 161
True
WWN>
```

This experiment suggests a general rule, for you to consider ...

Does it follow from the fact that a and b are co-prime with $a < b$ that b and $a + b$ are co-prime? Give a proof if your answer is 'yes' and a counterexample otherwise.

Theorem 8.4 For all positive $a, b \in \mathbb{N}$ there are integers m, n with $ma + nb = \text{GCD}(a, b)$.

Proof. Consider the pairs $(a_0, b_0), (a_1, b_1), \ldots, (a_k, b_k)$ generated by Euclid's algorithm. We know that $(a_0, b_0) = (a, b)$ and that $a_k = b_k = \text{GCD}(a, b)$.

a_0 satisfies $a_0 = ma + nb$ for $m = 1, n = 0$, b_0 satisfies $ma + nb = 1$ for $m = 0, n = 1$.

Suppose a_i satisfies $a_i = m_1 a + n_1 b$ and b_i satisfies $b_i = m_2 a + n_2 b$. If $a_i > b_i$, then a_{i+1} satisfies $a_{i+1} = (m_1 - m_2)a + (n_1 - n_2)b$ and b_{i+1} satisfies $b_{i+1} = m_2 a + n_2 b$. If $a_i < b_i$, then a_{i+1} satisfies $a_{i+1} = m_1 a + n_1 b$ and b_{i+1} satisfies $b_{i+1} = (m_2 - m_1)a + (n_2 - n_1)b$. Thus, every iteration through the loop of Euclid's algorithm preserves the fact that a_i and b_i are integral linear combinations $ma + nb$ of a and b.

This shows that there are integers m, n with $a_k = ma + nb$, hence that $ma + nb = \text{GCD}(a, b)$. ∎

Theorem 8.5 If p is a prime number that divides ab then p divides a or b.

Proof. Suppose p divides ab and p does not divide a. Then $\text{GCD}(a, p) = 1$. By the previous theorem there are integers m, n with $ma + np = 1$. Multiplying both sides by b gives:

$$mab + nbp = b.$$

By the fact that p divides ab we know that p divides both mab and nbp. Hence p divides $mab + nbp$. Hence p divides b. ∎

Theorem 8.5 is the tool for proving the fundamental theorem of arithmetic.

8.3. INTEGERS

Theorem 8.6 (Fundamental Theorem of Arithmetic)
Every natural number greater than 1 has a unique prime factorization.

Proof. We established in Section 1.7 that every natural number greater than 1 has at least one prime factorization.

To show that every natural number has at most one prime factorization, assume to the contrary that there is a natural number $N > 1$ with at least two different prime factorizations. Thus,

$$N = p_1 \cdots p_r = q_1 \cdots q_s,$$

with all of $p_1, \ldots, p_r, q_1, \ldots, q_s$ prime. Divide out common factors if necessary. This gives a p_i that is not among the q's. But this is a contradiction with Theorem 8.5, because p_i divides $N = q_1 \cdots q_s$ but p_i does not divide any of $q_1, \ldots, q_s$, since these are all prime numbers different from p_i. ∎

8.3 Integers

Suppose n, m, k are natural numbers. When $n = m + k$ we can view k as the difference of n and m, and it is tempting to write this as $k = n - m$ for the operation of subtraction. Addition and subtraction are called *inverse operations*, for if the addition of n to m is followed by the subtraction of n, the end result is the original natural number m. In other words, we have:

$$(m + n) - n = m.$$

But we have to be careful here. The operation of subtracting a natural number n from a natural number m will only result in a natural number if $m \geqslant n$. To make subtraction possible between any pair of numbers, we have to introduce negative numbers. This gives the domain of integers:

$$\mathbb{Z} = \{\ldots, -3, -2, -1, 0, 1, 2, 3, \ldots\}.$$

The symbol $\mathbb{Z}$ derives from *Zahl*, the German word for number.

In fact, we need not consider the integers as given by God, but we can view them as constructed from the natural numbers. This can be done in several ways. The following Haskell code illustrates one possibility:

```
data Sgn = P | N deriving (Eq,Show)
type MyInt = (Sgn,Natural)

myplus :: MyInt -> MyInt -> MyInt
myplus (s1,m) (s2,n) | s1 == s2              = (s1,m+n)
                     | s1 == P && n <= m = (P,m-n)
                     | s1 == P && n >  m = (N,n-m)
                     | otherwise             = myplus (s2,n) (s1,m)
```

Another way is as follows. We represent every integer as a 'difference pair' of two natural numbers. For example, $(0,3)$ represents -3, but the same number -3 is also represented by $(1,4)$ or $(2,5)$.

In general, for all $m_1, m_2 \in \mathbb{N}$, if $m_1 \geqslant m_2$ then there is a $k \in \mathbb{N}$ with $m_2 + k = m_1$, and (m_1, m_2) represents k. If, on the other hand, $m_1 < m_2$ then there is a $k \in \mathbb{N}$ with $m_1 + k = m_2$, and (m_1, m_2) represents $-k$. Thus, let $R \subseteq \mathbb{N}^2$ be defined as follows:

$$(m_1, m_2) R (n_1, n_2) :\equiv m_1 + n_2 = m_2 + n_1.$$

It is easy to see that R is an equivalence on $\mathbb{N}^2$. If $m \in \mathbb{N}$, then the integer $-m$ is represented by $(0, m)$, but also by $(k, k+m)$, for any $k \in \mathbb{N}$.

Intuitively (m_1, m_2) and (n_1, n_2) are equivalent modulo R when their differences are the same. This is the case precisely when $m_1 + n_2 = m_2 + n_1$. In this case we say that (m_1, m_2) and (n_1, n_2) *represent the same number*. In Section 6.8 we saw that the equivalence relation R is a congruence for addition on $\mathbb{N}^2$, where addition on $\mathbb{N}^2$ is given by:

$$(m_1, m_2) + (n_1, n_2) := (m_1 + n_1, m_2 + n_2),$$

Denoting the equivalence class of (m_1, m_2) as $[m_1 - m_2]$, we get:

$$[m_1 - m_2] := \{(n_1, n_2) \in \mathbb{N}^2 \mid m_1 + n_2 = m_2 + n_1\}.$$

We identify the integers with the equivalence classes $[m_1 - m_2]$, for $m_1, m_2 \in \mathbb{N}$. Note that in this notation $-$ is *not* an operator. Call the equivalence classes *difference classes*.

If an integer $m = [m_1 - m_2]$, then we can swap the sign by swapping the order of m_1 and m_2, and put $-m = [m_2 - m_1]$. Swapping the order twice gets us back to the original equivalence class. This reflects the familiar rule of sign $m = -(-m)$.

8.3. INTEGERS

We define addition and multiplication on difference classes as follows:

$$[m_1 - m_2] + [n_1 - n_2] := [(m_1 + n_1) - (m_2 + n_2)]$$
$$[m_1 - m_2] \cdot [n_1 - n_2] := [(m_1 n_1 + m_2 n_2) - (m_1 n_2 + n_1 m_2)].$$

The purpose of this definition is to extend the fundamental laws of arithmetic to the new domain. To see whether we have succeeded in this we have to perform a verification, as a justification of the definition.

In the verification that integers satisfy the law of commutativity for addition, we make use of the definition of addition for difference classes, and of laws of commutativity for addition on $\mathbb{N}$:

Proposition 8.7 For all $m, n \in \mathbb{Z}$: $m + n = n + m$.

Proof. Representing m, n as difference classes we get:

$$\begin{aligned}
[m_1 - m_2] + [n_1 - n_2] &= \quad [\text{definition of } + \text{ for difference classes}] \\
[(m_1 + n_1) - (m_2 + n_2)] &= \quad [\text{commutativity of } + \text{ for } \mathbb{N}] \\
[(n_1 + m_1) - (n_2 + m_2)] &= \quad [\text{definition of } + \text{ for difference classes}] \\
[(n_1 - n_2) + [m_1 - m_2)].
\end{aligned}$$

∎

Note that the proof uses just the definition of + for difference classes and the commutativity of + on $\mathbb{N}$. Note also that every equality statement is *justified* in the proof by a reference to a definition or to a fundamental law of arithmetic, where these laws are in turn justified by inductive proofs based on the recursive definitions of the natural number operations. Thus, the fundamental laws of arithmetic and the definition of a new kind of object (difference classes) in terms of a familiar one (natural numbers) are our starting point for the investigation of that new kind of object.

Exercise 8.8 Show (using the definition of integers as difference classes and the definition of addition for difference classes) that the associative law for addition holds for the domain of integers.

In a similar way it can be shown from the definition of integers by means of difference classes and the definition of multiplication for difference classes that the associative and commutative laws for multiplication continue to hold for the domain of integers.

As a further example of reasoning about this representation for the integers, we show that the distributive law continues to hold in the domain of integers (viewed as difference classes). In the verification we make use of the definition of addition and multiplication for difference classes, and of the fundamental laws for addition and multiplication on the natural numbers. Again, we justify every step by means of a reference to the definition of an operation on difference classes or to one of the laws of arithmetic for the natural numbers.

Proposition 8.9 For all $m, n, k \in \mathbb{Z} : m(n + k) = mn + mk$.

Proof. Representing m, n, k as difference classes we get:

$[m_1 - m_2]([n_1 - n_2] + [k_1 - k_2])$

$=$ [definition of $+$ for difference classes]

$[m_1 - m_2][(n_1 + k_1) - (n_2 + k_2)]$

$=$ [definition of $\cdot$ for difference classes]

$[(m_1(n_1 + k_1) + m_2(n_2 + k_2)) - (m_1(n_2 + k_2) + (n_1 + k_1)m_2)]$

$=$ [distribution of $\cdot$ over $+$ for $\mathbb{N}$]

$[(m_1 n_1 + m_1 k_1 + m_2 n_2 + m_2 k_2) - (m_1 n_2 + m_1 k_2 + m_2 n_1 + m_2 k_1)]$

$=$ [commutativity of $+$ for $\mathbb{N}$]

$[(m_1 n_1 + m_2 n_2 + m_1 k_1 + m_2 k_2) - (m_1 n_2 + m_2 n_1 + m_1 k_2 + m_2 k_1)]$

$=$ [definition of $+$ for difference classes]

$[(m_1 n_1 + m_2 n_2) - (m_1 n_2 + m_2 n_1)]$
$\quad + \quad [(m_1 k_1 + m_2 k_2) - (m_1 k_2 + m_2 k_1)]$

$=$ [definition of $\cdot$ for difference classes]

$[m_1 - m_2] \cdot [n_1 - n_2] + [m_1 - m_2] \cdot [k_1 - k_2]$.

∎

Once the integers and the operations of addition and multiplication on it are defined, and we have checked that the definitions are correct, we can forget of course about the representation by means of difference classes. It simply takes too much mental energy to keep such details of representation in mind. Mathematicians make friends with mathematical objects by forgetting about irrelevant details of their definition.

The natural numbers have representations as difference classes too: a natural number m is represented by $[m - 0]$. In fact, the original numbers

8.4. IMPLEMENTING INTEGER ARITHMETIC

and their representations in the new domain $\mathbb{Z}$ behave exactly the same, in a very precise sense: the function that maps $m \in \mathbb{N}$ to $[m - 0]$ is one-to-one (it never maps different numbers to the same pair) and it preserves the structure of addition and multiplication, in the following sense (see also Section 6.8). If $m + n = k$ then

$$[m - 0] + [n - 0] = [(m + n) - 0] = [k - 0],$$

and if $mn = k$ then

$$[m - 0] \cdot [n - 0] = [mn - 0] = [k - 0].$$

Again we forget about the differences in representation and we say: $\mathbb{N} \subseteq \mathbb{Z}$.

8.4 Implementing Integer Arithmetic

If we represent natural numbers as type Natural, then the moral of the above is that integers can be represented as *pairs* of naturals. E.g., minus five is represented by the pair (S Z,S(S(S(S(S Z))))), or by the pair (Z,S(S(S(S(S Z))))), and so on. Here is an appropriate data type declaration:

```
type NatPair = (Natural,Natural)
```

The gist of the previous section can be nicely illustrated by means of implementations of the integer operations. For suppose we have the natural number operations plus for addition and times for multiplication available for the type Natural). Then addition for integer pairs is implemented in Haskell as:

```
plus1 :: NatPair -> NatPair -> NatPair
plus1 (m1, m2) (n1, n2) = (m1+n1, m2+n2)
```

Subtraction is just addition, but with the sign of the second operand reversed. Sign reversal is done by swapping the elements of a pair, as we have seen. Thus, the implementation of subtraction can look like this:

```
subtr1 :: NatPair -> NatPair -> NatPair
subtr1 (m1, m2) (n1, n2) = plus1 (m1, m2) (n2, n1)
```

Here is the implementation of multiplication:

```
mult1 :: NatPair -> NatPair -> NatPair
mult1 (m1, m2) (n1, n2) = (m1*n1 + m2*n2, m1*n2 + m2*n1)
```

The implementation of equality for pairs of naturals is also straightforward:

```
eq1 :: NatPair -> NatPair -> Bool
eq1 (m1, m2) (n1, n2) = (m1+n2) == (m2+n1)
```

Finally, it is useful to be able to reduce a naturals pair to its simplest form (such a simplest form is often called a *canonical representation*). The simplest form of a naturals pair is a pair which has either its first or its second member equal to Z. E.g., the simplest form of the pair (S(S Z),S(S(S(S Z)))) is (Z,S(S Z)). Reduction to simplest form can again be done by recursion.

```
reduce1 :: NatPair -> NatPair
reduce1 (m1,Z) = (m1,Z)
reduce1 (Z,m2) = (Z,m2)
reduce1 (S m1, S m2) = reduce1 (m1, m2)
```

Exercise 8.10 Define and implement relations leq1 for $\leq$ and gt1 for $>$ for difference classes of naturals.

8.5 Rational Numbers

A further assumption that we would like to make is that any integer can be divided by any non-zero integer to form a fraction, or rational number, and that any rational number m/n can be 'canceled down' to its lowest form by dividing m and n by the same number. E.g., 12/39 cancels down to 4/13.

Again we can view the rational numbers as constructed by means of pairs, in this case pairs (m, n) with $m, n \in \mathbb{Z}$, $n \neq 0$. In this case the pairs are 'ratio pairs'. One and the same rational number can be represented in many different ways: $(1, 2)$, $(2, 4)$, $(13, 26)$ all represent the rational number $1/2$. Or in other words: the rational number m/n is nothing but the class of all (p, q) with $p, q \in \mathbb{Z}$, $q \neq 0$, and the property that $mq = np$. In such a case we say that m/n and p/q represent the same number.

As in the case of the representation of integers by means of difference classes we have an underlying notion of equivalence. Let S be the equivalence relation on $\mathbb{Z} \times (\mathbb{Z} - \{0\})$ given by $(m,n)S(p,q) :\equiv mq = np$. Note that (m, n) and (km, kn) are equivalent modulo S (provided $k \neq 0$), which justifies the simplification of km/kn to m/n by means of canceling down. The set $\mathbb{Q}$ of rational numbers (or fractional numbers) can be defined as:

$$\mathbb{Q} := (\mathbb{Z} \times (\mathbb{Z} - \{0\}))/S.$$

We write a class $[(m/n)]_S$ as $[m/n]$. Addition, multiplication and equality of rational numbers are now defined in terms of addition and multiplication of integers, as follows:

$$[m/n] + [p/q] := [(mq + pn)/nq]$$
$$[m/n] \cdot [p/q] := [mp/nq]$$
$$[m/n] = [p/q] :\equiv mq = np$$

Again, we have to check that these definitions make sense. It is easy to see that the sum, the difference, and the product of two ratio classes are again ratio classes. For instance, if $x \in \mathbb{Q}$ and $y \in \mathbb{Q}$, then there are integers $m, n, p, q \in \mathbb{Z}$ with $n \neq 0$, $q \neq 0$, and $x = [m/n]$ and $y = [p/q]$. Then $x + y = [m/n] + [p/q] = [(mq + pn)/nq]$, and from $mq + pn \in \mathbb{Z}$, $nq \in \mathbb{Z}$, and $nq \neq 0$ it follows that $x + y \in \mathbb{Q}$.

Proposition 8.11 The law of associativity for $+$ holds for the rationals.

Proof. If $x, y, z \in \mathbb{Q}$ then there are $m, n, p, q, r, s \in \mathbb{Z}$ with $x = [m/n]$, $y = [p/q]$, $z = [r/s]$, $n \neq 0$, $q \neq 0$ and $s \neq 0$. Now:

$$x + (y + z)$$
$$= \qquad [\text{definitions of } x, y, z]$$
$$[\tfrac{m}{n}] + ([\tfrac{p}{q}] + [\tfrac{r}{s}])$$
$$= \qquad [\text{definition of } + \text{ for } \mathbb{Q}]$$
$$[\tfrac{m}{n}] + [\tfrac{ps+rq}{qs}]$$
$$= \qquad [\text{definition of } + \text{ for } \mathbb{Q}]$$
$$[\tfrac{mqs+(ps+rq)n}{nqs}]$$
$$= \qquad [\text{distribution law for } \mathbb{Z}]$$
$$[\tfrac{mqs+psn+rqn}{nqs}]$$
$$= \qquad [\text{assoc of } \cdot \text{ for } \mathbb{Z}, \text{ dist law for } \mathbb{Z}]$$
$$[\tfrac{(mq+pn)s+rqn}{nqs}]$$
$$= \qquad [\text{definition of } + \text{ for } \mathbb{Q}]$$
$$[\tfrac{mq+pn}{nq}] + [\tfrac{r}{s}]$$
$$= \qquad [\text{definition of } + \text{ for } \mathbb{Q}]$$
$$([\tfrac{m}{n}] + [\tfrac{p}{q}]) + [\tfrac{r}{s}]$$
$$= \qquad [\text{definitions of } x, y, z]$$
$$(x + y) + z.$$

∎

In a similar way, it is not difficult (but admittedly a bit tedious) to check that the law of commutativity for + holds for the rationals, that the laws of associativity and commutativity for · hold for the rationals, and that the law of distribution holds for the rationals.

A thing to note about Proposition 8.11 is that the statement about rational numbers that it makes also tells us something about our ways of doing things in an implementation. It tells us that any program that uses the procedure `add` which implements addition for the rationals need not distinguish between

`(add (add (m,n) (p,q)) (r,s))` and `(add (m,n) (add (p,q) (r,s)))`.

Wherever one expression fits the other will fit.

8.5. RATIONAL NUMBERS

Again forgetting about irrelevant differences in representation, we see that $\mathbb{Z} \subseteq \mathbb{Q}$, for every integer n has a representation in the rationals, namely $n/1$.

If $x, y \in \mathbb{Q}$, then $x + y \in \mathbb{Q}$, $x - y \in \mathbb{Q}$, $xy \in \mathbb{Q}$. If, in addition $y \neq 0$, we have $x/y \in \mathbb{Q}$. We say that the domain of rationals is closed under addition, subtraction, and multiplication and 'almost closed' under division. If $y \neq 0$, then $1/y$ exists, which shows that every rational number y except 0 has an *inverse*: a rational number that yields 1 when multiplied with y.

A domain of numbers that is closed under the four operations of addition, subtraction, multiplication and division (with the extra condition on division that the divisor should be $\neq 0$) is called a *field*. The domain of rational numbers is a field.

It is easy to see that each rational number m/n can be written in decimal form, by performing the process of long division. There are two possible outcomes:

1. the division process stops: m/n can be written as a finite decimal expansion.

2. the division process repeats: m/n can be written as a infinite decimal expansion with a tail part consisting of a digit or group of digits which repeats infinitely often.

Examples of the first outcome are $1/4 = 0.25$, $-1/4 = -0.25$, $22/5 = 4.4$. Examples of the second outcome are $1/3 = 0.3333...$, $1/6 = 0.166666...$, $29/7 = 4.142857142857142857...$, $23/14 = 1.6428571428571428571....$ A handy notation for this is $1/3 = 0.\overline{3}$, $1/6 = 0.\overline{6}$, $29/7 = 4.\overline{142857}$, $23/14 = 1.6\overline{428571}$. The part under the line is the part that repeats infinitely often.

If the division process repeats for m/n then at each stage of the process there must be a non-zero remainder. This means that at each stage, the remainder must be in the range $1, \ldots, n-1$. But then after at most n steps, some remainder k is bound to reoccur, and after the reappearance of k the process runs through exactly the same stages as after the first appearance of k. The same remainders turn up, in the same order. We are in a loop.

Conversely, it is not very difficult to see that every repeating decimal corresponds to a rational number. First an example. Suppose $M = 0.133133133\ldots = 0.\overline{133}$. Then: $1000 * M = 133.\overline{133}$, and $(1000 * M) - M = 133.\overline{133} - 0.\overline{133} = 133$, so $M = \frac{133}{999}$.

Now for the general case.

Theorem 8.12 Every repeating decimal corresponds to a rational number.

Proof. A repeating decimal can always be written in our over-line notation in the form

$$M \pm 0.a_1 a_2 \cdots a_m \overline{b_1 b_2 \cdots b_n},$$

where M is an integer and the a_i and b_j are decimal digits. For example, $-37/14 = -2.6\overline{428571} = -2 - 0.6\overline{428571}$. If we can prove that the part

$$p = 0.a_1 a_2 \cdots a_m \overline{b_1 b_2 \cdots b_n}$$

is rational, then we are done, for if p is rational and M is an integer then there are integers n, k with $p = n/k$, so $M + p = M + n/k = \frac{kM+n}{k}$, i.e., $M + p$ is rational, and similarly for $M - p$. Setting

$$A = 0.a_1 a_2 \cdots a_m, \quad B = 0.b_1 b_2 \cdots b_n, \quad C = 0.\overline{b_1 b_2 \cdots b_n},$$

we get

$$p = 0.a_1 a_2 \cdots a_m \overline{b_1 b_2 \cdots b_n} = A + 10^{-m} C.$$

If we can write this in terms of rational operations on A and B we are done.

$$C = 0.\overline{b_1 b_2 \cdots b_n} = 0.b_1 b_2 \cdots b_n \overline{b_1 b_2 \cdots b_n}.$$

Thus,

$$\begin{aligned} C - 10^{-n} C &= 0.b_1 b_2 \cdots b_n \overline{b_1 b_2 \cdots b_n} - 0.0_1 \cdots 0_n \overline{b_1 b_2 \cdots b_n} = \\ &= 0.b_1 b_2 \cdots b_n = B. \end{aligned}$$

Therefore, $C = \frac{B}{1-10^{-n}}$, and we get:

$$p = A + \frac{10^{-m} B}{1 - 10^{-n}},$$

so we have proved that p is rational. ∎

There is a nice geometrical interpretation of the process of constructing rational numbers. Assume that we have been given the line of integers:

8.5. RATIONAL NUMBERS

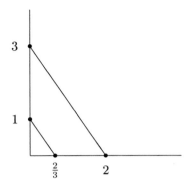

Figure 8.2: Constructing the fraction 2/3.

Place these on a (horizontal) x-axis and a (vertical) y-axis:

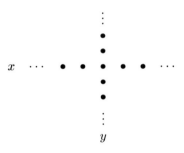

Now construct further points on the x-axis by first drawing a straight line l through a point m on the x-axis and a point n on the y-axis, and next drawing a line l' parallel to l through the point $(0,1)$. The intersection of l' and the x-axis is the rational point m/n. (Use congruence reasoning for triangles to establish that the ratio between m and n equals the ratio between the intersection point m/n and 1.) Figure 8.2 gives the construction of the fraction 2/3 on the x-axis.

Figures 8.3, 8.4, 8.5 and 8.6 give the geometrical interpretations of addition, negation, multiplication and reciprocal. Subtraction is addition of a negated number, division is multiplication with a reciprocal, so these give all the rational operations. Note that these constructions can all be performed by a process of (i) connecting previously constructed points by a straight line, and (ii) drawing lines parallel to previously constructed lines. These are the so-called *linear constructions*: the constructions that can be performed by using a ruler but no compass. In particular, use of a compass to construct line segments of equal lengths is forbidden.

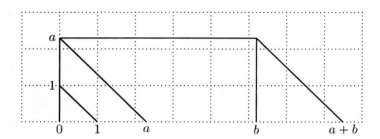

Figure 8.3: Geometrical Interpretation of Addition.

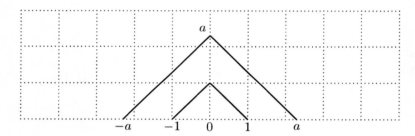

Figure 8.4: Geometrical Interpretation of Negation.

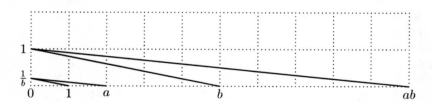

Figure 8.5: Geometrical Interpretation of Multiplication.

8.6. IMPLEMENTING RATIONAL ARITHMETIC

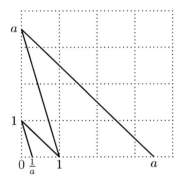

Figure 8.6: Geometrical Interpretation of Reciprocal.

8.6 Implementing Rational Arithmetic

Haskell has a standard implementation of the above, a type `Rational`, predefined as follows:

```
data Integral a => Ratio a = a :% a deriving (Eq)
type Rational              = Ratio Integer
```

To reduce a fraction to its simplest form, we first make sure that the denominator is non-negative. If (x,y) represents a fraction with numerator x and denominator y, then

   (x * signum y) (abs y)

is an equivalent representation with positive denominator. Here `abs` gives the absolute value, and `signum` the sign (1 for positive integers, 0 for 0, −1 for negative integers). The reduction to canonical form is performed by:

```
(%)                        :: Integral a => a -> a -> Ratio a
x % y                      = reduce (x * signum y) (abs y)

reduce                     :: Integral a => a -> a -> Ratio a
reduce x y | y == 0        = error "Ratio.%: zero denominator"
           | otherwise     = (x 'quot' d) :% (y 'quot' d)
                             where d = gcd x y
```

Functions for extracting the numerator and the denominator are provided:

```
numerator, denominator     :: Integral a => Ratio a -> a
numerator (x :% y)         = x
denominator (x :% y)       = y
```

Note that the numerator of (x % y) need not be equal to x and the denominator need not be equal to y:

```
Prelude> numerator (2 % 4)
1
Prelude> denominator (2 % 10)
5
```

A total order on the rationals is implemented by:

```
instance Integral a => Ord (Ratio a) where
    compare (x:%y) (x':%y') = compare (x*y') (x'*y)
```

The standard numerical operations from the class Num are implemented by:

```
instance Integral a => Num (Ratio a) where
    (x:%y) + (x':%y') = reduce (x*y' + x'*y) (y*y')
    (x:%y) * (x':%y') = reduce (x*x') (y*y')
    negate (x :% y)   = negate x :% y
    abs (x :% y)      = abs x :% y
    signum (x :% y)   = signum x :% 1
```

8.6. IMPLEMENTING RATIONAL ARITHMETIC

The rationals are also closed under the division operation $\lambda xy.\frac{x}{y}$ and the reciprocal operation $\lambda x.\frac{1}{x}$. These are implemented as follows:

```
instance Integral a => Fractional (Ratio a) where
    (x:%y) / (x':%y')    = (x*y') % (y*x')
    recip (x:%y)         = if x < 0 then (-y) :% (-x) else y :% x
```

If you want to try out decimal expansions of fractions on a computer, here is a Haskell program that generates the decimal expansion of a fraction.

```
decExpand :: Rational -> [Integer]
decExpand x | x < 0     = error "negative argument"
            | r == 0    = [q]
            | otherwise = q : decExpand ((r*10) % d)
    where
    (q,r) = quotRem n d
    n     = numerator x
    d     = denominator x
```

If the decimal expansion repeats, you will have to interrupt the process by typing *control c*:

WWN> decExpand (1 % 7)
[0, 1, 4, 2, 8, 5, 7, 1, 4, 2, 8, 5, 7, 1, 4, 2, 8, 5, 7, 1, 4, 2, 8,
5, 7, 1, 4, 2, 8, 5, 7, 1, 4, 2, 8, 5, 7, 1, 4, 2, 8, 5, 7, 1, 4, 2,
8, 5, 7, 1, 4, 2, 8, 5, 7, 1, 4, 2, 8, 5, 7, 1, 4, 2, 8, 5, 7, 1, 4,
2, 8, 5, 7, 1, 4, 2, 8, 5, 7, 1, 4, 2, 8, 5, 7, 1,
4, 2, 8, 5,{Interrupted!}

This problem can be remedied by checking every new quotient remainder pair against a list of quotient-remainder pairs, to spot a repetition. The main function decForm produces the integer part, and relegates the task of calculating the lists of non-repeating and repeating decimals to an auxiliary function decF.

```
decForm :: Rational -> (Integer,[Int],[Int])
decForm x | x < 0     = error "negative argument"
          | otherwise = (q,ys,zs)
  where
  (q,r)   = quotRem n d
  n       = numerator x
  d       = denominator x
  (ys,zs) = decF (r*10) d []
```

The function decF has a parameter for the list of quotient remainder pairs that have to be checked for repetition. The code for dForm uses elemIndex from the module List.hs to find the index of the first repeating digit, and splitAt to split a list at an index.

```
decF :: Integer -> Integer -> [(Int,Integer)] -> ([Int],[Int])
decF n d xs | r == 0         = (reverse (q: (map fst xs)),[])
            | elem (q,r) xs  = (ys,zs)
            | otherwise      = decF (r*10) d ((q,r):xs)
  where
  (q',r)  = quotRem n d
  q       = toInt q'
  xs'     = reverse xs
  Just k  = elemIndex (q,r) xs'
  (ys,zs) = splitAt k (map fst xs')
```

Here are a few examples:

```
WWN> decForm (133 % 999)
(0,[],[1,3,3])
WWN> decForm (1 % 7)
(0,[],[1,4,2,8,5,7])
WWN> decForm (2 % 7)
(0,[],[2,8,5,7,1,4])
WWN> decForm (3 % 7)
(0,[],[4,2,8,5,7,1])
WWN> decForm (4 % 7)
(0,[],[5,7,1,4,2,8])
WWN> decForm (5 % 7)
(0,[],[7,1,4,2,8,5])
WWN> decForm (6 % 7)
(0,[],[8,5,7,1,4,2])
```

There is no upper limit to the length of the period. Here is an example with a period length of 99:

```
WWN> decForm (468 % 199)
(2,[],[3,5,1,7,5,8,7,9,3,9,6,9,8,4,9,2,4,6,2,3,1,1,5,5,7,7,8,8,9,4,4,
7,2,3,6,1,8,0,9,0,4,5,2,2,6,1,3,0,6,5,3,2,6,6,3,3,1,6,5,8,2,9,1,4,5,
7,2,8,6,4,3,2,1,6,0,8,0,4,0,2,0,1,0,0,5,0,2,5,1,2,5,6,2,8,1,4,0,7,0])
```

Exercise 8.13 Write a Haskell program to find the longest period that occurs in decimal expansions of fractions with numerator and denominator taken from the set $\{1, \ldots, 999\}$.

8.7 Irrational Numbers

The Ancients discovered to their dismay that with just ruler and compass it is possible to construct line segments whose lengths do not form rational fractions. In other words, it is possible to construct a length q such that for no $m, n \in \mathbb{N}$ it holds that $q = \frac{m}{n}$. Here is the famous theorem from Antiquity stating this disturbing fact, with its proof.

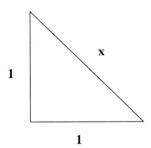

Theorem 8.14 There is no rational number x with $x^2 = 2$.

Proof. Assume there is a number $x \in \mathbb{Q}$ with $x^2 = 2$. Then there are $m, n \in \mathbb{Z}$, $n \neq 0$ with $(m/n)^2 = 2$. We can further assume that m/n is canceled down to its lowest form, i.e., there are no $k, p, q \in \mathbb{Z}$ with $k \neq 1$, $m = kp$ and $n = kq$.

We have: $2 = (m/n)^2 = m^2/n^2$, and multiplying both sides by n^2 we find $2n^2 = m^2$. In other words, m^2 is even, and since squares of odd numbers are always odd, m must be even, i.e., there is a p with $m = 2p$. Substitution in $2n^2 = m^2$ gives $2n^2 = (2p)^2 = 4p^2$, and we find that $n^2 = 2p^2$, which leads to the conclusion that n is also even. But this means that there is a q with $n = 2q$, and we have a contradiction with the assumption that m/n was in lowest form. It follows that there is no number $x \in \mathbb{Q}$ with $x^2 = 2$. The square root of 2 is not rational. ∎

Of course, we all use $\sqrt{2}$ for the square root of 2. The theorem tells us that $\sqrt{2} \notin \mathbb{Q}$. The collection of numbers that $\sqrt{2}$ does belong to is called the collection of real numbers, $\mathbb{R}$. It is possible to give a formal construction of the reals from the rationals, but we will not do so here. Instead, we just mention that $\mathbb{Q} \subseteq \mathbb{R}$, and we informally introduce the reals as the set of all signed (finite or infinite) decimal expansions. The domain of real numbers is closed under the four operations of addition, subtraction, multiplication and division, i.e., just like the rationals, the reals form a *field*.

Exercise 8.15 Use the method from the proof of Theorem 8.14 to show that $\sqrt{3}$ is irrational.

Exercise 8.16 Show that if p is prime, then $\sqrt{p}$ is irrational.

Exercise 8.17 Show that if n is a natural number with $\sqrt{n}$ not a natural number, then $\sqrt{n}$ is irrational.

Using the fundamental theorem of arithmetic, we can give the following alternative proof of Theorem 8.14:

Proof. If $\sqrt{2} = (p/q)$, then $2q^2 = p^2$. In the representation of p^2 as a product of prime factors, every prime factor has an even number of occurrences (for the square of p equals the product of the squares of p's prime factors). In the representation of $2q^2$ as a product of prime factors, the prime factor 2 has an *odd* number of occurrences. Contradiction with Theorem 8.6. ∎

In Section 8.8 we will discuss an algorithm, the mechanic's rule, for approaching the square root of any positive fraction p with arbitrary precision.

The irrational numbers are defined here informally as the infinite non-periodic decimal expansions. This is not a very neat definition, for its relies on the use of decimals, and there is nothing special about using the number *ten* as a basis for doing calculations. Other bases would serve just as well. In any case, looking at $\mathbb{Q}$ and $\mathbb{R}$ in terms of expansions provides a neat perspective on the relation between rational and irrational numbers. What, e.g., are the odds against constructing a rational by means of a process of tossing coins to get an infinite binary expansion?

8.7. IRRATIONAL NUMBERS

Exercise 8.18 Does it follow from the fact that $x + y$ is rational that x is rational or y is rational? If so, give a proof, if not, give a refutation.

We have seen that writing out the decimal expansion of a real number like $\sqrt{2}$ does not give a finite representation. In fact, since there are uncountably many reals (see Chapter 11), no finite representation scheme for arbitrary reals is possible.

In implementations it is customary to use floating point representation (or: scientific representation) of approximations of reals. E.g., the decimal fraction 1424213.56 is an approximation of $\sqrt{2} \times 10^6$, and gets represented as $1.42421356E + 5$. The decimal fraction 0.00000142421356 is an approximation of $\frac{\sqrt{2}}{10^6} = \sqrt{2} \times 10^{-6}$, and gets represented as $1.42421356E - 6$. The general form is $x.xxxEm$, where $x.xxx$ is a decimal fraction called the *mantissa* and m is an integer called the *exponent*. Here are the Hugs versions:

```
Prelude> sqrt 2 * 10^6
1.41421e+06
Prelude> sqrt 2 / 10^6
1.41421e-06
```

Haskell has a predefined type `Float` for single precision floating point numbers, and a type `Double` for double precision floating point numbers. Together these two types form the class `Floating`:

```
Prelude> :t sqrt 2
sqrt 2 :: Floating a => a
```

Floating point numbers are stored as pairs (m, n), where m is the matrix and n the exponent of the base used for the encoding. If x is a floating point number, the base of its representation is given by `floatRadix` and its matrix and exponent, as a value of type `(Integer,Int)`, by `decodeFloat`.

Thus, if `floatRadix` x equals b and `decodeFloat` x equals (m, n), then x is the number $m \cdot b^n$. `floatDigits` gives the number of digits of m in base b representation. In particular, if (m, n) is the value of `decodeFloat` x, and d the value of `floatDigits`, then either m and n are both zero, or $b^{d-1} \leqslant m < b^d$. Here is an example:

```
Prelude> floatRadix (sqrt 2)
2
Prelude> decodeFloat (sqrt 2)
(11863283,-23)
Prelude> 11863283 * 2^^(-23)
1.41421
Prelude> floatDigits (sqrt 2)
24
Prelude> 2^23 <= 11863283 && 11863283 < 2^24
True
```

The inverse to the function decodeFloat is encodeFloat:

```
Prelude> sqrt 2
1.41421
Prelude> encodeFloat 11863283 (-23)
1.41421
```

Scaling a floating point number is done by 'moving the point':

```
Prelude> scaleFloat 4 (sqrt 2)
22.6274
Prelude> 2^4 * sqrt 2
22.6274
```

A floating point number can always be scaled in such a way that its matrix is in the interval $(-1, 1)$. The matrix in this representation is given by the function significand, the exponent by exponent:

```
Prelude> significand (sqrt 2)
0.707107
Prelude> exponent (sqrt 2)
1
Prelude> 0.707107 * 2^1
1.41421
Prelude> scaleFloat 1 0.707107
1.41421
```

The definitions of exponent, significand and scaleFloat (from the Prelude):

```
exponent x       = if m==0 then 0 else n + floatDigits x
                   where (m,n) = decodeFloat x
significand x    = encodeFloat m (- floatDigits x)
                   where (m,_) = decodeFloat x
scaleFloat k x   = encodeFloat m (n+k)
                   where (m,n) = decodeFloat x
```

8.8 The Mechanic's Rule

Sequences of fractions can be used to find approximations to real numbers that themselves are *not* fractions (see Section 8.7). A well known algorithm for generating such sequences is the so-called *mechanic's rule* (also known

8.8. THE MECHANIC'S RULE

as *Newton's method*, a bit misleadingly, for the algorithm was already in use centuries before Newton):

$$p > 0, a_0 > 0, a_{n+1} = \frac{1}{2}(a_n + \frac{p}{a_n}).$$

In Exercise 8.19 you are asked to prove that this can be used to approximate the square root of any positive fraction p to any degree of accuracy. The Haskell implementation uses some fresh ingredients. The function recip takes the reciprocal of a fraction, the operation iterate iterates a function by applying it again to the result of the previous application, and takeWhile takes a property and a list and constructs the largest prefix of the list consisting of objects satisfying the property. In the present case, the property is (\ m -> m^2 <= p), having a square $\leqslant p$, and the list is the list of positive naturals. You should look up the implementations in Prelude.hs and make sure you understand.

```
mechanicsRule :: Rational -> Rational -> Rational
mechanicsRule p x = (1 % 2) * (x + (p * (recip x)))

mechanics :: Rational -> Rational -> [Rational]
mechanics p x = iterate (mechanicsRule p) x
```

```
sqrtM :: Rational -> [Rational]
sqrtM p | p < 0     = error "negative argument"
        | otherwise = mechanics p s
   where
   s = if xs == [] then 1 else last xs
   xs = takeWhile (\ m -> m^2 <= p) [1..]
```

As a demonstration, here are the first five steps in the approximation to $\sqrt{2}$, the first seven steps in the approximation to $\sqrt{4}$, and the first, second, third, fourth and fifth step in the approximation to $\sqrt{50}$, respectively. This already gives greater accuracy than is needed in most applications; the algorithm converges very fast.

```
WWN> take 5 (sqrtM 2)
[1 % 1,3 % 2,17 % 12,577 % 408,665857 % 470832]
```

```
WWN> take 7 (sqrtM 4)
[2 % 1,2 % 1,2 % 1,2 % 1,2 % 1,2 % 1,2 % 1]
WWN> sqrtM 50 !! 0
7 % 1
WWN> sqrtM 50 !! 1
99 % 14
WWN> sqrtM 50 !! 2
19601 % 2772
WWN> sqrtM 50 !! 3
768398401 % 108667944
WWN> sqrtM 50 !! 4
1180872205318713601 % 167000548819115088
```

Exercise 8.19 1. Prove that for every n,

$$\frac{a_{n+1} - \sqrt{p}}{a_{n+1} + \sqrt{p}} = \frac{(a_n - \sqrt{p})^2}{(a_n + \sqrt{p})^2}.$$

2. From the first item it follows (by induction on n) that

$$\frac{a_{n+1} - \sqrt{p}}{a_{n+1} + \sqrt{p}} = \left(\frac{a_0 - \sqrt{p}}{a_0 + \sqrt{p}}\right)^{2^n}.$$

Derive from this that sqrtM p converges to $\sqrt{p}$ for any positive rational number p.

3. Show that the approximation is from above, i.e., show that $n \geq 1$ implies that $a_n \geq \sqrt{p}$.

Exercise 8.20 1. Find a rule to estimate the number of correct decimal places in approximation a_n of $\sqrt{p}$. Use the result of Exercise 8.19. (Hint: try to find an inequality of the form $a_n - \sqrt{p} \leq t$, with t an expression that employs a_0 and a_1.)

2. Use the previous item to give an estimate of the number of correct decimal places in the successive approximations to $\sqrt{2}$.

8.9 Reasoning about Reals

Suppose one wants to describe the behaviour of moving objects by plotting their position or speed as a function of time. Moving objects do not suddenly disappear and reappear somewhere else (outside the Bermuda triangle, at least), so the path of a moving object does not have holes or

8.9. REASONING ABOUT REALS

gaps in it. This is where the notions of continuity and limit arise naturally. Analysis, with its emphasis on continuity and limit, is a rich source of examples where skill in quantifier reasoning is called for.

The following Lemma illustrates that many common functions are continuous. Its proof is given as an example: it is a nice illustration of the use of the logical rules. In fact, logic is all there is to this proof: properties of real numbers are not needed at all.

Lemma. *The composition of two continuous functions is continuous.*

I.e., if f and g are continuous functions from reals to reals, then the function h defined by $h(x) = g(f(x))$ (cf. Definition 6.29 p. 217) is continuous as well.

Proof. Given: f and g are continuous, i.e., ("ε-δ-definition" p. 64; for clarity we use different variables for the arguments):

$$\forall x \, \forall \varepsilon > 0 \, \exists \delta > 0 \, \forall y \, (|x - y| < \delta \implies |f(x) - f(y)| < \varepsilon), \qquad (8.1)$$

$$\forall a \, \forall \varepsilon > 0 \, \exists \delta > 0 \, \forall b \, (|a - b| < \delta \implies |g(a) - g(b)| < \varepsilon). \qquad (8.2)$$

To be proved: $\forall x \, \forall \varepsilon > 0 \, \exists \delta > 0 \, \forall y \, (|x-y| < \delta \implies |g(f(x)) - g(f(y))| < \varepsilon)$.
Proof (detailed version): Note that what we are asked to prove begins with two universal quantifiers $\forall x$ and $\forall \varepsilon > 0$. Therefore, the proof *has* to start (recall the *obligatory* opening that goes with $\forall$-introduction!) with choosing two arbitrary values $x \in \mathbb{R}$ and $\varepsilon > 0$. We now have to show that

$$\exists \delta > 0 \, \forall y \, (|x - y| < \delta \implies |g(f(x)) - g(f(y))| < \varepsilon). \qquad (8.3)$$

The next quantifier asks us to supply an example-δ that is positive. There appears to be no immediate way to reduce the proof problem further, so we start looking at what is given in order to obtain such an example.

It turns out that we have to use the *second* given (8.2) *first*. It can be used ($\forall$-elimination) if we specify values for a and ε. Later on it will become clear that $a = f(x)$ is a useful choice.

Thus, (8.2) delivers ($\exists$-elimination) some $\delta_1 >$ such that

$$\forall b \, (|f(x) - b| < \delta_1 \implies |g(f(x)) - g(b)| < \varepsilon). \qquad (8.4)$$

Applying the given (8.1) to our x and $\varepsilon = \delta_1$ ($\forall$-elimination) we obtain ($\exists$-elimination) $\delta > 0$ such that

$$\forall y \, (|x - y| < \delta \implies |f(x) - f(y)| < \delta_1). \qquad (8.5)$$

This is the example-δ we are looking for, i.e.:

Claim: $\forall y\, (|x - y| < \delta \implies |g(f(x)) - g(f(y))| < \varepsilon)$.
(From this follows what we have to show using $\exists$-introduction.)
Proof: Suppose that (to prepare for $\forall$-introduction and Deduction Rule — cf. Exercise 3.18 p. 91) y is such that $|x - y| < \delta$.
From (8.5) (using *this* y — $\forall$-elimination and Modus Ponens) you find:

$$|f(x) - f(y)| < \delta_1.$$

Finally, from (8.4) (with $b = f(y)$ — $\forall$-elimination) you find:

$$|g(f(x)) - g(f(y))| < \varepsilon.$$

The claim follows. ∎

Making up the score: the proof applies one rule for each of the fifteen (!) occurrences of logical symbols in what is given and what is to be proved. Make sure you understand every detail.

Of course, the amount of detail is excessive and there is a more common concise version as well. The following version is the kind of argument you will find in analysis textbooks. As is usual, it leaves to the reader to fill in which rules have been applied, and where.

Proof. Assume that $x \in \mathbb{R}$ and $\varepsilon > 0$.
From (8.2), obtain $\delta_1 > 0$ such that

$$\forall b\, (|f(x) - b| < \delta_1 \implies |g(f(x)) - g(b)| < \varepsilon). \tag{8.6}$$

Applying (8.1) to x and δ_1, obtain $\delta > 0$ such that

$$\forall y\, (|x - y| < \delta \implies |f(x) - f(y)| < \delta_1). \tag{8.7}$$

Then if $|x - y| < \delta$, by (8.7) we get that $|f(x) - f(y)| < \delta_1$, and from (8.6) it follows that $|g(f(x)) - g(f(y))| < \varepsilon$. ∎

This version of the proof will be considered very complete by every mathematician. Nevertheless, the compression attained is approximately 4 : 1.

The remaining examples of this section are about sequences of reals.

Limits. Assume that $a_0, a_1, a_2, \ldots$ is a sequence of reals and that $a \in \mathbb{R}$. The expression $\lim_{i \to \infty} a_i = a$ ("the sequence *converges* to a", "a is *limit* of the sequence") by definition means that

$$\forall \varepsilon > 0\, \exists n\, \forall i \geqslant n\, (|a - a_i| < \varepsilon).$$

8.9. REASONING ABOUT REALS

Generating concrete examples of converging sequences in Haskell is easy, as was demonstrated in Section 8.8. We have seen that the sequences produced by the mechanic's rule converge. We will prove some results about convergence.

Theorem. *Every sequence of reals has at most one limit.*

Proof. The situation should be analyzed as follows.
Given: $\lim_{i \to \infty} a_i = a$, $\lim_{i \to \infty} a_i = b$.
To be proved: $a = b$.
Proof: Proof by Contradiction is not a bad idea here, since the new given it provides, nl., that $a \neq b$, is equivalent with the positive $|a - b| > 0$. Thus, assume this. Proof by Contradiction now asks to look for something false.

In order to use the old given, you need to choose ($\forall$-elimination!) values for ε. As you'll see later, it is useful to choose $\varepsilon = \frac{1}{2}|a - b|$. Note that $\varepsilon > 0$.

From the given $\lim_{i \to \infty} a_i = a$ we obtain now, that $\exists n \forall i \geqslant n (|a - a_i| < \varepsilon)$. Thus ($\exists$-elimination) some n_1 exists such that $\forall i \geqslant n_1 (|a - a_i| < \varepsilon)$.

From the given $\lim_{i \to \infty} a_i = b$ we obtain, similarly, some n_2 such that $\forall i \geqslant n_2 (|b - a_i| < \varepsilon)$.

Define $n = \max(n_1, n_2)$. Since $n \geqslant n_1, n_2$, by $\forall$-elimination we now get from these facts that both $|a - a_n| < \varepsilon$ and $|b - a_n| < \varepsilon$.

Lastly, using the *triangle-inequality*

$$|x + y| \leqslant |x| + |y|,$$

it follows that $|a - b| = |a - a_n + a_n - b| \leqslant |a - a_n| + |b - a_n| < 2\varepsilon = |a - b|$ — and this is the falsity looked for. ∎

Exercise 8.21 Write a concise version of the above proof.

Exercise 8.22 Assume that $\lim_{i \to \infty} a_i = a$.

1. Show that $\lim_{i \to \infty} a_{2i} = a$.

2. Assume that $f : \mathbb{N} \to \mathbb{N}$ is a function such that $\forall n \exists m \forall i \geqslant m \, f(i) \geqslant n$. Show that $\lim_{i \to \infty} a_{f(i)} = a$.

Exercise 8.23 Assume that the sequences of reals $\{a_n\}_{n=0}^{\infty}$ and $\{b_n\}_{n=0}^{\infty}$ have limits a resp. b, and that $a < b$. Show that a number n exists such that $\forall m \geqslant n \, (a_m < b_m)$.

Exercise 8.24 Assume that $\lim_{i \to \infty} a_i = a$ and that $\lim_{i \to \infty} b_i = b$. Show that $\lim_{i \to \infty} (a_i + b_i) = a + b$.

Exercise 8.25 Show that a function $f : \mathbb{R} \to \mathbb{R}$ is continuous iff $\lim_{i \to \infty} f(a_i) = f(a)$ whenever $\lim_{i \to \infty} a_i = a$.

Cauchy. A sequence of reals $\{a_n\}_{n=0}^{\infty}$ is called *Cauchy* if

$$\forall \varepsilon > 0 \; \exists n \; \forall i, j \geqslant n \; (|a_i - a_j| < \varepsilon).$$

Exercise 8.26 Assume that the sequence $\{a_n\}_{n=0}^{\infty}$ is Cauchy.

1. Show that the sequence is *bounded*. I.e., that numbers b and c exist such that $\forall i \; (b < a_i < c)$.

2. Assume that $a \in \mathbb{R}$ is such that $\forall \varepsilon > 0 \forall n \exists i \geqslant n \; (|a - a_i| < \varepsilon)$. (The existence of such an a follows from the sequence being bounded, but you are not asked to prove this.) Show that $\lim_{i \to \infty} a_i = a$.

It follows immediately from Exercise 8.19 that the sequences of rationals produced by the mechanic's rule are Cauchy. Thanks to that we can implement a program for calculating square roots as follows (for still greater precision, change the value of ε in apprx):

```
approximate :: Rational -> [Rational] -> Rational
approximate eps (x:y:zs)
  | abs (y-x) < eps = y
  | otherwise       = approximate eps (y:zs)

apprx :: [Rational] -> Rational
apprx = approximate (1/10^6)

mySqrt :: Rational -> Rational
mySqrt p = apprx (sqrtM p)
```

Exercise 8.27 Just as we defined the integers from the naturals and the rationals from the integers by means of quotient sets generated from suitable equivalence classes, we construct the reals from the rationals, by defining the set of real numbers $\mathbb{R}$ as the set of all Cauchy sequences in $\mathbb{Q}$ modulo an appropriate equivalence relation. Define that equivalence relation and show that it is indeed an equivalence.

8.10 Complex Numbers

In the domain of rational numbers we cannot solve the equation $x^2 - 2 = 0$, but in the domain of real numbers we can: its roots are $x = \sqrt{2}$ and $x = -\sqrt{2}$. What about solving $x^2 + 1 = 0$? There are no real number solutions, for the square root of -1 does not exist in the realm of real numbers. The field of real numbers is not closed under the operation of taking square roots. To remedy this, we follow the by now familiar recipe of extending the number domain, by introducing numbers of a new kind. We extend the domain of reals to a domain of complex numbers, $\mathbb{C}$, by introducing an entity i called 'the imaginary unit', and postulating $i^2 = -1$.

We do not want to lose closure under addition, subtraction, multiplication and division, so we should be able to make sense of $\sqrt{2} + i$, $2i$, $-i$, and so on. In general, we want to be able to make sense of $x + iy$, where x and y are arbitrary real numbers, and we need rules for adding and multiplying complex numbers, in such a way that the laws of commutativity and associativity of $+$ and $\cdot$, and the law of distribution of $\cdot$ over $+$ continue to hold. x is called the real part of $x + iy$, iy its imaginary part. Adding complex numbers boils down to adding real and imaginary parts separately:

$$(x + iy) + (u + iw) = (x + u) + i(y + w).$$

This also gives a recipe for subtraction:

$$(x + iy) - (u + iw) = (x + iy) + (-u + -iw) = (x - u) + i(y - w).$$

For multiplication, we use the fact that $i^2 = -1$:

$$(x + iy)(u + iw) = xu + iyu + ixw + i^2yw = (xu - yw) + i(yu + xw).$$

Division uses the fact that $(x + iy)(x - iy) = x^2 + y^2$. It is given by:

$$\frac{x + iy}{u + iw} = \frac{x + iy}{u + iw} \cdot \frac{u - iw}{u - iw} = \frac{xu + yw}{u^2 + w^2} + i\frac{yu - xw}{u^2 + w^2}.$$

We see that like the rationals and the reals, the complex numbers are closed under the four operations addition, subtraction, multiplication and division, so $\mathbb{C}$ is a field. Moreover, any real number a can be viewed as a complex number of the form $a + 0i$, so we have that $\mathbb{R} \subseteq \mathbb{C}$.

Solving the equation $x^2 + 1 = 0$ in the domain $\mathbb{C}$ gives two roots, viz., $x = i$ and $x = -i$. In general, solving the equation $x^n + a_{n-1}x^{n-1} + \ldots + a_1x + a_0 = 0$, where $a_0, \ldots, a_{n-1}$ may be either real or complex, gives n complex roots, for the fundamental theorem of algebra (which we will not prove here) states that every *polynomial of degree n*,

$$f(x) = x^n + a_{n-1}x^{n-1} + \ldots + a_1 x + a_0,$$

can be factored into a product of exactly n factors,

$$(x - b_1)(x - b_2) \cdots (x - b_n).$$

Exercise 8.28 Check that the commutative and associative laws and the distributive law hold for $\mathbb{C}$.

There is a standard Haskell module *Complex.hs* with an implementation of complex numbers.

```
infix  6  :+

data (RealFloat a) => Complex a = !a :+ !a
                       deriving (Eq,Read,Show)
```

The exclamation marks in the typing `!a :+ !a` indicate that the real and imaginary parts of type `RealFloat` are evaluated in a strict way.

The real part of a complex number $x + iy$ is the real number x, the imaginary part the real number y. Notation for the real part of z: $\text{Re}(z)$. Notation for the imaginary part of z: $\text{Im}(z)$. The Haskell implementations are:

```
realPart, imagPart :: (RealFloat a) => Complex a -> a
realPart (x:+y)     = x
imagPart (x:+y)     = y
```

The complex number $z = x + iy$ can be represented geometrically in either of two ways:

1. Associate with z the point with coordinates (x, y) in the plane $\mathbb{R}^2$. In this way, we view the plane $\mathbb{R}^2$ as the *complex plane*.

2. Associate with z the two-dimensional vector with components x and y. Think of this vector as a *free vector*, i.e., a vector that may be moved around freely as long as its direction remains unchanged.

8.10. COMPLEX NUMBERS

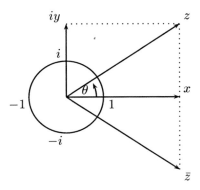

Figure 8.7: Geometrical Representation of Complex Numbers.

The two representations can be combined by attaching the vector z to the origin of the plane. We then get the picture of Figure 8.7. Call the horizontal axis through the origin the *real axis* and the vertical axis through the origin the *imaginary axis*.

The *conjugate* of a complex number $z = x + iy$ is the number $\bar{z} = x - iy$. Its geometrical representation is the reflection of z in the real axis (see again Figure 8.7). Its implementation is given by:

```
conjugate :: (RealFloat a) => Complex a -> Complex a
conjugate (x:+y)    = x :+ (-y)
```

The *magnitude* or *modulus* or *absolute value* of a complex number z is the length r of the z vector. Notation $|z|$. The magnitude of $z = x + iy$ is given by $\sqrt{x^2 + y^2}$. Its Haskell implementation:

```
magnitude :: (RealFloat a) => Complex a -> a
magnitude (x:+y) =
   scaleFloat k (sqrt ((scaleFloat mk x)^2 + (scaleFloat mk y)^2))
   where k  = max (exponent x) (exponent y)
         mk = - k
```

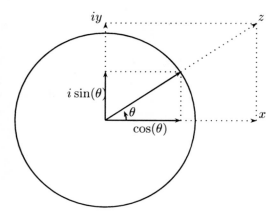

Figure 8.8: $\angle\theta = \cos(\theta) + i\sin(\theta)$.

The *phase* or *argument* of a complex number z is the angle θ of the vector z. Notation: $\arg(z)$. The phase θ of a vector of magnitude 1 is given by the vector $\cos(\theta) + i\sin(\theta)$ (see Figure 8.8), so the phase of $z = x + iy$ is given by $\arctan\frac{y}{x}$ for $x > 0$, and $\arctan\frac{y}{x} + \pi$ for $x < 0$. In case $x = 0$, the phase is 0. This computation is taken care of by the Haskell function atan2. Here is the implementation:

```
phase                :: (RealFloat a) => Complex a -> a
phase (0:+0)         = 0
phase (x:+y)         = atan2 y x
```

The *polar representation* of a complex number $z = x + iy$ is $r\angle\theta$, where $r = |z|$ and $\theta = \arg(z)$. The advantage of polar representation is that complex multiplication looks much more natural: just multiply the magnitudes and add the phases.

$$(R\angle\varphi)(r\angle\theta) = Rr\angle(\varphi + \theta).$$

The polar representation of a complex number is given by:

8.10. COMPLEX NUMBERS

```
polar                  :: (RealFloat a) => Complex a -> (a,a)
polar z                = (magnitude z, phase z)
```

Note that polar representations are not unique, for we have:

$$\ldots = r\angle(\theta - 2\pi) = r\angle\theta = r\angle(\theta + 2\pi) = r\angle(\theta + 4\pi) = \ldots$$

To get from representation $r\angle\theta$ to the representation as a vector sum of real and imaginary parts, use $r\angle\theta = r(\cos(\theta) + i\sin(\theta))$ (see again Figure 8.8). Implementation:

```
mkPolar                :: (RealFloat a) => a -> a -> Complex a
mkPolar r theta        = r * cos theta :+ r * sin theta
```

Converting a phase θ to a vector in the unit circle is done by:

```
cis                    :: (RealFloat a) => a -> Complex a
cis theta              = cos theta :+ sin theta
```

The implementations of the arithmetical operations of addition, subtraction, multiplication, negation, absolute value and signum (plus typecasts from integers and ints) are given by:

```
instance (RealFloat a) => Num (Complex a) where
    (x:+y) + (x':+y')  = (x+x') :+ (y+y')
    (x:+y) - (x':+y')  = (x-x') :+ (y-y')
    (x:+y) * (x':+y')  = (x*x'-y*y') :+ (x*y'+y*x')
    negate (x:+y)      = negate x :+ negate y
    abs z              = magnitude z :+ 0
    signum 0           = 0
    signum z@(x:+y)    = x/r :+ y/r where r = magnitude z
    fromInteger n      = fromInteger n :+ 0
    fromInt n          = fromInt n :+ 0
```

Note that the signum of a complex number z is in fact the vector representation of $\arg(z)$. This suggests that $\arg(z)$ is a generalization of the $+$ or $-$ sign for real numbers. That this is indeed the case can be seen when we perform complex multiplication on real numbers: positive real numbers, represented in the complex plane, have phase 0, or in general, phase $2k\pi$. Negative real number, represented in the complex plane, have phase π, or in general, $(2k+1)\pi$. Multiplying two negative real numbers means multiplying their values and adding their phases, so we get phase 2π, which is the same as phase 0, modulo $2k\pi$.

Complex division is the inverse of multiplication, so it boils down to performing division on the magnitudes, and subtraction on the phases. The implementation in terms of the vector representations looks slightly more involved than this, however:

```
instance (RealFloat a) => Fractional (Complex a) where
    (x:+y) / (x':+y')  = (x*x''+y*y'') / d :+ (y*x''-x*y'') / d
            where x'' = scaleFloat k x'
                  y'' = scaleFloat k y'
                  k   = - max (exponent x') (exponent y')
                  d   = x'*x'' + y'*y''
```

Square root is given by:

```
instance (RealFloat a) => Floating (Complex a) where
    sqrt 0       = 0
    sqrt z@(x:+y) = u :+ (if y < 0 then -v else v)
        where (u,v) = if x < 0 then (v',u') else (u',v')
    v'   = abs y / (u'*2)
    u'   = sqrt ((magnitude z + abs x) / 2)
```

We can use this to generalize the code from page 58 for solve quadratic equations with the 'abc'-formula $x = \frac{-b \pm \sqrt{b^2 - 4ac}}{2a}$:

8.10. COMPLEX NUMBERS

```
solveQ :: (Complex Float, Complex Float, Complex Float)
                              -> (Complex Float, Complex Float)
solveQ = \ (a,b,c) -> if a == 0 then error "not quadratic"
                          else let d = b^2 - 4*a*c in
                             ((- b + sqrt d) / 2*a,
                              (- b - sqrt d) / 2*a)
```

This gives:

```
WWN> solveQ (1,0,-1)
(1.0 :+ 0.0,(-1.0) :+ 0.0)
WWN> solveQ (1,0,1)
(0.0 :+ 1.0,0.0 :+ (-1.0))
WWN> solveQ (1,0,-2)
(1.41421 :+ 0.0,(-1.41421) :+ 0.0)
WWN> solveQ (1,0,2)
(0.0 :+ 1.41421,0.0 :+ (-1.41421))
```

For the definition of further numerical operations on complex numbers we refer to the library file *Complex.hs*.

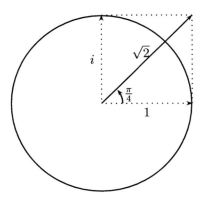

Figure 8.9: The number $1 + i$.

Complex numbers lose their mystery when one gets well acquainted with their geometrical representations. Here are some examples, for getting the feel of them. The number $1 + i$ has magnitude $\sqrt{2}$ and phase $\frac{\pi}{4}$: see Figure 8.9. Squaring this number involves squaring the magnitude and doubling the phase, so $(1 + i)^2$ has magnitude 2 and phase $\frac{\pi}{2}$: see

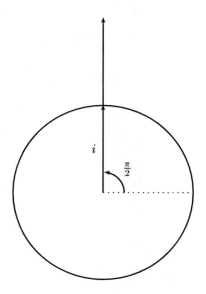

Figure 8.10: The number $(1+i)^2$.

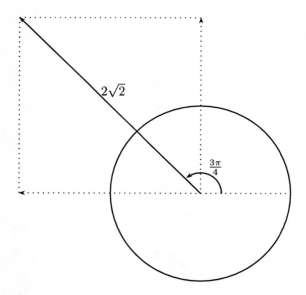

Figure 8.11: The number $(1+i)^3$.

8.10. COMPLEX NUMBERS

Figure 8.10. Raising $1+i$ to the third power involves multiplying the magnitudes 2 and $\sqrt{2}$ and adding the phases $\frac{\pi}{4}$ and $\frac{\pi}{2}$. This gives magnitude $2\sqrt{2}$ and phase $\frac{3\pi}{4}$. See Figure 8.11 for a picture of the number $(1+i)^3$. Raising $1+i$ to the fourth power involves squaring $(1+i)^2$, so the magnitude 2 is squared and the phase $\frac{\pi}{2}$ is doubled, which gives magnitude 4 and phase π. Translating all of this back into vector sum notation, we get $(1+i)^2 = 2i$, $(1+i)^3 = -2+2i$, and $(1+i)^4 = -4$. Sure enough, the Haskell library *Complex.hs* confirms these findings:

```
WWN> (1 :+ 1)^2
0.0 :+ 2.0
WWN> (1 :+ 1)^2
0.0 :+ 2.0
WWN> (1 :+ 1)^3
(-2.0) :+ 2.0
WWN> (1 :+ 1)^4
(-4.0) :+ 0.0
```

Similarly, we see that multiplying i and $-i$ involves multiplying the magnitudes 1 and adding the phases $\frac{\pi}{2}$ and $\frac{3\pi}{2}$, with result the number with magnitude 1 and phase $2\pi = 0$, i.e., the number 1. Here is the Haskell confirmation:

```
WWN> (0 :+ 1) * (0 :+ (-1))
1.0 :+ 0.0
```

Exercise 8.29 You are encouraged to familiarize yourself further with complex numbers by checking the following by means of pictures:

1. $\text{Re}(z) = \frac{1}{2}(z + \bar{z})$.

2. $\text{Im}(z) = \frac{1}{2i}(z - \bar{z})$.

3. $\tan(\arg(z)) = \frac{\text{Im}(z)}{\text{Re}(z)}$.

4. $\frac{R\angle\varphi}{r\angle\theta} = \frac{R}{r}\angle(\varphi - \theta)$.

5. $\arg(\frac{z_1}{z_2}) = \arg(z_1) - \arg(z_2)$.

6. $\arg(\frac{1}{z}) = -\arg(z)$.

Exercise 8.30 1. Use induction on n to prove *De Moivre's formula* for $n \in \mathbb{N}$:
$$(\cos(\varphi) + i\sin(\varphi))^n = \cos(n\varphi) + i\sin(n\varphi).$$
Draw a picture to see what is happening!

2. Prove *De Moivre's formula* for exponents in $\mathbb{Z}$ by using the previous item, plus:

$$(\cos(\varphi) + i\sin(\varphi))^{-m} = \frac{1}{(\cos(\varphi) + i\sin(\varphi))^m}.$$

8.11 Further Reading

A classic overview of the ideas and methods of mathematics, beautifully written, and a book everyone with an interest in mathematics should possess and read is Courant and Robbins [CR78, CrbIS96]. Here is praise for this book by Albert Einstein:

> A lucid representation of the fundamental concepts and methods of the whole field of mathematics. It is an easily understandable introduction for the layman and helps to give the mathematical student a general view of the basic principles and methods.

Another beautiful book of numbers is [CG96]. For number theory and its history see [Ore88].

Chapter 9

Polynomials

Preview

Polynomials or integral rational functions are functions that can be represented by a finite number of additions, subtractions, and multiplications with one independent variable. The closed forms that we found and proved by induction for the sums of evens, sums of odds, sums of squares, sums of cubes, and so on, in Chapter 7, are all polynomial functions. In this chapter we will first study the process of automating the search for polynomial functions that produce given sequences of integers. Next, we establish the connection with the binomial theorem, we implement a datatype for the polynomials themselves, and we use this datatype for the study of combinatorial problems.

```
module POL

where

import Polynomials
```

9.1 Difference Analysis of Polynomial Sequences

Suppose $\{a_n\}$ is a sequence of natural numbers, i.e., $f = \lambda n.a_n$ is a function in $\mathbb{N} \to \mathbb{N}$. The function f is a *polynomial function of degree* k if

f can be presented in the form

$$c_k n^k + c_{k-1} n^{k-1} + \cdots + c_1 n + c_0,$$

with $c_i \in \mathbb{Q}$ and $c_k \neq 0$.

Example 9.1 The sequence

$$[1, 4, 11, 22, 37, 56, 79, 106, 137, 172, 211, 254, 301, 352, \ldots]$$

is given by the polynomial function $f = \lambda n.(2n^2 + n + 1)$. This is a function of the second degree.

Here is the Haskell check:

```
Prelude> take 15 (map (\ n -> 2*n^2 + n + 1) [0..])
[1,4,11,22,37,56,79,106,137,172,211,254,301,352,407]
```

Consider the *difference sequence* given by the function

$$d(f) = \lambda n. a_{n+1} - a_n.$$

The Haskell implementation looks like this:

```
difs :: [Integer] -> [Integer]
difs [] = []
difs [n] = []
difs (n:m:ks) = m-n : difs (m:ks)
```

This gives:

```
POL> difs [1,4,11,22,37,56,79,106,137,172,211,254,301]
[3,7,11,15,19,23,27,31,35,39,43,47]
```

The difference function $d(f)$ of a polynomial function f is itself a polynomial function. E.g., if $f = \lambda n.(2n^2 + n + 1)$, then:

$$\begin{aligned} d(f) &= \lambda n. 2(n+1)^2 + (n+1) + 1 - (2n^2 + n + 1) \\ &= \lambda n. 4n + 3. \end{aligned}$$

The Haskell check:

```
POL> take 15 (map (\n -> 4*n + 3) [0..])
[3,7,11,15,19,23,27,31,35,39,43,47,51,55,59]
POL> take 15 (difs (map (\ n -> 2*n^2 + n + 1) [0..]))
[3,7,11,15,19,23,27,31,35,39,43,47,51,55,59]
```

9.1. DIFFERENCE ANALYSIS OF POLYNOMIAL SEQUENCES

Proposition 9.2 If f is a polynomial function of degree k then $d(f)$ is a polynomial function of degree $k-1$.

Proof. Suppose $f(n)$ is given by

$$c_k n^k + c_{k-1} n^{k-1} + \cdots + c_1 n + c_0.$$

Then $d(f)(n)$ is given by

$$c_k(n+1)^k + c_{k-1}(n+1)^{k-1} + \cdots + c_1(n+1) + c_0$$
$$- (c_k n^k + c_{k-1} n^{k-1} + \cdots + c_1 n + c_0).$$

It is not hard to see that $f(n+1)$ has the form $c_k n^k + g(n)$, with g a polynomial of degree $k-1$. Since $f(n)$ also is of the form $c_k n^k + h(n)$, with h a polynomial of degree $k-1$, $d(f)(n)$ has the form $g(n) - h(n)$, so $d(f)$ is itself a polynomial of degree $k-1$. ∎

It follows from Proposition 9.2 that if f is a polynomial function of degree k, then $d^k(f)$ will be a constant function (a polynomial function of degree 0).

Here is a concrete example of computing difference sequences until we hit at a constant sequence:

```
-12     -11      6      45     112     213     354     541
    1       17      39      67     101     141     187
       16      22      28      34      40      46
           6       6       6       6       6
```

We find that the sequence of third differences is constant, which means that the form of the original sequence is a polynomial of degree 3. To find the next number in the sequence, just take the sum of the last elements of the rows. This gives $6 + 46 + 187 + 541 = 780$.

Charles Babbage (1791–1871), one of the founding fathers of computer science, used these observations in the design of his *difference engine*. We will give a Haskell version of the machine.

According to Proposition 9.2, if a given input list has a polynomial form of degree k, then after k steps of taking differences the list is reduced to a constant list:

```
POL> difs [-12,-11,6,45,112,213,354,541,780,1077]
[1,17,39,67,101,141,187,239,297]
POL> difs [1,17,39,67,101,141,187,239,297]
[16,22,28,34,40,46,52,58]
POL> difs [16,22,28,34,40,46,52,58]
[6,6,6,6,6,6,6]
```

The following function keeps generating difference lists until the differences get constant:

```
difLists :: [[Integer]]->[[Integer]]
difLists [] = []
difLists lists@(xs:xss) =
  if constant xs then lists else difLists ((difs xs):lists)
  where
  constant (n:m:ms) = all (==n) (m:ms)
  constant _        = error "lack of data or not a polynomial fct"
```

This gives the lists of all the difference lists that were generated from the initial sequence, with the constant list upfront.

```
POL> difLists [[-12,-11,6,45,112,213,354,541,780,1077]]
[[6,6,6,6,6,6,6],
 [16,22,28,34,40,46,52,58],
 [1,17,39,67,101,141,187,239,297],
 [-12,-11,6,45,112,213,354,541,780,1077]]
```

The list of differences can be used to generate the next element of the original sequence: just add the last elements of all the difference lists to the last element of the original sequence. In our example case, to get the next element of the list

$$[-12, -11, 6, 45, 112, 213, 354, 541, 780, 1077]$$

add the list of last elements of the difference lists (including the original list): $6 + 58 + 297 + 1077 = 1438$. To see that this is indeed the next element, note that the difference of 1438 and 1077 is 361, the difference of 361 and 297 is 64, and the difference of 64 and 58 is 6, so the number 1438 'fits' the difference analysis. The following function gets the list of last elements that we need (in our example case, the list [6,58,297,1077]):

```
genDifs :: [Integer] -> [Integer]
genDifs xs = map last (difLists [xs])
```

A new list of last elements of difference lists is computed from the current one by keeping the constant element d_1, and replacing each d_{i+1} by $d_i + d_{i+1}$.

9.1. DIFFERENCE ANALYSIS OF POLYNOMIAL SEQUENCES

```
nextD :: [Integer] -> [Integer]
nextD [] = error "no data"
nextD [n] = [n]
nextD (n:m:ks) = n : nextD (n+m : ks)
```

The next element of the original sequence is given by the last element of the new list of last elements of difference lists:

```
next :: [Integer] -> Integer
next = last . nextD . genDifs
```

In our example case, this gives:

```
POL> next [-12,-11,6,45,112,213,354,541,780,1077]
1438
```

All this can now be wrapped up in a function that continues any list of polynomial form, provided that enough initial elements are given as data:

```
continue :: [Integer] -> [Integer]
continue xs = map last (iterate nextD differences)
  where
  differences = nextD (genDifs xs)
```

This uses the predefined Haskell function iterate, that is given by:

```
iterate        :: (a -> a) -> a -> [a]
iterate f x    = x : iterate f (f x)
```

This is what we get:

```
POL> take 20 (continue [-12,-11,6,45,112,213,354,541,780,1077])
[1438,1869,2376,2965,3642,4413,5284,6261,7350,8557,9888,11349,
12946,14685,16572,18613,20814,23181,25720,28437]
```

If a given list is generated by a polynomial, then the degree of the polynomial can be computed by difference analysis, as follows:

```
degree :: [Integer] -> Int
degree xs = length (difLists [xs]) - 1
```

The difference engine is smart enough to be able to continue a list of sums of squares, or a list of sums of cubes:

```
POL> take 10 (continue [1,5,14,30,55])
[91,140,204,285,385,506,650,819,1015,1240]
POL> take 10 (continue [1,9,36,100,225,441])
[784,1296,2025,3025,4356,6084,8281,11025,14400,18496]
```

Exercise 9.3 What continuation do you get for [3,7,17,39,79,143]? Can you reconstruct the polynomial function that was used to generate the sequence?

Difference analysis yields an algorithm for continuing any finite sequence with a polynomial form. Is it also possible to give an algorithm for finding the form? This would solve the problem of how to guess the closed forms for the functions that calculate sums of squares, sums of cubes, and so on. The answer is 'yes', and the method is Gaussian elimination.

9.2 Gaussian Elimination

If we know that a sequence $a_0, a_1, a_2, a_3, \ldots$ has a polynomial form of degree 3, then we know that the form is $a + bx + cx^2 + dx^3$ (listing the coefficients in increasing order). This means that we can find the form of the polynomial by solving the following quadruple of linear equations in a, b, c, d:

$$
\begin{aligned}
a &= a_0 \\
a + b + c + d &= a_1 \\
a + 2b + 4c + 8d &= a_2 \\
a + 3b + 9c + 27d &= a_3
\end{aligned}
$$

Since this is a set of four linear equations in four unknowns, where the equations are linearly independent (none of them can be written as a multiple of any of the others), this can be solved by eliminating the unknowns one by one.

9.2. GAUSSIAN ELIMINATION

Example 9.4 Find the appropriate set of equations for the sequence

$$[-7, -2, 15, 50, 109, 198, 323]$$

and solve it.

Difference analysis yields that the sequence is generated by a polynomial of the third degree, so the sequence leads to the following set of equations:

$$\begin{aligned} a &= -7 \\ a + b + c + d &= -2 \\ a + 2b + 4c + 8d &= 15 \\ a + 3b + 9c + 27d &= 50 \end{aligned}$$

Eliminating a and rewriting gives:

$$\begin{aligned} b + c + d &= 5 \\ 2b + 4c + 8d &= 22 \\ 3b + 9c + 27d &= 57 \end{aligned}$$

Next, eliminate the summand with factor d from the second and third equation. Elimination from the second equation is done by subtracting the second equation from the 8-fold product of the first equation. This gives $6b + 4c = 18$, which can be simplified to $3b + 2c = 9$. Elimination from the third equation is done by subtracting the third equation from the 27-fold product of the first, with result $24b + 18c = 78$. We get the following pair of equations:

$$\begin{aligned} 3b + 2c &= 9 \\ 24b + 18c &= 78 \end{aligned}$$

Elimination of c from this pair is done by subtracting the second equation from the 9-fold product of the first. This gives $3b = 3$, whence $b = 1$. Together with $3b + 2c = 9$ we get $c = 3$. Together with $b + c + d = 5$ we get $d = 1$. Thus, the polynomial we are looking for has the form $\lambda n.(n^3 + 3n^2 + n - 7)$.

Exercise 9.5 Find the appropriate set of equations for the sequence

$$[13, 21, 35, 55, 81, 113, 151]$$

and solve it.

Solving sets of linear equations can be viewed as manipulation of matrices of coefficients. E.g., the quadruple of linear equations in a, b, c, d for a polynomial of the third degree gives the following matrix:

$$\begin{pmatrix} 1 & 0 & 0 & 0 & a_0 \\ 1 & 1 & 1 & 1 & a_1 \\ 1 & 2 & 4 & 8 & a_2 \\ 1 & 3 & 9 & 27 & a_3 \end{pmatrix}$$

To solve this, we transform it to an equivalent matrix in so-called *echelon form* or *left triangular form*, i.e., a matrix of the form:

$$\begin{pmatrix} a_{00} & a_{01} & a_{02} & a_{03} & b_0 \\ 0 & a_{11} & a_{12} & a_{13} & b_1 \\ 0 & 0 & a_{22} & a_{23} & b_2 \\ 0 & 0 & 0 & a_{33} & b_3 \end{pmatrix}$$

From this form, compute the value of variable d from the last row, next eliminate this variable from the third row, and find the value of c. Then use the values of d and c to find the value of b from the second row, and finally, use the values of b, c, d to find the value of a from the first row.

To handle matrices, the following type declarations are convenient.

```
type Matrix = [Row]
type Row    = [Integer]
```

It is also convenient to be able to have functions for the numbers of rows and columns of a matrix.

```
rows, cols :: Matrix -> Int
rows m = length m
cols m | m == []    = 0
       | otherwise = length (head m)
```

The function `genMatrix` produces the appropriate matrix for a list generated by a polynomial:

9.2. GAUSSIAN ELIMINATION

```
genMatrix :: [Integer] -> Matrix
genMatrix xs = zipWith (++) (genM d) [ [x] | x <- xs ]
  where
  d      = degree xs
  genM n = [ [ (toInteger x^m) | m <- [0..n] ] | x <- [0..n] ]
```

`zipWith` is predefined in the Haskell prelude as follows:

```
zipWith              :: (a->b->c) -> [a]->[b]->[c]
zipWith z (a:as) (b:bs) = z a b : zipWith z as bs
zipWith _ _      _      = []
```

In a picture:

$$[(z\ x_0\ y_0), (z\ x_1\ y_1), \ldots, (z\ x_n\ y_n), \left\{ \begin{array}{l} [x_{n+1}, x_{n+2}, \ldots \\ [y_{n+1}, y_{n+2}, \ldots \end{array} \right.$$

`genMatrix` gives, e.g.:

```
POL> genMatrix [-7,-2,15,50,109,198,323]
[[1,0,0,0,-7],[1,1,1,1,-2],[1,2,4,8,15],[1,3,9,27,50]]
```

The process of transforming the matrix to echelon form is done by so-called *forward elimination*: use one row to eliminate the first coefficient from the other rows by means of the following process of adjustment (the first row is used to adjust the second one):

```
adjustWith :: Row -> Row -> Row
adjustWith (m:ms) (n:ns) = zipWith (-) (map (n*) ms) (map (m*) ns)
```

To transform a matrix into echelon form, proceed as follows:

1. If the number of rows or the number of columns of the matrix is 0, then the matrix is already in echelon form.

2. If every row of `rs` begins with a 0 then the echelon form of `rs` can be found by putting 0's in front of the echelon form of `map tail rs`.

3. If `rs` has rows that do not start with a 0, then take the first one of these, `piv`, and use it to eliminate the leading coefficients from the other rows. This gives a matrix of the form

$$\begin{pmatrix} a_{00} & a_{01} & a_{02} & a_{03} & \cdots & b_0 \\ 0 & a_{11} & a_{12} & a_{13} & \cdots & b_1 \\ 0 & a_{21} & a_{22} & a_{23} & \cdots & b_2 \\ 0 & a_{31} & a_{32} & a_{33} & \cdots & b_3 \\ \vdots & \vdots & \vdots & \vdots & \vdots & \vdots \\ 0 & a_{n1} & a_{n2} & a_{n3} & \cdots & b_n \end{pmatrix}$$

where the first row is the pivot row. All that remains to be done in this case is to put the following sub matrix in echelon form:

$$\begin{pmatrix} a_{11} & a_{12} & a_{13} & \cdots & b_1 \\ a_{21} & a_{22} & a_{23} & \cdots & b_2 \\ a_{31} & a_{32} & a_{33} & \cdots & b_3 \\ \vdots & \vdots & \vdots & \vdots & \vdots \\ a_{n1} & a_{n2} & a_{n3} & \cdots & b_n \end{pmatrix}$$

The code for this can be found in the Haskell demo file *Matrix.hs* (part of the *Hugs* system):

```
echelon     :: Matrix -> Matrix
echelon rs
    | null rs || null (head rs) = rs
    | null rs2                = map (0:) (echelon (map tail rs))
    | otherwise               = piv : map (0:) (echelon rs')
    where rs'             = map (adjustWith piv) (rs1++rs3)
          (rs1,rs2)       = span leadZero rs
          leadZero (n:_)  = n==0
          (piv:rs3)       = rs2
```

Here is an example:

POL> echelon [[1,0,0,0,-7],[1,1,1,1,-2],[1,2,4,8,15],[1,3,9,27,50]]
[[1,0,0,0,-7],[0,-1,-1,-1,-5],[0,0,-2,-6,-12],[0,0,0,-12,-12]]

Backward Gaussian elimination, or computing the values of the variables from a matrix in echelon form, is done by computing the value of the

9.2. GAUSSIAN ELIMINATION

variable in the last row, eliminate that variable from the other rows to get a smaller matrix in echelon form, and repeating that process until the values of all variables are found. If we know that $ax = c$ (we may assume $a \neq 0$), and the coordinate of x is the coordinate $a_{in} = b$ of $a_{i1} \mid \cdots \mid a_{in-1} \mid a_{in} \mid d$, then we can eliminate this coordinate by replacing $a_{i1} \mid \cdots \mid a_{in-1} \mid a_{in} \mid d$ with the following:

$$a \cdot a_{i1} \mid \cdots \mid a \cdot a_{in-1} \mid ad - bc.$$

It does make sense to first reduce $x = c/a$ to its simplest form by dividing out common factors of c and a. The implementation of rational numbers does this for us if we express x as a number of type `Rational` (see Section 8.6 for further details). Note that an elimination step transforms a matrix in echelon form, minus its last row, into a smaller matrix in echelon form. Here is the implementation:

```
eliminate :: Rational -> Matrix -> Matrix
eliminate p rs = map (simplify c a) rs
  where
  c = numerator   p
  a = denominator p
  simplify c a row = init (init row') ++ [a*d - b*c]
    where
    d    = last row
    b    = last (init row)
    row' = map (*a) row
```

The implementation of backward substitution runs like this:

```
backsubst :: Matrix -> [Rational]
backsubst rs = backsubst' rs []
  where
  backsubst' [] ps = ps
  backsubst' rs ps = backsubst' rs' (p:ps)
    where
    a   = (last rs) !! ((cols rs) - 2)
    c   = (last rs) !! ((cols rs) - 1)
    p   = c % a
    rs' = eliminate p (init rs)
```

We get:

```
POL> backsubst [[1,0,0,0,-7],[0,-1,-1,-1,-5],[0,0,-2,-6,-12],[0,0,0,-12,-12]]
[-7 % 1,1 % 1,3 % 1,1 % 1]
```

To use all this to analyze a polynomial sequence, generate the appropriate matrix (appropriate for the degree of the polynomial that we get from difference analysis of the sequence), put it in echelon form, and compute the values of the unknowns by backward substitution.

```
solveSeq :: [Integer] -> [Rational]
solveSeq = backsubst . echelon . genMatrix
```

Recall that the sequence of sums of squares starts as follows: $0, 1, 5, 14, 30, \ldots$. Solving this sequence with `solveSeq` gives:

```
POL> solveSeq [0,1,5,14,30]
[0 % 1,1 % 6,1 % 2,1 % 3]
```

This gives the form

$$\frac{1}{3}n^3 + \frac{1}{2}n^2 + \frac{1}{6}n = \frac{2n^3 + 3n^2 + n}{6} = \frac{n(n+1)(2n+1)}{6}.$$

Here is a Haskell check (the use of the / operator creates a list of Fractionals):

```
POL> map (\ n -> (1/3)*n^3 + (1/2)*n^2 + (1/6)*n) [0..4]
[0.0,1.0,5.0,14.0,30.0]
```

Similarly, $0, 1, 9, 36, 100, 225$ is the start of the sequence of sums of cubes. Solving this with `solveSeq` gives:

```
POL> solveSeq [0, 1, 9, 36, 100, 225]
[0 % 1,0 % 1,1 % 4,1 % 2,1 % 4]
```

This gives the form

$$\frac{1}{4}n^4 + \frac{1}{2}n^3 + \frac{1}{4}n^2 = \frac{n^4 + 2n^3 + n^2}{4} = \frac{n^2(n+1)^2}{4} = \left(\frac{n(n+1)}{2}\right)^2.$$

The running example from the previous section is solved as follows:

```
POL> solveSeq [-12,-11,6,45,112,213,354,541,780,1077]
[-12 % 1,-5 % 1,5 % 1,1 % 1,1 % 1]
```

9.3. POLYNOMIALS AND THE BINOMIAL THEOREM

Thus, the sequence has the form $n^3 + 5n^2 - 5n - 12$.

Before we look at the confirmation, let us note that we are now in fact using *representations* of polynomial functions as lists of their coefficients, starting from the constant coefficient. It is easy to implement a conversion from these representations to the polynomial functions that they represent:

```
p2fct :: Num a => [a] -> a -> a
p2fct [] x = 0
p2fct (a:as) x = a + (x * p2fct as x)
```

We can use this in the confirmations, as follows:

```
POL> [ n^3 + 5 * n^2 - 5 * n - 12 | n <- [0..9] ]
[-12,-11,6,45,112,213,354,541,780,1077]
POL> map (p2fct [-12,-5,5,1]) [0..9]
[-12,-11,6,45,112,213,354,541,780,1077]
```

Finally, here is the automated solution of Exercise 9.3:

```
POL> solveSeq [3,7,17,39,79]
[3 % 1,3 % 1,0 % 1,1 % 1]
```

This represents the form $n^3 + 3n + 3$.

Exercise 9.6 Suppose you want to find a closed form for the number of pieces you can cut a pie into by making n straight cuts. After some experimentation it becomes clear that to obtain the maximum number of pieces no cut should be parallel to a previous cut, and no cut should pass through an intersection point of previous cuts. Under these conditions you find that the n-th cut can be made to intersect all the $n-1$ previous cuts, and can thereby made to split n of the old regions. This gives the recurrence $C_0 = 1$ and $C_n = C_{n-1} + n$, which yields the sequence $1, 2, 4, 7, 11, \ldots$ Next, you use `solveSeq` to obtain:

```
POL> solveSeq [1,2,4,7,11]
[1 % 1,1 % 2,1 % 2]
```

You conclude that n cuts can divide a pie into $\frac{1}{2}n^2 + \frac{1}{2}n + 1 = \frac{n(n+1)}{2} + 1$ pieces. Is there still need for an inductive proof to show that this answer is correct?

$$\begin{aligned}
a' &= 0, \\
(a \cdot f(x))' &= a \cdot f'(x), \\
(x^n)' &= nx^{n-1}, \\
(f(x) \pm g(x))' &= f'(x) \pm g'(x), \\
(f(x) \cdot g(x))' &= f'(x)g(x) + f(x)g'(x)b, \\
(f(g(x)))' &= g'(x) \cdot f'(g(x)).
\end{aligned}$$

Figure 9.1: Differentiation Rules.

9.3 Polynomials and the Binomial Theorem

In this section we will establish a connection between polynomials and lists, namely lists of coefficients of a polynomial. Let $f(x)$ be a polynomial of degree n, i.e., let $f(x)$ be a function

$$f(x) = a_n x^n + a_{n-1} x^{n-1} + \cdots + a_1 x + a_0,$$

with a_i constants, and $a_n \neq 0$. Let c be a constant and consider the case $x = y + c$. Substitution of $y + c$ for x in $f(x)$ gives a new polynomial of degree n in y, say

$$f(x) = f(y+c) = b_n y^n + b_{n-1} y^{n-1} + \cdots + b_1 y + b_0.$$

To consider an example, take $f(x) = 3x^4 - x + 2$, and let $c = -1$. Substituting $y + c$ for x we get:

$$\begin{aligned}
f(x) &= f(y+c) = f(y-1) \\
&= 3(y-1)^4 - (y-1) + 2 \\
&= (3y^4 - 12y^3 + 18y^2 - 12y + 3) - (y-1) + 2 \\
&= 3y^4 - 12y^3 + 18y^2 - 13y + 6.
\end{aligned}$$

We will see shortly that the coefficients b_i can be computed in a very simple way. Substitution of $x - c$ for y in in $f(y + c)$ gives

$$f(x) = b_n(x-c)^n + b_{n-1}(x-c)^{n-1} + \cdots + b_1(x-c) + b_0.$$

9.3. POLYNOMIALS AND THE BINOMIAL THEOREM

Calculation of $f'(x), f''(x), \ldots, f^{(n)}(x)$ (the first, second, ..., n-th derivative of f), is done with the familiar rules of Figure 9.1.[1] In particular, this gives $(b(x-c)^k)' = kb(x-c)^{k-1}$, and we get:

$$f'(x) = b_1 + 2b_2(x-c) + \cdots + nb_n(x-c)^{n-1}$$
$$f''(x) = 2b_2 + 3 \cdot 2b_3(x-c) + \cdots + n(n-1)b_n(x-c)^{n-2}$$
$$\vdots$$
$$f^{(n)}(x) = n(n-1)(n-2)\cdots 3 \cdot 2b_n.$$

Substitution of $x = c$ gives:

$$f(c) = b_0, f'(c) = b_1, f''(c) = 2b_2, \ldots, f^{(n)}(c) = n!b_n.$$

This yields the following instruction for calculating the b_k:

$$b_0 = f(c), b_1 = f'(c), b_2 = \frac{f''(c)}{2}, \ldots, b_n = \frac{f^{(n)}(c)}{n!}.$$

In general:

$$b_k = \frac{f^{(k)}(c)}{k!}.$$

Applying this to the example $f(x) = 3x^4 - x + 2$, with $c = -1$, we see that $b_0 = f(-1) = 3(-1)^4 + 1 + 2 = 6$, $b_1 = f'(-1) = 12(-1)^3 - 1 = -13$, $b_2 = \frac{f''(-1)}{2} = \frac{36(-1)^2}{2} = 18$, $b_3 = \frac{f^{(3)}(-1)}{6} = \frac{-72}{6} = -12$, $b_4 = \frac{f^{(4)}(-1)}{24} = 3$.

Another example is the expansion of $(z+1)^n$. Using the calculation method with derivatives, we get, for $c = 0$:

$$f(z) = (z+1)^n = b_n z^n + b_{n-1} z^{n-1} + \cdots + b_1 z + b_0,$$

with the following derivatives:

$$f'(z) = n(z+1)^{n-1}, f''(z) = n(n-1)(z+1)^{n-2}, \ldots, f^{(n)}(z) = n!.$$

Substituting $z = 0$ this gives:

$$b_0 = 1, b_1 = n, b_2 = \frac{n(n-1)}{2}, \ldots, b_k = \frac{n(n-1)\cdots(n-k+1)}{k!}, \ldots, b_n = \frac{n!}{n!} = 1.$$

The general coefficient b_k has the form

$$b_k = \frac{n(n-1)\cdots(n-k+1)}{k!} = \frac{n(n-1)\cdots(n-k+1)}{k!} \cdot \frac{(n-k)!}{(n-k)!}$$
$$= \frac{n!}{k!\,(n-k)!}.$$

---
[1] If these rules are unfamiliar to you, or if you need to brush up your knowledge of analysis, you should consult a book like [Bry93].

Define:
$$\binom{n}{k} := \frac{n!}{k!\,(n-k)!}.$$

Pronounce $\binom{n}{k}$ as 'n choose k' or 'n over k'. We have derived:

Theorem 9.7 (Newton's binomial theorem)

$$(z+1)^n = \sum_{k=0}^{n} \binom{n}{k} z^k.$$

If A is a set of n objects, then there are $\binom{n}{k}$ ways to pick a subset B from A with $|B| = k$. Thus, $\binom{n}{k}$ is also the number of k-sized subsets of an n-sized set. To see why this is so, note that the number of k-sequences picked from A, without repetitions, equals

$$n \cdot (n-1) \cdots (n-k+1),$$

for there are n ways to pick the first element in the sequence, $n-1$ ways to pick the second element in the sequence, ..., and $n - (k-1)$ ways to pick the k-th element in the sequence. The number $n \cdot (n-1) \cdots (n-k+1)$ is equal to $\frac{n!}{(n-k)!}$.

For picking k-sized subsets from A, order does not matter. There are $k!$ ways of arranging sequences of size k without repetition. These are all equivalent. This gives $\frac{n!}{k!\,(n-k)!}$ for the number of k-sized subsets of a set of size n. This connection explains the phrasing 'n choose k'.

Note the following:

$$\binom{n}{0} = 1, \quad \binom{n}{1} = n, \quad \binom{n}{2} = \frac{n(n-1)}{2}, \quad \binom{n}{3} = \frac{n(n-1)(n-2)}{6}.$$

Here is a straightforward implementation of $\binom{n}{k}$.

```
choose n k = (product [(n-k+1)..n]) `div` (product [1..k])
```

The more general version of Newton's binomial theorem runs:

Theorem 9.8 (Newton's binomial theorem, general version)

$$(x+y)^n = \sum_{k=0}^{n} \binom{n}{k} x^k y^{n-k}.$$

9.3. POLYNOMIALS AND THE BINOMIAL THEOREM

Note: a *binomial* is the sum of two terms, so $(x+y)$ is a binomial.

Proof: To get this from the special case $(z+1)^n = \sum_{k=0}^{n} \binom{n}{k} z^k$ derived above, set $z = \frac{x}{y}$ to get $(\frac{x}{y}+1)^n = \sum_{k=0}^{n} \binom{n}{k} \frac{x^k}{y^k}$, and multiply by y^n:

$$\begin{aligned}
(x+y)^n &= \left(\frac{x}{y}+1\right)^n \cdot y^n \\
&= \left(\sum_{k=0}^{n} \binom{n}{k} \frac{x^k}{y^k}\right) \cdot y^n = \sum_{k=0}^{n} \binom{n}{k} \frac{x^k \cdot y^n}{y^k} = \sum_{k=0}^{n} \binom{n}{k} x^k y^{n-k}.
\end{aligned}$$

Because of their use in the binomial theorem, the numbers $\binom{n}{k}$ are called *binomial coefficients*. What the binomial theorem gives us is:

$$\begin{aligned}
(x+y)^0 &= 1x^0 y^0 \\
(x+y)^1 &= 1x^1 y^0 + 1x^0 y^1 \\
(x+y)^2 &= 1x^2 y^0 + 2x^1 y^1 + 1x^0 y^2 \\
(x+y)^3 &= 1x^3 y^0 + 3x^2 y^1 + 3x^1 y^2 + 1x^0 y^3 \\
(x+y)^4 &= 1x^4 y^0 + 4x^3 y^1 + 6x^2 y^2 + 4x^1 y^3 + 1x^0 y^4 \\
(x+y)^5 &= 1x^5 y^0 + 5x^4 y^1 + 10x^3 y^2 + 10x^2 y^3 + 5x^1 y^4 + 1x^0 y^5
\end{aligned}$$

$\vdots$

To see how this pattern arises, look at what happens when we raise $x+y$ to the n-th power by performing the n multiplication steps to work out the product of $\underbrace{(x+y)(x+y)\cdots(x+y)}_{n \text{ factors}}$:

$$\begin{array}{r}
x+y \\
x+y \quad \times \\ \hline
x^2 + xy + xy + y^2 \\
x+y \quad \times \\ \hline
x^3 + x^2y + x^2y + x^2y + xy^2 + xy^2 + xy^2 + y^3 \\
x+y \quad \times \\ \hline
\vdots
\end{array}$$

Every term in this expansion is itself the product of x-factors and y-factors, with a total number of factors always n, so that each term has the form $x^k y^{n-k}$. Every binomial $(x+y)$ in $(x+n)^n$ either contributes an x-factor or a y-factor to $x^k y^{n-k}$. The number of ways to get at the term $x^k y^{n-k}$ equals the number of k-sized subsets from a set of size n (pick any subset of k binomials from the set of n binomial factors). Thus, this term occurs exactly $\binom{n}{k}$ times.

We can arrange the binomial coefficients in the well-known triangle of Pascal.

$$\begin{array}{ccccccccc} & & & & \binom{0}{0} & & & & \\ & & & \binom{1}{0} & & \binom{1}{1} & & & \\ & & \binom{2}{0} & & \binom{2}{1} & & \binom{2}{2} & & \\ & \binom{3}{0} & & \binom{3}{1} & & \binom{3}{2} & & \binom{3}{3} & \\ \binom{4}{0} & & \binom{4}{1} & & \binom{4}{2} & & \binom{4}{3} & & \binom{4}{4} \\ & & & & \vdots & & & & \end{array}$$

Working this out, we get:

$$\begin{array}{ccccccccc} & & & & 1 & & & & \\ & & & 1 & & 1 & & & \\ & & 1 & & 2 & & 1 & & \\ & 1 & & 3 & & 3 & & 1 & \\ 1 & & 4 & & 6 & & 4 & & 1 \\ & & & & \vdots & & & & \end{array}$$

Studying the pattern of Pascal's triangle, we see that that it is built according to the following law:

$$\binom{n}{k} = \binom{n-1}{k-1} + \binom{n-1}{k}.$$

This is called the addition law for binomial coefficients. To see that this law is correct, consider a set A of size n, and single out one of its objects, a. To count the number of ways of picking a k-sized subset B from A, consider the two cases (i) $a \in B$ and (ii) $a \notin B$. The number of ways of picking a k-sized subset B from A with $a \in B$ is equal to the number of ways of picking a $k-1$-sized subset from $A - \{a\}$, i.e., $\binom{n-1}{k-1}$. The number of ways of picking a k-sized subset B from A with $a \notin B$ is equal to the number of ways of picking a k-sized subset from $A - \{a\}$, i.e., $\binom{n-1}{k}$. Thus, there are $\binom{n-1}{k-1} + \binom{n-1}{k}$ ways of picking a k-sized subset from an n-sized set.

It is of course also possible to prove the addition law directly from the definition of $\binom{n}{k}$. Assume $k > 0$. Then:

$$\begin{aligned} \binom{n-1}{k-1} + \binom{n-1}{k} &= \frac{(n-1)!}{(k-1)!\,(n-k)!} + \frac{(n-1)!}{k!\,(n-1-k)!} \\ &= \frac{(n-1)!\,k}{k!\,(n-k)!} + \frac{(n-1)!\,(n-k)}{k!\,(n-k)!} \\ &= \frac{(n-1)!\,n}{k!\,(n-k)!} = \frac{n!}{k!\,(n-k)!} = \binom{n}{k}. \end{aligned}$$

9.3. POLYNOMIALS AND THE BINOMIAL THEOREM

We can use the addition law for an implementation of $\binom{n}{k}$. In accordance with the interpretation of $\binom{n}{k}$ as the number of k-sized subset of an n-sized set, we will put $\binom{n}{0} = 1$ (there is just one way to pick a 0-sized subset from any set) and $\binom{n}{k} = 0$ for $n < k$ (no ways to pick a k-sized subset from a set that has less than k elements). A further look at Pascal's triangle reveals the following law of symmetry:

$$\binom{n}{k} = \binom{n}{n-k}.$$

This makes sense under the interpretation of $\binom{n}{k}$ as the number of k-sized subsets of a set of size n, for the number of k-sized subsets equals the number of their complements. There is just one way to pick an n-sized subset from an n-sized set (pick the whole set), so $\binom{n}{n} = 1$. This leads to the following implementation:

```
choose' n 0 = 1
choose' n k | n < k      = 0
            | n == k     = 1
            | otherwise =
                choose' (n-1) (k-1) + (choose' (n-1) (k))
```

Exercise 9.9 Which implementation is more efficient, choose or choose'? Why?

Exercise 9.10 Derive the symmetry law for binomial coefficients directly from the definition.

We will now give an inductive proof of Newton's binomial theorem. The proof uses the addition law for binomials, in the form $\binom{n}{k-1} + \binom{n}{k} = \binom{n+1}{k}$.

Theorem 9.11 (Newton's binomial theorem again)

$$(x+y)^n = \sum_{k=0}^{n} \binom{n}{k} x^k y^{n-k}.$$

Proof. Induction on n.

Basis:
$$(x+y)^0 = 1 = \binom{0}{0}x^0y^0 = \sum_{k=0}^{0}\binom{0}{k}x^ky^{0-k}.$$

Induction step: Assume
$$(x+y)^n = \sum_{k=0}^{n}\binom{n}{k}x^ky^{n-k}.$$

Then:

$$
\begin{aligned}
(x+y)^{n+1} &= (x+y)(x+y)^n \\
&\stackrel{\text{ih}}{=} (x+y)\sum_{k=0}^{n}\binom{n}{k}x^ky^{n-k} \\
&= x\sum_{k=0}^{n}\binom{n}{k}x^ky^{n-k} + y\sum_{k=0}^{n}\binom{n}{k}x^ky^{n-k} \\
&= \sum_{k=0}^{n}\binom{n}{k}x^{k+1}y^{n-k} + \sum_{k=0}^{n}\binom{n}{k}x^ky^{(n+1)-k} \\
&= x^{n+1} + \sum_{k=0}^{n-1}\binom{n}{k}x^{k+1}y^{n-k} + \sum_{k=1}^{n}\binom{n}{k}x^ky^{(n+1)-k} + y^{n+1} \\
&= x^{n+1} + \sum_{k=1}^{n}\binom{n}{k-1}x^ky^{(n+1)-k} + \sum_{k=1}^{n}\binom{n}{k}x^ky^{(n+1)-k} + y^{n+1} \\
&= x^{n+1} + \sum_{k=1}^{n}\left(\binom{n}{k-1}x^ky^{(n+1)-k} + \binom{n}{k}x^ky^{(n+1)-k}\right) + y^{n+1} \\
&\stackrel{\text{add}}{=} x^{n+1} + \sum_{k=1}^{n}\binom{n+1}{k}x^ky^{(n+1)-k} + y^{n+1} \\
&= \sum_{k=1}^{n+1}\binom{n+1}{k}x^ky^{(n+1)-k} + y^{n+1} = \sum_{k=0}^{n+1}\binom{n+1}{k}x^ky^{(n+1)-k}.
\end{aligned}
$$

∎

Exercise 9.12 Show from the definition that if $0 < k \leqslant n$ then:
$$\binom{n}{k} = \frac{n}{k}\cdot\binom{n-1}{k-1}.$$

9.4. POLYNOMIALS FOR COMBINATORIAL REASONING

The law from Exercise 9.12 is the so-called absorption law for binomial coefficients. It allows for an alternative implementation of a function for binomial coefficients, for we have the following recursion:

$$\binom{n}{0} = 1, \quad \binom{n}{k} = 0 \text{ for } n < k, \quad \binom{n}{k} = \frac{n}{k} \cdot \binom{n-1}{k-1} \text{ for } 0 < k \leqslant n.$$

Thus we get a more efficient function for $\binom{n}{k}$:

```
binom n 0 = 1
binom n k | n < k     = 0
          | otherwise = (n * binom (n-1) (k-1)) 'div' k
```

Exercise 9.13 Use a combinatorial argument (an argument in terms of sizes the subsets of a set) to prove Newton's law:

$$\binom{n}{m} \cdot \binom{m}{k} = \binom{n}{k} \cdot \binom{n-k}{m-k}.$$

Exercise 9.14 Prove:

$$\binom{n}{n} + \binom{n+1}{n} + \binom{n+2}{n} + \cdots + \binom{n+k}{n} = \binom{n+k+1}{n+1}.$$

9.4 Polynomials for Combinatorial Reasoning

To implement the polynomial functions in a variable z, we will represent a polynomial

$$f(z) = f_0 + f_1 z + f_2 z^2 + \cdots + f_{n-1} z^{n-1} + f_n z^n$$

as a list of its coefficients:

$$[f_0, f_1, \ldots, f_n].$$

As we have seen, the function p2fct maps such lists to the corresponding functions.

The constant zero polynomial has the form $f(z) = 0$. In general we will avoid trailing zeros in the coefficient list, i.e., we will assume that if $n > 0$ then $f_n \neq 0$. The constant function $\lambda z.c$ will get represented as $[c]$, so there is a map from integers to polynomial representations, given by $\lambda c.[c]$. We will also allow rationals as coefficients, to there is also a map from rationals to polynomial representations, given by $\lambda r.[r]$.

We need some conventions for switching back and forth between a polynomial and its list of coefficients. If $f(z)$ is a polynomial, then we use f for its coefficient list. If this list of coefficients is non-empty then, as before, we will indicate the tail of f as $\overline{f}$. Thus, if $f = [f_0, f_1, \ldots, f_n]$, then $\overline{f} = [f_1, \ldots, f_n]$, and we have the identity $f = f_0 : \overline{f}$. Moreover, if $f(z) = f_0 + f_1 z + f_2 z^2 + \cdots + f_{n-1} z^{n-1} + f_n z^n$, then we use $\overline{f}(z)$ for $f_1 + f_2 z + \cdots + f_{n-1} z^{n-2} + f_n z^{n-1}$. This convention yields the following important equality:

$$f(z) = f_0 + z\overline{f}(z).$$

The identity function $\lambda z.z$ will get represented as $[0, 1]$, for this function is of the form $\lambda z.f_0 + f_1 z$, with $f_0 = 0$ and $f_1 = 1$. This gives:

```
z :: Num a => [a]
z = [0,1]
```

To negate a polynomial, simply negate each term in its term expansion. For if $f(z) = f_0 + f_1 z + f_2 z^2 + \cdots$, then $-f(z) = -f_0 - f_1 z - f_2 z^2 - \cdots$. This gives:

```
negate []      = []
negate (f:fs)  = (negate f) : (negate fs)
```

To add two polynomials $f(z)$ and $g(z)$, just add their coefficients, for clearly, if $f(z) = f_0 + f_1 z + f_2 z^2 + \cdots + f_k z^k$ and $g(z) = b_0 + b_1 z + g_2 z^2 + \cdots = g_m z^m$, then

$$f(z) + g(z) = (f_0 + g_0) + (f_1 + g_1)z + (f_2 + g_2)z^2 + \cdots$$

This translates into Haskell as follows:

```
fs      + []     = fs
[]      + gs     = gs
(f:fs)  + (g:gs) = f+g : fs+gs
```

Note that this uses overloading of the + sign: in f+g we have addition of numbers, in fs+gs addition of polynomial coefficient sequences.

9.4. POLYNOMIALS FOR COMBINATORIAL REASONING

The product of $f(z) = f_0 + f_1 z + f_2 z^2 + \cdots + f_k z^k$ and $g(z) = g_0 + g_1 z + g_2 z^2 + \cdots + g_m z^m$ looks like this:

$$f(z) \cdot g(z) = (f_0 + f_1 z + f_2 z^2 + \cdots + f_k z^k) \cdot (g_0 + g_1 z + g_2 z^2 + \cdots + g_m z^m)$$
$$= f_0 g_0 + (f_0 g_1 + f_1 g_0) z + (f_0 g_2 + f_1 g_1 + f_2 g_0) z^2 + \cdots$$
$$= f_0 g_0 + z(f_0 \overline{g}(z) + g_0 \overline{f}(z) + z\overline{f}(z)\overline{g}(z))$$
$$= f_0 g_0 + z(f_0 \overline{g}(z) + \overline{f}(z) g(z))$$

Here $\overline{f}(z) = f_1 + f_2 z + \cdots + f_k z^{k-1}$ and $\overline{g}(z) = g_1 + g_2 z + \cdots + g_m z^{m-1}$, i.e., $\overline{f}(z)$ and $\overline{g}(z)$ are the polynomials that get represented as $[f_1, \ldots, f_k]$ and $[g_1, \ldots, g_m]$, respectively.

If $f(z)$ and $g(z)$ are polynomials of degree k, then for all $n \leqslant k$, in the list of coefficients for $f(z)g(z)$, the n-th coefficient has the form $f_0 g_n + f_1 g_{n-1} + \cdots + f_{n-1} g_1 + f_n g_0$. The list of coefficients of the product is called the **convolution** of the lists of coefficients f and g.

Multiplying a polynomial by z boils down to shifting its sequence of coefficients one place to the right. This leads to the following Haskell implementation, where (.*) is an auxiliary multiplication operator for multiplying a polynomial by a numerical constant. Note that (*) is overloaded: f*g multiplies two numbers, but fs * (g:gs) multiplies two lists of coefficients. We cannot extend this overloading to multiplication of numbers with coefficient sequences, since Haskell insists on operands of the same type for (*). Hence the use of (.*).

```
infixl 7 .*
(.*) :: Num a => a -> [a] -> [a]
c .* []     = []
c .* (f:fs) = c*f : c .* fs

fs     * []     = []
[]     * gs     = []
(f:fs) * (g:gs) = f*g : (f .* gs + fs * (g:gs))
```

Example 9.15 In Figure 9.2 the polynomials are declared as a data type in class Num. This entails that all Haskell operations for types in this class are available. We get:

```
POL> (z + 1)^0
[1]
```

```
POL> (z + 1)
[1,1]
POL> (z + 1)^2
[1,2,1]
POL> (z + 1)^3
[1,3,3,1]
POL> (z + 1)^4
[1,4,6,4,1]
POL> (z + 1)^5
[1,5,10,10,5,1]
POL> (z + 1)^6
[1,6,15,20,15,6,1]
```

This gives yet another way to get at the binomial coefficients.

Now suppose we have a polynomial $f(z)$. We are interested in the difference list of its coefficients $[f_0, f_1 - f_0, f_2 - f_1, \ldots]$. It is easy to see that this difference list is the list of coefficients of the polynomial $(1 - z)f(z)$:

$$\begin{array}{rcl} f(z) & \rightsquigarrow & [\quad f_0, \quad\quad f_1, \quad\quad f_2, \quad\quad f_3, \quad \cdots \quad] \\ -zf(z) & \rightsquigarrow & [\quad 0, \quad\quad -f_0, \quad\quad -f_1, \quad\quad -f_2, \quad \cdots \quad] \\ (1-z)f(z) & \rightsquigarrow & [\quad f_0, \quad f_1 - f_0, \quad f_2 - f_1, \quad f_3 - f_2, \quad \cdots \quad] \end{array}$$

This is implemented by the following function:

```
delta :: Num a => [a] -> [a]
delta = ([1,-1] *)
```

This gives, for example:

```
POL> delta [2,4,6]
[2,2,2,-6]
```

Note that the coefficient of z^4 in $[2, 4, 6]$ is 0, so this is correct. Note also that we are now looking at difference lists of coefficients, not at different lists of the result of mapping the polynomial function to $[0..]$, as in Section 9.1.

9.4. POLYNOMIALS FOR COMBINATORIAL REASONING

```
module Polynomials

where

infixl 7 .*
(.*) :: Num a => a -> [a] -> [a]
c .* []      = []
c .* (f:fs)  = c*f : c .* fs

z :: Num a => [a]
z = [0,1]

instance Num a => Num [a] where
  fromInteger c  = [fromInteger c]
  negate []      = []
  negate (f:fs)  = (negate f) : (negate fs)
  fs      + []      = fs
  []      + gs      = gs
  (f:fs)  + (g:gs)  = f+g : fs+gs
  fs      * []      = []
  []      * gs      = []
  (f:fs)  * (g:gs)  = f*g : (f .* gs + fs * (g:gs))

delta :: Num a => [a] -> [a]
delta = ([1,-1] *)

shift :: [a] -> [a]
shift = tail

p2fct :: Num a => [a] -> a -> a
p2fct [] x = 0
p2fct (a:as) x = a + (x * p2fct as x)

comp :: Num a => [a] -> [a] -> [a]
comp _       []     = error ".."
comp []      _      = []
comp (f:fs) (0:gs) = f : gs * (comp fs (0:gs))
comp (f:fs) (g:gs) = ([f] + [g] * (comp fs (g:gs)))
                     + (0 : gs * (comp fs (g:gs)))

deriv :: Num a => [a] -> [a]
deriv []     = []
deriv (f:fs) = deriv1 fs 1 where
  deriv1 []     _ = []
  deriv1 (g:gs) n = n*g : deriv1 gs (n+1)
```

Figure 9.2: A Module for Polynomials.

The composition of two polynomials $f(z)$ and $g(z)$ is again a polynomial $f(g(z))$. It is given by:

$$\begin{aligned} f(z) &= f_0 + f_1 z + f_2 z^2 + f_3 z^3 + \cdots \\ f(g(z)) &= f_0 + f_1 g(z) + f_2 (g(z))^2 + f_3 (g(z))^3 + \cdots \end{aligned}$$

We see from this:

$$f(g(z)) = f_0 + g(z) \cdot \overline{f}(g(z))$$

This leads immediately to the following implementation (the module of Figure 9.2 has a slightly more involved implementation comp that gets explained in the next chapter, on page 377):

```
comp1 :: Num a => [a] -> [a] -> [a]
comp1 _ [] = error ".."
comp1 [] _ = []
comp1 (f:fs) gs = [f] + (gs * comp1 fs gs)
```

Example 9.16 We can use this to pick an arbitrary layer in Pascal's triangle:

POL> comp1 (z^2) (z+1)
[1,2,1]
POL> comp1 (z^3) (z+1)
[1,3,3,1]
POL> comp1 (z^12) (z+1)
[1,12,66,220,495,792,924,792,495,220,66,12,1]

We can also use it to generate Pascal's triangle up to arbitrary depth:

POL> comp1 [1,1,1,1,1,1] [[0],[1,1]]
[[1],[1,1],[1,2,1],[1,3,3,1],[1,4,6,4,1],[1,5,10,10,5,1]]

Note that this uses the composition of $f(z) = 1 + z + z^2 + z^3 + z^4 + z^5 + z^6$ with $g(z) = (y+1)z + 0$. The result of this is $f(g(z)) = 1 + (y+1)z + (y+1)^2 z^2 + \cdots + (y+1)^6 z^6$.

If $f(z) = f_0 + f_1 z + f_2 z^2 + \cdots + f_k z^k$, the derivative of $f(z)$ is given by (Figure 9.1):

$$f'(z) = f_1 + 2f_2 z + \cdots + k f_k z^{k-1}.$$

This has a straightforward implementation, as follows:

9.4. POLYNOMIALS FOR COMBINATORIAL REASONING

```
deriv :: Num a => [a] -> [a]
deriv []     = []
deriv (f:fs) = deriv1 fs 1 where
   deriv1 []     _ = []
   deriv1 (g:gs) n = n*g : deriv1 gs (n+1)
```

The close link between binomial coefficients and combinatorial notions makes polynomial reasoning a very useful tool for finding solutions to combinatorial problems.

Example 9.17 How many ways are there of selecting ten red, blue or white marbles from a vase, in such a way that there are at least two of each color and at most five marbles have the same colour? The answer is given by the coefficient of z^{10} in the following polynomial:

$$(z^2 + z^3 + z^4 + z^5)^3.$$

This is easily encoded into a query in our implementation:

POL> ([0,0,1,1,1,1]^3) !! 10
12

How many ways are there of selecting ten red, blue or white marbles from a vase, in such manner that there is even number of marbles of each colour:

POL> ([1,0,1,0,1,0,1,0,1,0,1]^3) !! 10
21

We associate coefficient lists with combinatorial problems by saying that

$$[f_0, f_1, f_2, \ldots, f_n]$$

solves a combinatorial problem if f_r gives the number of solutions for that problem.

Example 9.18 The polynomial $(1+z)^{10}$ solves the problem of picking r elements from a set of 10. The finite list $[1, 10, 45, 120, 210, 252, 210, 120, 45, 10, 1]$ solves the problem. It is implemented by:

POL> (1 + z)^10
[1,10,45,120,210,252,210,120,45,10,1]

Example 9.19 The list $[1, 3, 6, 10, 15, 18, 19, 18, 15, 10, 6, 3, 1]$ is a solution for the problem of picking r marbles from a vase containing red, white or blue marbles, with a maximum of five of each colour. A polynomial for this problem is $(1 + z + z^2 + z^3 + z^4 + z^5)^3$. In the implementation:

```
POL> (1 + z + z^2 + z^3 + z^4 + z^5)^3
[1,3,6,10,15,21,25,27,27,25,21,15,10,6,3,1]
```

Exercise 9.20 Use polynomials to find out how many ways there are of selecting ten red, blue or white marbles from a vase, in such manner that the number of marbles from each colour is prime.

9.5 Further Reading

Charles Babbage's difference engine is described in [Lar34] (reprinted in [Bab61]), and by Babbage himself in his memoirs [Bab94]. The memoirs are very amusing:

> Among the various questions which have been asked respecting the Difference Engine, I will mention a few of the most remarkable: one gentleman addressed me thus: 'Pray, Mr Babbage, can you explain to me in two words what is the principle of your machine?' Had the querist possessed a moderate acquaintance with mathematics I might in four words have conveyed to him the required information by answering, 'The Method of Differences.'
> [...] On two occasions I have been asked - 'Pray, Mr Babbage, if you put into the machine wrong figures, will the right answers come out?' In one case a member of the Upper, and in the other a member of the Lower, House put this question.

There are many good textbooks on calculus, but [Bry93] is particularly enlightening. An excellent book on discrete mathematics and combinatorial reasoning is [Bal91].

Chapter 10

Corecursion

Preview

In this chapter we will look the construction of infinite objects and at proof methods suited to reasoning with infinite data structures. The most important kind of infinite data structures are streams (infinite lists), so the main topic of this chapter is the logic of stream processing. We will show how non-deterministic processes can be viewed as functions from random integer streams to streams. For the implementation of this we will use two functions from Random.hs, a module for random number generation and processing from the Haskell library. At the end of the chapter we will connect combinatorial reasoning with stream processing, via the study of power series and generating functions. Our Haskell treatment of power series is modeled after the beautiful [McI99, McI00].

```
module COR

where

import Random (mkStdGen,randomRs)
import Polynomials
import PowerSeries
```

The default for the display of fractional numbers in Haskell is floating point notation. As we are going to develop streams of integers and streams of fractions in this chapter, it is convenient to have them displayed with unlimited precision in integer or rational notation. The `default` command takes care of that.

```
default (Integer, Rational, Double)
```

10.1 Corecursive Definitions

As we have seen, it is easy to generate infinite lists in Haskell. Infinite lists are often called *streams*. Here is the code again for generating an infinite list (or a stream) of ones:

```
ones = 1 : ones
```

This looks like a recursive definition, but there is no base case. Here is a definition of a function that generates an infinite list of all natural numbers:

```
nats = 0 : map (+1) nats
```

Again, the definition of `nats` looks like a recursive definition, but there is no base case. Definitions like this are called *corecursive definitions*. Corecursive definitions always yield infinite objects. When you come to think of it, the funny explanation of the acronym *GNU* as *GNU's Not Unix* is also an example.

As we have seen in Section 3.7, generating the odd natural numbers can be done by corecursion.

```
odds = 1 : map (+2) odds
```

Exercise 10.1 Write a corecursive definition that generates the *even* natural numbers.

10.1. CORECURSIVE DEFINITIONS

We can make the corecursive definitions more explicit with the use of `iterate`. The definition of `iterate` in the Haskell prelude is itself an example of corecursion:

```
iterate           :: (a -> a) -> a -> [a]
iterate f x       = x : iterate f (f x)
```

Here are versions of the infinite lists above in terms of `iterate`:

```
theOnes = iterate id 1
theNats = iterate (+1) 0
theOdds = iterate (+2) 1
```

Exercise 10.2 Use `iterate` to define the infinite stream of even natural numbers.

The list `[0..]` can be defined corecursively from `ones` with `zipWith`.

Suppose n is a natural number. Then its successor can be got by adding 1 to n. 0 is the first natural number. The second natural number, 1, is got by adding 1 to 0. The third natural number, 2, is got by adding 1 to the second natural number, and so on:

```
theNats1 = 0 : zipWith (+) ones theNats1
```

The technique that produced `theNats1` can be used for generating the Fibonacci numbers:

```
theFibs = 0 : 1 : zipWith (+) theFibs (tail theFibs)
```

The process on Fibonacci numbers that was defined in Exercise 7.17 can be defined with corecursion, as follows:

```
pr (x1:x2:x3:xs) = x1*x3 - x2*x2 : pr (x2:x3:xs)
```

As we proved in Exercise 7.17, applying this process to theFibs gives the list $\lambda n.(-1)^{n+1}$:

```
COR> take 20 (pr theFibs)
[-1,1,-1,1,-1,1,-1,1,-1,1,-1,1,-1,1,-1,1,-1,1,-1,1]
```

The definition of the sieve of Eratosthenes (page 104) also uses corecursion:

```
sieve :: [Integer] -> [Integer]
sieve (0 : xs) = sieve xs
sieve (n : xs) = n : sieve (mark xs 1 n)
   where
   mark (y:ys) k m | k == m    = 0 : (mark ys 1     m)
                   | otherwise = y : (mark ys (k+1) m)
```

What these definitions have in common is that they generate infinite objects, and that they look like recursive definitions, except for the fact that there is no base case.

Here is a faster way to implement the Sieve of Eratosthenes. This time, we actually *remove* multiples of x from the list on encountering x in the sieve. The counting procedure now has to be replaced by a calculation, for the removals affect the distances in the list. The property of *not* being a multiple of n is implemented by the function (\ m -> (rem m n) /= 0). Removing all numbers that do not have this property is done by filtering the list with the property.

```
sieve' :: [Integer] -> [Integer]
sieve' (n:xs) = n : sieve' (filter (\ m -> (rem m n) /= 0) xs)

primes' :: [Integer]
primes' = sieve' [2..]
```

How does one prove things about corecursive programs? E.g., how does one prove that sieve and sieve' compute the same stream result for every stream argument? Proof by induction does not work here, for there is no base case.

10.2. PROCESSES AND LABELED TRANSITION SYSTEMS

Exercise 10.3* The Thue-Morse sequence is a stream of 0's and 1's that is produced as follows. First produce 0. Next, at any stage, swap everything that was produced so far (by interchanging 0's and 1's) and append that. The first few stages of producing this sequence look like this:

$$
\begin{array}{l}
0 \\
01 \\
0110 \\
01101001 \\
0110100110010110
\end{array}
$$

Thus, if A_k denotes the first 2^k symbols of the sequence, then A_{k+1} equals $A_k \mathbin{+\!\!+} B_k$, where B_k is obtained from A_k by interchanging 0's and 1's. Give a corecursive program for producing the Thue-Morse sequence as a stream.

10.2 Processes and Labeled Transition Systems

The notion of a nondeterministic sequential process is so general that it is impossible to give a rigorous definition. Informally we can say that processes are interacting procedures. Typical examples are (models of) mechanical devices such as clocks, protocols for traffic control, vending machines, operating systems, client-server computer systems, and so on. A formal notion for modeling processes that has turned out to be extremely fruitful is the following.

A **labeled transition system** (Q, A, T) consists of a set of **states** Q, a set of **action labels** A, and a ternary relation $T \subseteq Q \times A \times Q$, the **transition relation**. If $(q, a, q') \in T$ we write this as $q \xrightarrow{a} q'$.

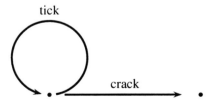

Figure 10.1: Ticking clock.

Example 10.4 Perhaps the simplest example of a labeled transition system is the system given by the two states c and c_0 and the two transitions $c \xrightarrow{\text{tick}} c$ and $c \xrightarrow{\text{crack}} c_0$ (Figure 10.1). This is a model of a clock that ticks until it gets unhinged.

Note that the process of the ticking clock is nondeterministic. The clock keeps ticking, until at some point, for no reason, it gets stuck.

To implement nondeterministic processes like the clock process from Example 10.4, we have to find a way of dealing with the nondeterminism. Nondeterministic behaviour is behaviour determined by random factors, so a simple way of modeling nondeterminism is by modeling a process as a map from a randomly generated list of integers to a stream of actions. The following function creates random streams of integers, within a specified bound $[0, .., b]$, and starting from a particular seed s. It uses `randomRs` and `mkStdGen` from the library module `Random.hs`.

```
randomInts :: Int -> Int -> [Int]
randomInts bound seed =
  tail (randomRs (0,bound) (mkStdGen seed))
```

Exercise 10.5 Note that `randomInts 1 seed` generates a random stream of 0's and 1's. In the long run, the proportion of 0's and 1's in such a stream will be 1 to 1. How would you implement a generator for streams of 0's and 1's with, in the long run, a proportion of 0's and 1's of 2 to 1?

We define a process as a map from streams of integers to streams of action labels. To start a process, create an appropriate random integer stream and feed it to the process.

```
type Process = [Int] -> [String]

start :: Process -> Int -> Int -> [String]
start process bound seed = process (randomInts bound seed)
```

The clock process can now be modeled by means of the following corecursion:

10.2. PROCESSES AND LABELED TRANSITION SYSTEMS

```
clock :: Process
clock (0:xs) = "tick"  : clock xs
clock (1:xs) = "crack" : []
```

This gives:
```
COR> start clock 1 1
["tick","crack"]
COR> start clock 1 2
["crack"]
COR> start clock 1 25
["tick","tick","tick","tick","crack"]
```

The parameter for the integer bound in the start function (the second argument of start function) should be set to 1, to ensure that we start out from a list of 0's and 1's.

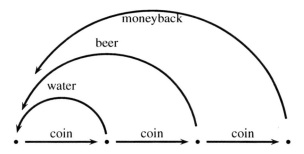

Figure 10.2: A simple vending machine.

Example 10.6 Consider a very simple vending machine that sells mineral water and beer. Water costs one euro, beer two euros. The machine has a coin slot and a button. It only accepts 1 euro coins. If a coin is inserted and the dispense button is pushed, it dispenses a can of mineral water. If instead of pushing the dispense button, another one euro coin is inserted and next the dispense button is pushed, it dispenses a can of beer. If, instead of pushing the button for beer, a third coin is inserted, the machine returns the inserted money (three 1 euro coins) and goes back to its initial state. This time we need four states, and the following transitions (Figure 10.2):
$q \xrightarrow{\text{coin}} q_1$, $q_1 \xrightarrow{\text{water}} q$, $q_1 \xrightarrow{\text{coin}} q_2$, $q_2 \xrightarrow{\text{beer}} q$, $q_2 \xrightarrow{\text{coin}} q_3$, $q_3 \xrightarrow{\text{moneyback}} q$.

Again, this is easily modeled. This time, the random stream is not needed for the transitions $q \xrightarrow{\text{coin}} q_1$ and $q_3 \xrightarrow{\text{moneyback}} q$, for insertion of a coin is the only possibility for the first of these, and return of all the inserted money for the second.

```
vending, vending1, vending2, vending3 :: Process
vending  (0:xs) = "coin"       : vending1 xs
vending  (1:xs) =                vending  xs
vending1 (0:xs) = "coin"       : vending2 xs
vending1 (1:xs) = "water"      : vending  xs
vending2 (0:xs) = "coin"       : vending3 xs
vending2 (1:xs) = "beer"       : vending  xs
vending3 (0:xs) = "moneyback"  : vending  xs
vending3 (1:xs) =                vending3 xs
```

This gives:

```
COR> take 9 (start vending 1 1)
["coin","water","coin","water","coin","water","coin","coin","beer"]
COR> take 8 (start vending 1 3)
["coin","water","coin","coin","coin","moneyback","coin","water"]
COR> take 8 (start vending 1 22)
["coin","water","coin","water","coin","coin","coin","moneyback"]
```

Example 10.7 A parking ticket dispenser works as follows. As long as pieces of 1 or 2 euro are inserted, the parking time is incremented by 20 minutes per euro. If the red button is pressed, all the inserted money is returned, and the machine returns to its initial state. If the green button is pressed, a parking ticket is printed indicating the amount of parking time, and the machine returns to its initial state. There are the following transitions: $q(i) \xrightarrow{\text{return}(i)} q(0)$, $q(i) \xrightarrow{\text{1euro}} q(i+1)$, $q(i) \xrightarrow{\text{2euro}} q(i+2)$, $q(0) \xrightarrow{\text{no time}} q(0)$, $q(i) \xrightarrow{\text{time } i*20 \text{ min}} q(0)$. Note that the number of states is infinite.

Here is an implementation of the parking ticket dispenser:

10.2. PROCESSES AND LABELED TRANSITION SYSTEMS

```
ptd :: Process
ptd = ptd0 0

ptd0 :: Int -> Process
ptd0 0 (0:xs) = ptd0 0 xs
ptd0 i (0:xs) = ("return " ++ show i ++ " euro") : ptd0 0 xs
ptd0 i (1:xs) = "1 euro" : ptd0 (i+1) xs
ptd0 i (2:xs) = "2 euro" : ptd0 (i+2) xs
ptd0 0 (3:xs) = ptd0 0 xs
ptd0 i (3:xs) = ("ticket " ++ show (i * 20) ++ " min") : ptd0 0 xs
```

This yields:

```
COR> take 6 (start ptd 3 457)
["1 euro","2 euro","2 euro","ticket 100 min","1 euro","ticket 20 min"]
```

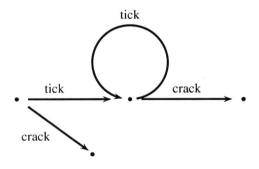

Figure 10.3: Another ticking clock.

Example 10.8 Intuitively, the clock process of Example 10.4 is the same as the clock process of the following example (Figure 10.3): $c_1 \xrightarrow{tick} c_2$, $c_1 \xrightarrow{crack} c_3$, $c_2 \xrightarrow{tick} c_2$ and $c_2 \xrightarrow{crack} c_4$. It is clear that this is also a clock that ticks until it gets stuck.

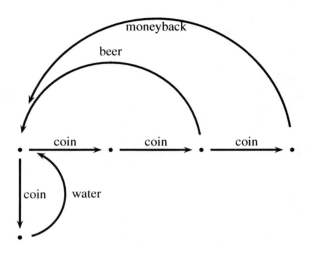

Figure 10.4: Another simple vending machine.

Exercise 10.9 Consider the vending machine given by the following transitions (Figure 10.4): $q \xrightarrow{\text{coin}} q_1$, $q \xrightarrow{\text{coin}} q_4$, $q_1 \xrightarrow{\text{coin}} q_2$, $q_2 \xrightarrow{\text{beer}} q$, $q_2 \xrightarrow{\text{coin}} q_3$, $q_3 \xrightarrow{\text{moneyback}} q$, $q_4 \xrightarrow{\text{water}} q$. Taking the machine from Example 10.6 and this machine to be black boxes, how can a user find out which of the two machines she is operating?

Exercise 10.10 Give a Haskell implementation of the vending machine from Exercise 10.9.

The key question about processes is the question of identity: How does one prove that two processes are the same? How does one prove that they are different? Proof methods for this will be developed in the course of this chapter.

Before we end this brief introduction to processes we give a simple example of process interaction.

Example 10.11 A user who continues to buy beer from the vending machine in Example 10.6 can be modeled by: $u \xrightarrow{\text{coin}} u_1$, $u_1 \xrightarrow{\text{coin}} u_2$, $u_2 \xrightarrow{\text{beer}} u$. It is clear how this should be implemented.

How about a beer drinker who interacts with a vending machine? It turns out that we can model this very elegantly as follows. We let the user

10.3. PROOF BY APPROXIMATION

start with buying his (her?) first beer. Next, we feed the stream of actions produced by the vending machine as input to the user, and the stream of actions produced by the user as input to the vending machine, and they keep each other busy. This interaction can be modeled corecursively, as follows:

```
actions   = user [0,0,1] responses
responses = vending actions

user acts ~(r:s:p:resps) = acts ++ user (proc [r,s,p]) resps
proc ["coin","coin","beer"] =   [0,0,1]
```

This gives:
```
COR> take 8 actions
[0,0,1,0,0,1,0,0]
COR> take 8 responses
["coin","coin","beer","coin","coin","beer","coin","coin"]
```

The user starts by inserting two coins and pressing the button, the machine responds with collecting the coins and issuing a can of beer, the user responds to this by inserting two more coins and pressing the button once more, and so on. One hairy detail: the pattern ~(r:s:p:resps) is a so-called **lazy pattern**. Lazy patterns always match, they are irrefutable. This allows the initial request to be submitted 'before' the list (r:s:p:resps) comes into existence by the response from the vending machine.

10.3 Proof by Approximation

One of the proof methods that work for corecursive programs is proof by approximation. For this, we have to extend each data type to a so-called **domain** with a partial ordering $\sqsubseteq$ (the approximation order). Every data type gets extended with an element $\bot$. This is the lowest element in the approximation order.

Let $(D, \sqsubseteq)$ be a set D with a partial order $\sqsubseteq$ on it, and let $A \subseteq D$.

An element $x \in A$ is the **greatest element** of A if $a \sqsubseteq x$ for all $a \in A$. $x \in A$ is the **least element** of A if $x \sqsubseteq a$ for all $a \in A$. Note that there are D with $A \subseteq D$ for which such least and greatest elements do not exist.

Exercise 10.12 Give an example of a set D, a partial order $\sqsubseteq$ on D, and a subset A of D such that A has no greatest and no least element.

An element $x \in D$ is an **upper bound** of A if $a \sqsubseteq x$ for all $a \in A$. Use A^u for the set of all upper bounds of A, i.e., $A^u := \{x \in D \mid \forall a \in A \; a \sqsubseteq x\}$.

E.g., consider $\mathbb{N}$ with the usual order $\leqslant$. Take $\{2,4,6\} \subseteq \mathbb{N}$. Then $\{2,4,6\}^u = \{n \in \mathbb{N} \mid 6 \leqslant n\}$. But the set of even numbers has no upper bounds in $\mathbb{N}$.

An element $x \in D$ is a **lower bound** of A if $x \sqsubseteq a$ for all $a \in A$. Use A^l for the set of all upper bounds of A, i.e., $A^l := \{x \in D \mid \forall a \in A \; x \sqsubseteq a\}$.

An element $x \in D$ is the **lub** or **least upper bound** of A if x is the least element of A^u. The lub of A is also called the supremum of A. Notation $\sqcup A$.

E.g., consider $\mathbb{R}$ with the usual order $\leqslant$. Take $A = \{\frac{n}{n+1} \mid n \in \mathbb{N}\} \subseteq \mathbb{R}$. Then $A = \{0, \frac{1}{2}, \frac{2}{3}, \frac{3}{4}, \ldots\}$, $A^u = \{r \in \mathbb{R} \mid r \geqslant 1\}$, and $\sqcup A = 1$.

An element $x \in D$ is the **glb** or **greatest lower bound** of A if x is the greatest element of A^l. The glb of A is also called the infimum of A. Notation $\sqcap A$. Caution: there may be $A \subseteq D$ for which $\sqcup A$ and $\sqcap A$ do not exist.

Exercise 10.13 Give an example of a set D, a partial order $\sqsubseteq$ on D, and a subset A of D such that A has no lub and no glb.

A subset A of D is called a **chain** in D if the ordering on A is total, i.e., if for all $x, y \in A$ either $x \sqsubseteq y$ or $y \sqsubseteq x$. E.g., the set $\mathcal{A} = \{\{k \in \mathbb{N} \mid k < n\} \mid n \in \mathbb{N}\}$ is a chain in $\wp(\mathbb{N})$ under $\subseteq$.

A set D with a partial order $\sqsubseteq$ is called a **domain** if D has a least element $\bot$ and $\sqcup A$ exists for every chain A in D. E.g., $\wp(\mathbb{N})$ with $\subseteq$ is a domain: the least element is given by $\emptyset$, $\sqcup \mathcal{A}$ is given by $\bigcup \mathcal{A}$.

Exercise 10.14 Show that $\mathbb{N}$, with the usual ordering $\leqslant$, is not a domain. Can you extend $\mathbb{N}$ to a domain?

We will now define approximation orders that make each data type into a domain. For basic data types A the approximation order is given by:

$$x \sqsubseteq y \; :\equiv \; x = \bot \vee x = y.$$

To see that the result is indeed a domain, observe that $\bot$ is the least element of the data type, and that the only chains in basic data types are $\{\bot\}$, $\{x\}$ and $\{\bot, x\}$. Obviously, we have $\sqcup\{\bot\} = \bot$, $\sqcup\{x\} = x$, $\sqcup\{\bot, x\} = x$. Thus, every chain has a lub.

10.3. PROOF BY APPROXIMATION

For pair data types $A \times B$ — represented in Haskell as (a,b) — the approximation order is given by:

$$\bot \sqsubseteq (x,y)$$
$$(x,y) \sqsubseteq (u,v) \; :\equiv \; x \sqsubseteq u \wedge y \sqsubseteq v.$$

Again, it is not difficult to see that every chain in a pair data type has a lub.

For functional data types $A \to B$ the approximation order is given by:

$$f \sqsubseteq g \; :\equiv \; \forall x \in A \; (fx \sqsubseteq gx).$$

Here it is assumed that A and B are domains.

If A is a basic data type, f, g can be viewed as partial functions, with $f \sqsubseteq g$ indicating that g is defined wherever f is defined, and f and g agreeing on every x for which they are both defined.

Exercise 10.15 Show that functional data types $A \to B$ under the given approximation order form a domain. Assume that A and B are domains.

For list data types $[A]$ the approximation order is given by:

$$\bot \sqsubseteq \text{xs}$$
$$[] \sqsubseteq \text{xs} \; :\equiv \; \text{xs} = []$$
$$x : \text{xs} \sqsubseteq y : \text{ys} \; :\equiv \; x \sqsubseteq y \wedge \text{xs} \sqsubseteq \text{ys}$$

Exercise 10.16 Show that list data types $[a]$ under the given approximation order form a domain. Assume that a is a domain.

Using $\bot$ one can create partial lists, where a partial list is a list of the form $x_0 : x_1 : \cdots : \bot$. The Haskell guise of $\bot$ is a program execution error or a program that diverges. The value $\bot$ shows up in the Haskell prelude in the following weird definition of the undefined object.

```
undefined           :: a
undefined | False = undefined
```

A call to `undefined` always gives rise to an error due to case exhaustion. An example of a partial list would be '1':'2':undefined. Partial lists can be used to approximate finite and infinite lists. It is easy to check that

$$\bot \sqsubseteq x_0 : \bot \sqsubseteq x_0 : x_1 : \bot \sqsubseteq \cdots \sqsubseteq x_0 : x_1 : \cdots : x_n : \bot \sqsubseteq x_0 : x_1 : \cdots : x_n : [].$$

This finite set of approximations is a chain, and we have:

$$\bigsqcup \{\bot, x_0 : \bot, x_0 : x_1 : \bot, \ldots, x_0 : x_1 : \cdots : x_n : \bot, x_0 : x_1 : \cdots : x_n : []\}$$

$$= x_0 : x_1 : \cdots : x_n : [].$$

Also, for infinite lists, we can form infinite sets of approximations:

$$\bot \sqsubseteq x_0 : \bot \sqsubseteq x_0 : x_1 : \bot \sqsubseteq x_0 : x_1 : x_2 : \bot \sqsubseteq \cdots$$

This infinite set of approximations is a chain. We will show (in Lemma 10.17) that for infinite lists $\text{xs} = x_0 : x_1 : x_2 : \cdots$ we have:

$$\bigsqcup \{\bot, x_0 : \bot, x_0 : x_1 : \bot, x_0 : x_1 : x_2 : \bot, \ldots\} = \text{xs}.$$

The function `approx` can be used to give partial approximations to any list:

```
approx :: Integer -> [a] -> [a]
approx (n+1) []     = []
approx (n+1) (x:xs) = x : approx n xs
```

Since $n + 1$ matches only positive integers, the call `approx n xs`, with n less than or equal to the length of the list, will cause an error (by case exhaustion) after generation of n elements of the list, i.e., it will generate $x_0 : x_1 : \cdots : x_{n-1} : \bot$. If n is greater than the length of xs, the call `approx n xs` will generate the whole list xs. We can now write

$$\bigsqcup \{\bot, x_0 : \bot, x_0 : x_1 : \bot, x_0 : x_1 : x_2 : \bot, \ldots\}$$

as

$$\bigsqcup_{n=0}^{\infty} \text{approx } n \text{ xs}.$$

10.3. PROOF BY APPROXIMATION

Lemma 10.17 (Approximation Lemma) For any list xs:

$$\bigsqcup_{n=0}^{\infty} \text{approx } n \text{ xs} = \text{xs}.$$

Proof. We have to show two things:

1. xs $\in \{\text{approx } n \text{ xs} \mid n \in \mathbb{N}\}^u$.

2. xs is the least element of $\{\text{approx } n \text{ xs} \mid n \in \mathbb{N}\}^u$.

To show (1), we have to establish that for every n and every xs, approx n xs $\sqsubseteq$ xs. We prove this with induction on n, for arbitrary lists.

Basis: approx 0 xs $= \bot \sqsubseteq$ xs for any list xs.

Induction step: Assume that for every xs, approx n xs $\sqsubseteq$ xs. We have to show that for any list xs, approx $(n+1)$ xs $\sqsubseteq$ xs.

If xs $= \bot$ or xs $= []$, then the result trivially holds by the definition of approx, so we assume xs $=$ x:xs'. Then:

$$\begin{array}{rl}
& \text{approx } (n+1) \text{ x : xs'} \\
= & \{\text{ def approx }\} \\
& x : \text{approx } n \text{ xs'} \\
\sqsubseteq & \{\text{ induction hypothesis }\} \\
& \text{x : xs'}.
\end{array}$$

To show (2), we have to show that for any list ys, if ys $\in \{\text{approx } n \text{ xs} \mid n \in \mathbb{N}\}^u$, then xs $\sqsubseteq$ ys, i.e., we have to show that xs is the least element of $\{\text{approx } n \text{ xs} \mid n \in \mathbb{N}\}^u$.

Assume ys $\in \{\text{approx } n \text{ xs} \mid n \in \mathbb{N}\}^u$. This means that for all $n \in \mathbb{N}$, approx n xs $\sqsubseteq$ ys. We have to show that xs $\sqsubseteq$ ys.

Suppose xs $\not\sqsubseteq$ ys. Then there has to be a k with (approx $(k+1)$ xs) !! $k \not\sqsubseteq$ ys !! k. But then approx $(k+1)$ xs $\not\sqsubseteq$ ys , and contradiction with ys $\in \{\text{approx } n \text{ xs} \mid n \in \mathbb{N}\}^u$. ∎

Theorem 10.18 (Approximation Theorem)

$$\text{xs} = \text{ys} \Leftrightarrow \forall n \text{ (approx } n \text{ xs} = \text{approx } n \text{ ys)}.$$

Proof. $\Rightarrow$: Immediate from the fact that xs !! n = ys !! n, for every $n \in \mathbb{N}$.
$\Leftarrow$:

$$\begin{aligned}
& \forall n (\text{approx } n \text{ xs} = \text{approx } n \text{ ys}) \\
\Longrightarrow \quad & \{ \text{ property of lub } \} \\
& \bigsqcup_{n=0}^{\infty} (\text{approx } n \text{ xs}) = \bigsqcup_{n=0}^{\infty} (\text{approx } n \text{ ys}) \\
\Longleftrightarrow \quad & \{ \text{ Lemma 10.17 } \} \\
& \text{xs} = \text{ys}.
\end{aligned}$$

∎

The approximation theorem is one of our tools for proving properties of streams.

Example 10.19 Suppose we want to prove the following:

$$\text{map } f \text{ (iterate } f \text{ } x) = \text{iterate } f \text{ } (f \text{ } x).$$

This equality cannot be proved by list induction, as the lists on both sides of the equality sign are infinite. We can prove an equivalent property by induction on n, as follows.

$$\forall n (\text{approx } n \text{ (map } f \text{ (iterate } f \text{ } x)) = \text{approx } n \text{ (iterate } f \text{ } (f \text{ } x))).$$

Proof by induction on n.

Basis. For $n = 0$, the property holds by the fact that approx 0 xs = $\bot$ for all lists xs.

Induction step. Assume (for arbitrary x):

$$\text{approx } n \text{ (map } f \text{ (iterate } f \text{ } x)) = \text{approx } n \text{ (iterate } f \text{ } (f \text{ } x)).$$

10.3. PROOF BY APPROXIMATION

We have:

$$\begin{aligned}
&\quad \text{approx } (n+1) \text{ (map } f \text{ (iterate } f \text{ } x)) \\
&= \{ \text{ definition of iterate } \} \\
&\quad \text{approx } (n+1) \text{ (map } f \text{ } (x : \text{iterate } f \text{ } (f \text{ } x))) \\
&= \{ \text{ definition of map } \} \\
&\quad \text{approx } (n+1) \text{ } (f \text{ } x : \text{map } f \text{ (iterate } f \text{ } (f \text{ } x))) \\
&= \{ \text{ definition of approx } \} \\
&\quad f \text{ } x : \text{approx } n \text{ (map } f \text{ (iterate } f \text{ } (f \text{ } x))) \\
&= \{ \text{ induction hypothesis } \} \\
&\quad f \text{ } x : \text{approx } n \text{ (iterate } f \text{ } (f \text{ } (f \text{ } x))) \\
&= \{ \text{ definition of approx } \} \\
&\quad \text{approx } (n+1) \text{ } (f \text{ } x : \text{iterate } f \text{ } (f \text{ } (f \text{ } x))) \\
&= \{ \text{ definition of iterate } \} \\
&\quad \text{approx } (n+1) \text{ (iterate } f \text{ } (f \text{ } x)).
\end{aligned}$$

Exercise 10.20 In Exercise 7.52 you showed that for every finite list xs :: [a], every function f :: a -> b, every total predicate p :: b -> Bool the following holds:

$$\text{filter } p \text{ (map } f \text{ xs)} = \text{map } f \text{ (filter } (p \cdot f) \text{ xs)}.$$

Use proof by approximation to show that this also holds for infinite lists.

Example 10.21 To reason about sieve, we would like to show that mark n k has the same effect on a list of integers as mapping with the following function:

$$\lambda m. \text{if rem } m \text{ } n \neq 0 \text{ then } m \text{ else } 0.$$

This will only hold when mark n k is applied to sequences that derive from a sequence [q..] by replacement of certain numbers by zeros. Let us use [q..]• for the general form of such a sequence. Suppose xs equals [q..]• for some q, i.e., suppose that xs is the result of replacing some of the items in the infinite list $[q, q+1, q+2, \ldots]$ by zeros. We prove by approximation that if $q = an + k$, with $1 \leqslant k \leqslant n$, then:

$$\text{mark } n \text{ } k \text{ xs} = \text{map } (\lambda m. \text{if rem } m \text{ } n \neq 0 \text{ then } m \text{ else } 0) \text{ xs}.$$

Basis. For $n' = 0$, the property holds by the fact that approx 0 xs = $\bot$ for all lists xs.

Induction step. Assume (for arbitrary xs of the form [q..]$^\bullet$ with $q = an + k$ and $1 \leqslant k \leqslant n$):

approx n' (mark n k xs) = approx n' (map (λm.if rem m $n \neq 0$ then m else 0) xs)

Two cases: (i) $k = n$, and (ii) $k < n$.
Case (i) is the case where $n|q$, i.e., rem x $n = 0$. We get:

$\quad$ approx $(n' + 1)$ (mark n k x:xs)
$=$ { definition of mark }
$\quad$ approx $(n' + 1)$ $(0 :$ mark n 1 xs)
$=$ { definition of approx }
$\quad 0 :$ approx n' mark n 1 xs
$=$ { induction hypothesis }
$\quad 0 :$ approx n' (λm.if rem m $n \neq 0$ then m else 0) xs
$=$ { definition of approx }
$\quad$ approx $(n' + 1)$ $(0 : (\lambda m$.if rem m $n \neq 0$ then m else 0) xs)
$=$ { definition of map, plus the fact that rem x $n = 0$ }
$\quad$ approx $(n' + 1)$ map (λm.if rem m $n \neq 0$ then m else 0) x:xs.

Case (ii) is the case where $n \not| \ q$. The reasoning for this case uses the other case of the definition of `mark`.

A proof by approximation of the fact that `sieve` and `sieve'` define the same function on streams can now use the result from Example 10.21 as a lemma.

10.4 Proof by Coinduction

To compare two streams xs and ys, intuitively it is enough to compare their observational behaviour. The key observation on a stream is to inspect its head. If two streams have the same head, and their tails have the same observational behaviour, then they are equal. Similarly, to compare two infinite binary trees it is enough to observe that they have the same information at their root nodes, that their left daughters have the same observational behaviour, and that their right daughters have the same observational behaviour. And so on, for other infinite data structures.

To make this talk about observational behaviour rigorous, it is fruitful to consider the infinite data structures we are interested in as labeled transition systems. The observations are the action labels that mark a transition.

10.4. PROOF BY COINDUCTION

A **bisimulation** on a labeled transition system (Q, A, T) is a binary relation R on Q with the following properties. If qRp and $a \in A$ then:

1. If $q \xrightarrow{a} q'$ then there is a $p' \in Q$ with $p \xrightarrow{a} p'$ and $q'Rp'$.
2. If $p \xrightarrow{a} p'$ then there is an $q' \in Q$ with $p \xrightarrow{a} p'$ and $q'Rp'$.

Example 10.22 Take decimal representations of rational numbers with over-line notation to indicate repeating digits (Section 8.5). The following representations all denote the same number $\frac{1}{7}$:

$$0.\overline{142857}, 0.1\overline{428571}, 0.14\overline{285714}, 0.142857142\overline{857142}.$$

Why? Because if you check them against each other decimal by decimal, you will never hit at a difference. The relation 'having the same infinite expansion' is a bisimulation on decimal representations, and the representations for $\frac{1}{7}$ are all bisimilar.

Exercise 10.23 Show that the relation of equality on a set A is a bisimulation on A.

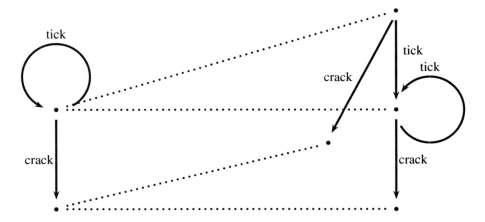

Figure 10.5: Bisimulation between ticking clocks.

Example 10.24 Collect the transition for the two clock processes of Examples 10.4 and 10.8 together in one transition system. Then it is easy to check that the relation R given by $cRc_1, cRc_2, c_0Rc_3, c_0Rc_4$ is a bisimulation on this transition system (Figure 10.5). The bisimulation connects the states c and c_1. The two clock processes are indeed indistinguishable.

Exercise 10.25 Show that there is no bisimulation that connects the starting states of the two vending machines of Examples 10.6 and 10.9.

Exercise 10.26 Show that the union of two bisimulations is again a bisimulation.

Exercise 10.27 Show that there is always a greatest bisimulation on a set A under the inclusion ordering. (Hint: show that the union of all bisimulations on A is itself a bisimulation.)

Exercise 10.28 The bisimulation relation given in Example 10.24 is not the greatest bisimulation on the transition system. Find the greatest bisimulation.

Use $\sim$ for the greatest bisimulation on a given set A (Exercise 10.27). Call two elements of A bisimilar when they are related by a bisimulation on A. Being bisimilar then coincides with being related by the greatest bisimulation:

$$a \sim b \Leftrightarrow \exists R (R \text{ is a bisimulation, and } aRb).$$

To show that two infinite objects x and y are equal, we show that they exhibit the same behaviour, i.e. we show that $x \sim y$. Such a proof is called a **proof by coinduction**.

In the particular case of comparing streams we can make the notion of a bisimulation more specific. There are only two kinds of observations on a stream: observing its head and observing its tail. Comparison of the heads should exhibit objects that are the same. Comparison of the tails should exhibit bisimilar streams.

When talking about streams $f = [f_0, f_1, f_2, f_3, \ldots]$ it is convenient to refer to the tail of the stream as $\overline{f}$. Thus, a stream $f = [f_0, f_1, f_2, f_3, \ldots]$ always has the form $f = f_0 : \overline{f}$, where f_0 is the head and $\overline{f}$ is the tail.

Viewing a stream f as a transition system, we get transitions $f \xrightarrow{f_0} \overline{f}$, and so on (Figure 10.6). A bisimulation between streams f and g that connects f and g is given in Figure 10.7. It is immediate from the picture that the notion of a bisimulation between streams boils down to the following.

10.4. PROOF BY COINDUCTION

$$f \bullet \xrightarrow{f_0} \bullet \overline{f}$$

Figure 10.6: A stream viewed as a transition system.

Figure 10.7: Bisimulation between streams.

A **stream bisimulation** on a set A is a relation R on $[A]$ with the following property. If $f, g \in [A]$ and fRg then both $f_0 = g_0$ and $\overline{f}R\overline{g}$.

The general pattern of a proof by coinduction using stream bisimulations of $f \sim g$, where $f, g \in [A]$, is simply this. Define a relation R on $[A]$. Next, show that R is a stream bisimulation, with fRg.

Example 10.29 As an example, we will show:

$$\text{map } f \text{ (iterate } f \ x) \sim \text{iterate } f \ (f \ x).$$

Let R be the following relation on $[A]$.

$$\{(\text{map } f \text{ (iterate } f \ x), \text{iterate } f \ (f \ x)) \mid f :: a \to a, x :: a\}.$$

Suppose:

$$(\text{map } f \text{ (iterate } f \ x)) \ R \ (\text{iterate } f \ (f \ x)).$$

> *Given:* A labeled transition system (Q, A, T) ...
> *To be proved:* $q \sim p$
> *Proof:*
> >   Let R be given by ... and suppose qRp.
> >   *To be proved:* R is a bisimulation.
> > >   Suppose $q \xrightarrow{a} q'$
> > >   *To be proved:* There is a $p' \in Q$ with
> > > $p \xrightarrow{a} p'$ and $q'Rp'$.
> > >   *Proof:* ...
> > >   Suppose $p \xrightarrow{a} p'$
> > >   *To be proved:* There is a $q' \in Q$ with
> > > $q \xrightarrow{a} q'$ and $q'Rp'$.
> > >   *Proof:* ...
> >   Thus R is a bisimulation with qRp.
> Thus $q \sim p$.

> *Given:* ...
> *To be proved:* $f \sim g$
> *Proof:*
> >   Let R be given by ... and suppose fRg.
> >   *To be proved:* R is a stream bisimulation.
> > >   *To be proved:* $f_0 = g_0$.
> > >   *Proof:* ...
> > >   *To be proved:* $\overline{f}R\overline{g}$.
> > >   *Proof:* ...
> >   Thus R is a stream bisimulation with fRg.
> Thus $f \sim g$.

Figure 10.8: Proof Recipes for Proofs by Coinduction.

We show that R is a bisimulation. We have:

$$\text{map } f \text{ (iterate } f \text{ } x) \stackrel{\text{iterate}}{=} \text{map } f \text{ } x : (\text{iterate } f \text{ } (f \text{ } x))$$
$$\stackrel{\text{map}}{=} (f \text{ } x) : \text{map } f \text{ (iterate } f \text{ } (f \text{ } x))$$

$$\text{iterate } f \text{ } (f \text{ } x) \stackrel{\text{iterate}}{=} (f \text{ } x) : \text{iterate } f \text{ } (f \text{ } (f \text{ } x)).$$

This shows:

$$\text{map } f \text{ (iterate } f \text{ } x) \xrightarrow{\text{head}} (f \text{ } x) \quad (1)$$
$$\text{map } f \text{ (iterate } f \text{ } x) \xrightarrow{\text{tail}} \text{map } f \text{ } x : (\text{iterate } f \text{ } (f \text{ } x)) \quad (2)$$
$$\text{iterate } f \text{ } (f \text{ } x) \xrightarrow{\text{head}} (f \text{ } x) \quad (3)$$
$$\text{iterate } f \text{ } (f \text{ } x) \xrightarrow{\text{tail}} \text{iterate } f \text{ } (f \text{ } (f \text{ } x)) \quad (4)$$

Clearly, map f (iterate f x) and iterate f $(f$ $x)$ have the same heads, and also, their tails are R-related. This shows that R is a bisimimulation that connects map f (iterate f x) and iterate f $(f$ $x)$. Hence

$$(\text{map } f \text{ (iterate } f \text{ } x)) \sim (\text{iterate } f \text{ } (f \text{ } x)).$$

Exercise 10.30 Show by means of a coinduction argument that for every **infinite** list xs :: [a], every f :: a -> b, and every **total** predicate p :: b -> Bool the following holds:

$$\text{filter } p \text{ (map } f \text{ xs)} = \text{map } f \text{ (filter } (p.f) \text{ xs)}.$$

10.5 Power Series and Generating Functions

In Chapter 9 we have seen how polynomials can get represented as finite lists of their coefficients, and how operations on polynomials can be given directly on these list representations. We will now generalize this to infinite sequences. A possible motivation for this (one of many) is the wish to define a division operation on polynomials.

Suppose we want to divide

$$f(z) = f_0 + f_1 z + f_2 z^2 + \cdots + f_k z^k$$

by

$$g(z) = g_0 + g_1 z + g_2 z^2 + \cdots + g_m z^m.$$

```
module PowerSeries

where

import Polynomials

instance Fractional a => Fractional [a] where
  fromRational c  = [fromRational c]
  fs      / []    = error "division by 0 attempted"
  []      / gs    = []
  (0:fs)  / (0:gs) = fs / gs
  (_:fs)  / (0:gs) = error "division by 0 attempted"
  (f:fs)  / (g:gs) = let q = f/g in
       q : (fs - q.*gs) / (g:gs)

int :: Fractional a => [a] -> [a]
int fs = 0 : int1 fs 1 where
  int1 []     _ = []
  int1 (g:gs) n = g/n : (int1 gs (n+1))

expz = 1 + (int expz)
```

Figure 10.9: A Module for Power Series.

10.5. POWER SERIES AND GENERATING FUNCTIONS

The outcome $h(z) = h_0 + h_1 z + h_2 z^2 + \cdots$ has to satisfy:

$$f(z) = h(z) \cdot g(z).$$

This gives:

$$\begin{aligned}
f_0 + z\overline{f}(z) &= (h_0 + z\overline{h}(z))g(z) \\
&= h_0 g(z) + z\overline{h}(z)g(z) \\
&= h_0(g_0 + z\overline{g}(z)) + z\overline{h}(z)g(z) \\
&= h_0 g_0 + z(h_0 \overline{g}(z) + \overline{h}(z)g(z)).
\end{aligned}$$

Thus, $f_0 = h_0 g_0$, hence $h_0 = \frac{f_0}{g_0}$, and $\overline{f}(z) = h_0 \overline{g}(z) + \overline{h}(z)g(z)$, so from this $\overline{h}(z) = \frac{\overline{f}(z) - h_0 \overline{g}(z)}{g(z)}$. We see from this that computing fractions can be done by a process of long division:

$$\frac{f(z)}{g(z)} = \frac{f_0}{g_0} + z\frac{\overline{f}(z) - (f_0/g_0)\overline{g}(z)}{g(z)}. \qquad \text{(div)}$$

An example case is $\frac{1}{1-z}$. Long division gives:

$$\begin{aligned}
\frac{1}{1-z} &= \frac{1-z}{1-z} + z\frac{1}{1-z} = 1 + z\frac{1}{1-z} \\
&= 1 + z(1 + z\frac{1}{1-z}) = 1 + z(1 + z(1 + z\frac{1}{1-z})) = \cdots
\end{aligned}$$

The representation of $1 - z$ is $[1, -1]$, so in terms of computation on sequences we get:

$$\begin{aligned}
\frac{[1]}{[1,-1]} &= \frac{1}{1} : \frac{[1]}{[1,-1]} \\
&= \frac{1}{1} : \frac{1}{1} : \frac{[1]}{[1,-1]} \\
&= \frac{1}{1} : \frac{1}{1} : \frac{1}{1} : \frac{[1]}{[1,-1]} \\
&= \frac{1}{1} : \frac{1}{1} : \frac{1}{1} : \frac{1}{1} : \frac{[1]}{[1,-1]} \\
&= \cdots
\end{aligned}$$

This shows that $\frac{1}{1-z}$ does not have finite degree. It is not a polynomial.

To get at a class of functions closed under division we define **power series** to be functions of the form

$$f(z) = f_0 + f_1 z + f_2 z^2 + \cdots = \sum_{k=0}^{\infty} f_k z^k.$$

A power series is represented by an infinite stream $[f_0, f_1, f_2, \ldots]$. Power series in a variable z can in fact be viewed as approximations to complex numbers (Section 8.10), but we will not pursue that connection here.

The implementation of division follows the formula (div). Since trailing zeros are suppressed, we have to take the fact into account that [] represents an infinite stream of zeros.

```
fs      / []      = error "division by 0 attempted"
[]      / gs      = []
(0:fs)  / (0:gs)  = fs / gs
(_:fs)  / (0:gs)  = error "division by 0 attempted"
(f:fs)  / (g:gs)  = let q = f/g in q : (fs - q.*gs)/(g:gs)
```

Here are some Haskell computations involving division:

```
COR> take 10 (1/(1-z))
[1 % 1,1 % 1,1 % 1,1 % 1,1 % 1,1 % 1,1 % 1,1 % 1,1 % 1,1 % 1]
COR> take 10 (1/(1+z))
[1 % 1,-1 % 1,1 % 1,-1 % 1,1 % 1,-1 % 1,1 % 1,-1 % 1,1 % 1,-1 % 1]
```

Example 10.31 To get a feel for 'long division with sequences', we compute $\frac{3}{3-z}$, as a calculation on sequences. The sequence representation of $3-z$ is $[3, -1]$, so we get:

$$\begin{aligned}
\frac{[3]}{[3,-1]} &= \frac{3}{3} : \frac{[1]}{[3,-1]} \\
&= \frac{3}{3} : \frac{1}{3} : \frac{[1/3]}{[3,-1]} \\
&= \frac{3}{3} : \frac{1}{3} : \frac{1}{9} : \frac{[1/9]}{[3,-1]} \\
&= \frac{3}{3} : \frac{1}{3} : \frac{1}{9} : \frac{1}{27} : \frac{[1/27]}{[3,-1]} \\
&= \cdots
\end{aligned}$$

This is indeed what Haskell calculates for us:

10.5. POWER SERIES AND GENERATING FUNCTIONS

```
COR> take 9 (3 /(3-z))
[1 % 1,1 % 3,1 % 9,1 % 27,1 % 81,1 % 243,1 % 729,1 % 2187,1 % 6561]
```

Integration involves division, for $\int_0^z f(t)dt$ is given by

$$0 + f_0 z + \frac{1}{2}f_1 z^2 + \frac{1}{3}f_2 z^3 + \cdots$$

To give an example, if $g(z) = \frac{1}{1-z}$, then $\int_0^z g(t)dt$ equals:

$$0 + z + \frac{1}{2}z^2 + \frac{1}{3}z^3 + \cdots$$

Integration is implemented by:

```
int :: Fractional a => [a] -> [a]
int [] = []
int fs = 0 : int1 fs 1 where
   int1 []       _ = []
   int1 (g:gs) n = g/n : (int1 gs (n+1))
```

We get:

```
COR> take 10 (int (1/(1-z)))
[0 % 1,1 % 1,1 % 2,1 % 3,1 % 4,1 % 5,1 % 6,1 % 7,1 % 8,1 % 9]
```

To extend composition to power series, we have to be careful, for the equation $f(g(z)) = f_0 + g(z) \cdot \overline{f}(g(z))$ has a snag: the first coefficient depends on all of f, which is a problem if f is a power series. Here is an example:

```
COR> comp1 ones [0,2] where ones = 1 : ones
```

ERROR - Garbage collection fails to reclaim sufficient space

To solve this, we must develop the equation for composition a bit further:

$$\begin{aligned} f(g(z)) &= f_0 + g(z) \cdot \overline{f}(g(z)) \\ &= f_0 + (g_0 + z\overline{g}(z)) \cdot \overline{f}(g(z)) \\ &= (f_0 + g_0 \cdot \overline{f}(g(z))) + z\overline{g}(z) \cdot \overline{f}(g(z)) \end{aligned}$$

In the special case where $g_0 = 0$ we can simplify this further:

$$f(z\overline{g}(z)) = f_0 + z\overline{g}(z) \cdot \overline{f}(z\overline{g}(z)).$$

This leads immediately to the following implementation (part of the module for Polynomials):

```
comp :: Num a => [a] -> [a] -> [a]
comp _       []      = error ".."
comp []      _       = []
comp (f:fs) (0:gs) = f : gs * (comp fs (0:gs))
comp (f:fs) (g:gs) = ([f] + [g] * (comp fs (g:gs)))
                     + (0 : gs * (comp fs (g:gs)))
```

This gives:

```
COR> take 15 (comp ones [0,2]) where ones = 1 : ones
[1,2,4,8,16,32,64,128,256,512,1024,2048,4096,8192,16384]
```

Figure 10.9 gives a module for power series. We will now show how this is can be used as a tool for combinatorial calculations. We are interested in finite lists $[f_0, \ldots, f_n]$ or infinite lists $[f_0, f_1, f_2, \ldots]$ that can be viewed as solutions to combinatorial problems.

We associate sequences with combinatorial problems by saying that $[f_0, f_1, f_2, \ldots]$ **solves** a combinatorial problem if f_r gives the number of solutions for that problem (cf. the use of polynomials for solving combinatorial problems in Section 9.4, which is now generalized to power series).

We call a power series $f(z)$ a **generating function** for a combinatorial problem if the list of coefficients of $f(z)$ solves that problem. If $f(z)$ has power series expansion $f_0 + f_1 z + f_2 z^2 + f_3 z^3 + \cdots$, then $f(z)$ is a generating function for $[f_0, f_1, f_2, f_3, \ldots]$.

Example 10.32 Find a generating function for $[1, 1, 1, 1, \ldots]$. Solution: as we saw above $\frac{1}{1-z}$ has power series expansion $1 + z + z^2 + z^3 + z^4 + \cdots$. The list of coefficients is $[1, 1, 1, 1, \ldots]$.

The generating function $\frac{1}{1-z}$ is an important building block for constructing other generating functions, so we should give it a name. In fact, we already did: ones names the list [1,1,1,1,..]. As long as we don't use division, this is OK.

Example 10.33 Find a generating function for the list of natural numbers. Solution: recall the corecursive definition of the natural numbers in terms of zipWith (+). This can now be expressed succinctly in terms of addition of power series:

```
COR> take 20 nats where nats = 0 : (nats + ones)
[0,1,2,3,4,5,6,7,8,9,10,11,12,13,14,15,16,17,18,19]
```

10.5. POWER SERIES AND GENERATING FUNCTIONS

This gives the following specification for the generating function:

$$g(z) = z(g(z) + \frac{1}{1-z})$$
$$g(z) - zg(z) = \frac{z}{1-z}$$
$$g(z) = \frac{z}{(1-z)^2}$$

Here is the check:

```
COR> take 20 (z * ones^2)
[0,1,2,3,4,5,6,7,8,9,10,11,12,13,14,15,16,17,18,19]
```

Another way of looking at this: differentiation of $1 + z + z^2 + z^3 + z^4 + \cdots$ gives $1 + 2z + 3z^2 + 4z^3 + \cdots$. This means that $z\left(\frac{1}{1-z}\right)' = \frac{z}{(1-z)^2}$ is a generating function for the naturals.

Example 10.34 Find the generating function for $[0, 0, 0, 1, 2, 3, 4, \ldots]$. Solution: multiplication by z has the effect of shifting the sequence of coefficients one place to the right and inserting a 0 at the front. Thus the generating function can be got from that of the previous example through multiplying by z^2. The generating function is: $\frac{z^3}{(1-z)^2}$. Here is the check:

```
COR> take 20 (z^3 * ones^2)
[0,0,0,1,2,3,4,5,6,7,8,9,10,11,12,13,14,15,16,17]
```

Example 10.35 Find the generating function for the sequence of powers of two (the sequence $\lambda n.2^n$). Solution: start out from a corecursive program for the powers of two:

```
COR> take 15 powers2 where powers2 = 1 : 2 * powers2
[1,2,4,8,16,32,64,128,256,512,1024,2048,4096,8192,16384]
```

This immediately leads to the specification $g(z) = 1 + 2zg(z)$ (the factor z shifts one place to the right, the summand 1 puts 1 in first position). It follows that $g(z) = \frac{1}{1-2z}$. Here is the confirmation:

```
COR> take 10 (1/(1-2*z))
[1 % 1,2 % 1,4 % 1,8 % 1,16 % 1,32 % 1,64 % 1,128 % 1,256 % 1,512 % 1]
```

Example 10.36 If $g(z)$ is the generating function for $\lambda n.g_n$, then $\lambda n.g_{n+1}$ is generated by $\frac{g(z)}{z}$, for multiplication by $\frac{1}{z}$ shifts the coefficients of the power series one place to the left. Thus, the generating function for the sequence $\lambda n.n + 1$ is $\frac{1}{(1-z)^2}$. This sequence can also be got by adding the sequence of naturals and the sequence of ones, i.e., it is also generated by $\frac{z}{(1-z)^2} + \frac{1}{1-z}$. This identification makes algebraic sense:

$$\frac{z}{(1-z)^2} + \frac{1}{1-z} = \frac{z}{(1-z)^2} + \frac{1-z}{(1-z)^2} = \frac{1}{(1-z)^2}.$$

Exercise 10.37 Find the generating functions for

$$[0,0,0,1,1,1,\ldots],$$

for

$$[1,1,1,0,0,0,\ldots],$$

and for

$$[1, \frac{1}{2}, \frac{1}{4}, \frac{1}{8}, \ldots].$$

Check your answers by computer.

Example 10.38 Find the generating function for $[0,1,0,1,0,1,\ldots]$. Solution: the sequence is generated by the following corecursive program.

```
COR> take 20 things where things = 0 : 1 : things
[0,1,0,1,0,1,0,1,0,1,0,1,0,1,0,1,0,1,0,1]
```

This leads to the specification $g(z) = 0 + z(1 + zg(z))$, which reduces to $g(z) = z + z^2 g(z)$. From this, $g(z) = \frac{z}{1-z^2}$. In a similar way we can derive the generating function for $[1,0,1,0,1,0,\ldots]$, which turns out to be $g(z) = \frac{1}{1-z^2}$. Alternatively, observe that $[1,0,1,0,1,0,\ldots]$ is the result of shifting $[0,1,0,1,0,1,\ldots]$ one place to the left, so division by z does the trick.

Theorem 10.39 summarizes a number of recipes for finding generating functions.

Theorem 10.39 Suppose $f(z)$ is a generating function for $\lambda n.f_n$ and $g(z)$ is the generating function for $\lambda n.g_n$. Then:

1. $\frac{c}{1-z}$ is the generating function for $\lambda n.c$.

2. $\frac{z}{(1-z)^2}$ is the generating function for $\lambda n.n$.

3. $\frac{f(z)}{z}$ is the generating function for $\lambda n.f_{n+1}$.

4. $cf(z) + dg(z)$ is the generating function for $\lambda n.cf_n + dg_n$.

10.5. POWER SERIES AND GENERATING FUNCTIONS 381

5. $(1-z)f(z)$ is the generating function for the difference sequence $\lambda n.f_n - f_{n-1}$.

6. $\frac{1-z}{z}f(z)$ is the generating function for the difference sequence $\lambda n.f_{n+1} - f_n$.

7. $\frac{1}{1-z}f(z)$ is the generating function for $\lambda n.f_0 + f_1 + \cdots + f_n$.

8. $f(z)g(z)$ is the generating function for $\lambda n.f_0 g_n + f_1 g_{n-1} + \cdots + f_{n-1} g_1 + f_n g_0$ (the convolution of f and g).

9. $zf'(z)$ is the generating function for $\lambda n.n f_n$.

10. $\frac{1}{z}\int_0^z f(t)dt$ is the generating function for $\lambda n.\frac{f_n}{n+1}$.

Exercise 10.40 Prove the statements of Theorem 10.39 that were not proved in the examples.

Example 10.41 Find the generating function for the sequence $\lambda n.2n+1$.
Solution: the odd natural numbers are generated by the following corecursive program:

```
COR> take 20 (ones + 2 * nats)
[1,3,5,7,9,11,13,15,17,19,21,23,25,27,29,31,33,35,37,39]
```

This immediately gives $g(z) = \frac{1}{1-z} + \frac{2}{(1-z)^2}$.

Example 10.42 Find the generating function for the sequence $\lambda n.(n+1)^2$.
Solution: the generating function is the sum of the generating functions for $\lambda n.n^2$, $\lambda n.2n$ and $\lambda n.1$. One way to proceed would be to find these generating functions and take their sum. But there is an easier method. Recall that $(n+1)^2$ is the sum of the first n odd numbers, and use the recipe from Theorem 10.39 for forming the sequence $\lambda n.a_0 + \cdots + a_n$, by means of dividing the generating function from Example 10.41 by $1-z$. This immediately gives $\frac{1}{(1-z)^2} + \frac{2z}{(1-z)^3}$.

Alternatively, Theorem 10.39 tells us that $\lambda n.(n+1)^2$ can also be defined from the naturals with the help of differentiation:

```
COR> take 20 (deriv nats)
[1,4,9,16,25,36,49,64,81,100,121,144,169,196,225,256,289,324,361,400]
```

It follows that the generating function $g(z)$, for $g(z)$ should satisfy:

$$g(z) = \left(\frac{z}{(1-z)^2}\right)'.$$

Working this out we find that $g(z) = \frac{1}{(1-z)^2} + \frac{2z}{(1-z)^3}$.
Finally, here is computational confirmation:

```
COR> take 20 (ones^2 + 2 * z * ones^3)
[1,4,9,16,25,36,49,64,81,100,121,144,169,196,225,256,289,324,361,400]
```

Example 10.43 Find the generating function for $\lambda n.n^2$. Solution: shift the solution for the previous example to the right by multiplication by z. This gives:

```
COR> take 20 (z * ones^2 + 2 * z^2 * ones^3)
[0,1,4,9,16,25,36,49,64,81,100,121,144,169,196,225,256,289,324,361]
```

So the generating function is $g(z) = \frac{z}{(1-z)^2} + \frac{2z^2}{(1-z)^3}$. Sure enough, we can also get the desired sequence from the naturals by shift and differentiation:

```
COR> take 20 (z * deriv nats)
[0,1,4,9,16,25,36,49,64,81,100,121,144,169,196,225,256,289,324,361]
```

Example 10.44 Find the generating function for the sequence of Fibonacci numbers. Solution: consider again the corecursive definition of the Fibonacci numbers. We can express this succinctly in terms of addition of power series, as follows:

```
COR> take 20 fibs where fibs = 0 : 1 : (fibs + (tail fibs))
[0,1,1,2,3,5,8,13,21,34,55,89,144,233,377,610,987,1597,2584,4181]
```

This is easily translated into an instruction for a generating function:

$$g(z) = z^2 \left(g(z) + \frac{g(z)}{z} \right) + z.$$

Explanation: multiplying by z^2 inserts two 0's in front of the fibs sequence, and adding z changes the second of these to 1 (the meaning is: the second coefficient is obtained by add 1 to the previous coefficient). $\frac{g(z)}{z}$ gives the tail of the fibs sequence. From this we get:

$$g(z) - zg(z) - z^2 g(z) = z$$
$$g(z) = \frac{z}{1 - z - z^2}$$

Lo and behold:

```
COR> take 10 (z/(1-z-z^2))
[0 % 1,1 % 1,1 % 1,2 % 1,3 % 1,5 % 1,8 % 1,13 % 1,21 % 1,34 % 1]
```

10.5. POWER SERIES AND GENERATING FUNCTIONS

Exercise 10.45 The Lucas numbers are given by the following recursion:

$$L_0 = 2, \quad L_1 = 1, \quad L_{n+2} = L_n + L_{n+1}.$$

This gives:

$$[2, 1, 3, 4, 7, 11, 18, 29, 47, 76, 123, 199, 322, 521, 843, 1364, 2207, 3571, \ldots]$$

The recipe is that for the Fibonacci numbers, but the initial element of the list is 2 instead of 0. Find a generating function for the Lucas numbers, starting out from an appropriate corecursive definition.

Example 10.46 Recall the definition of the Catalan numbers from Exercise 7.20:

$$C_0 = 1, \quad C_{n+1} = C_0 C_n + C_1 C_{n-1} + \cdots + C_{n-1} C_1 + C_n C_0.$$

Clearly, C_{n+1} is obtained from $[C_0, \ldots, C_n]$ by convolution of that list with itself. This leads to the following corecursive definition of the Catalan numbers:

```
COR> take 15 cats where cats = 1 : cats * cats
[1,1,2,5,14,42,132,429,1430,4862,16796,58786,208012,742900,2674440]
```

From this, we get at the following specification for the generating function:

$$g(z) = 1 + z \cdot (g(z))^2$$
$$z(g(z))^2 - g(z) + 1 = 0$$

Considering $g(z)$ as the unknown, we can solve this as a quadratic equation $ax^2 + bx + c = 0$, using the formula $x = \frac{-b \pm \sqrt{b^2 - 4ac}}{2a}$. We get the following generating functions:

$$g(z) = \frac{1 \pm \sqrt{1 - 4z}}{2z}.$$

Consider the following power series.

$$g(z) = 1 + z + \frac{z^2}{2!} + \frac{z^3}{3!} + \cdots = \sum_{n=0}^{\infty} \frac{z^n}{n!}$$

Note that by the definition of the derivative operation for power series, $g(z)' = g(z)$. The function with this property is the function e^z, where e

is the base of the natural logarithm, Napier's number e. The number e is defined by the following infinite sum:

$$1 + \frac{1}{1!} + \frac{1}{2!} + \frac{1}{3!} + \cdots = \sum_{n=0}^{\infty} \frac{1}{n!}$$

Therefore we call $g(z)$ the exponential function e^z. Note that by the rules for integration for power series,

$$\int_0^z e^t dt = 0 + z + \frac{z^2}{2!} + \frac{z^3}{3!} + \cdots = e^z - 1,$$

and therefore, $e^z = 1 + \int_0^z e^t dt$. This gives us a means of defining e^z by corecursion, as follows:

```
expz = 1 + (int expz)
```

We get:

```
COR> take 9 expz
[1 % 1,1 % 1,1 % 2,1 % 6,1 % 24,1 % 120,1 % 720,1 % 5040,1 % 40320]
```

Since we know (Theorem 10.39) that $\frac{1}{1-z} f(z)$ gives incremental sums of the sequence generated by $f(z)$, we now have an easy means to approximate Napier's number (the number e):

```
COR> take 20 (1/(1-z) * expz)
[1 % 1,2 % 1,5 % 2,8 % 3,65 % 24,163 % 60,1957 % 720,685 % 252,
109601 % 40320,98641 % 36288,9864101 % 3628800,13563139 % 4989600,
260412269 % 95800320,8463398743 % 3113510400,
47395032961 % 17435658240,888656868019 % 326918592000,
56874039553217 % 20922789888000,7437374403113 % 2736057139200,
17403456103284421 % 6402373705728000,82666416490601 % 30411275102208]
```

10.6 Exponential Generating Functions

Up until now, the generating functions were used for solving combinatorial problems in which order was irrelevant. These generating functions are sometimes called **ordinary generating functions**. To tackle combinatorial problems where order of selection plays a role, it is convenient to use

10.6. EXPONENTIAL GENERATING FUNCTIONS

generating functions in a slightly different way. The **exponential generating function** for a sequence $\lambda n.f_n$ is the function

$$f(z) = f_0 + f_1 z + \frac{f_2 z^2}{2!} + \frac{f_3 z^3}{3!} + \cdots = \sum_{n=0}^{\infty} \frac{f_n z^n}{n!}$$

Example 10.47 The (ordinary) generating function of $[1,1,1,\ldots]$ is $\frac{1}{1-z}$, the exponential generating function of $[1,1,1,\ldots]$ is e^z. If e^z is taken as an ordinary generating function, it generates the sequence $[1, \frac{1}{1!}, \frac{1}{2!}, \frac{1}{3!}, \frac{1}{4!}, \ldots]$.

Example 10.48 $(1+z)^r$ is the ordinary generating function for the problem of picking an n-sized subset from an r-sized set, and it is the exponential generating function for the problem of picking a sequence of n distinct objects from an r-sized set.

We see from Examples 10.47 and 10.48 that the same function $g(z)$ can be used for generating different sequences, depending on whether $g(z)$ is used as an ordinary or an exponential generating function.

It makes sense, therefore, to define an operation that maps ordinarily generated sequences to exponentially generated sequences.

```
o2e :: Num a => [a] -> [a]
o2e []     = []
o2e (f:fs) = f : o2e (deriv (f:fs))
```

With this we get:

COR> take 10 (o2e expz)
[1 % 1,1 % 1,1 % 1,1 % 1,1 % 1,1 % 1,1 % 1,1 % 1,1 % 1,1 % 1]

A function for converting from exponentially generated sequences to ordinarily generated sequences is the converse of this, so it uses integration:

```
e2o :: Fractional a => [a] -> [a]
e2o []     = []
e2o (f:fs) = [f] + (int (e2o (fs)))
```

This gives:

```
COR> take 9 (e2o (1/(1-z)))
[1 % 1,1 % 1,1 % 2,1 % 6,1 % 24,1 % 120,1 % 720,1 % 5040,1 % 40320]
```

Example 10.49 Here is how $(z+1)^{10}$ is used to solve the problem of finding the number of ways of picking subsets from a set of 10 elements:

```
COR> (1+z)^10
[1,10,45,120,210,252,210,120,45,10,1]
```

Here is how the same function is used to solve the problem of finding the number of ways to pick sequences from a set of 10 elements:

```
COR> o2e ((1+z)^10)
[1,10,90,720,5040,30240,151200,604800,1814400,3628800,3628800]
```

Example 10.50 Suppose a vase contains red, white and blue marbles, at least four of each kind. How many ways are there of arranging sequences of marbles from the vase, with at most four marbles of the same colour? Solution: the exponential generating function is $(1 + z + \frac{z^2}{2} + \frac{z^3}{6} + \frac{z^4}{24})^3$. The idea: if you pick n marbles of the same colour, then $n!$ of the marble orderings become indistinguishable. Here is the Haskell computation of the solution.

```
COR> o2e ([1,1,1/2,1/6,1/24]^3)
[1 % 1,3 % 1,9 % 1,27 % 1,81 % 1,240 % 1,690 % 1,1890 % 1,4830 % 1,
  11130 % 1,22050 % 1,34650 % 1,34650 % 1]
```

Example 10.51 Suppose a vase contains red, white and blue marbles, an unlimited supply of each colour. How many ways are there of arranging sequences of marbles from the vase, assume that at least two marbles are of the same colour? Solution: a suitable exponential generating function is $(e^z - z - 1)^3 = (\frac{z^2}{2} + \frac{z^3}{6} + \frac{z^4}{24} + \cdots)^3$:

```
COR> take 10 (o2e ((expz - z - 1)^3))
[0 % 1,0 % 1,0 % 1,0 % 1,0 % 1,0 % 1,90 % 1,630 % 1,2940 % 1,11508 % 1]
```

This gives the number of solutions for up to 9 marbles. For up to 5 marbles there are no solutions. There are $90 = \frac{6!}{2^3}$ sequences containing two marbles of each colour. And so on.

Exercise 10.52 Suppose a vase contains red, white and blue marbles, at least four of each kind. How many ways are there of arranging sequences of marbles from the vase, with at least two and at most four marbles of the same colour?

10.7 Further Reading

Domain theory and proof by approximation receive fuller treatment in [DP02]. Generating functions are a main focus in [GKP89]. The connection between generating functions and lazy semantics is the topic of [Kar97, McI99, McI00]. A coinductive calculus of streams and power series is presented in [Rut03]. Corecursion is intimately connected with circularity in definitions [BM96] and with the theory of processes and process communication [Fok00, Mil99].

Chapter 11

Finite and Infinite Sets

Preview

Some sets are bigger than others. For instance, finite sets such as $\emptyset$ and $\{0, 1, 2\}$, are smaller than infinite ones such as $\mathbb{N}$ and $\mathbb{R}$. But there are varieties of infinity: the infinite set $\mathbb{R}$ is bigger than the infinite set $\mathbb{N}$, in a sense to be made precise in this chapter.

This final chapter starts with a further discussion of the Principle of Mathematical Induction as the main instrument to fathom the infinite with, and explains why some infinite sets are incomparably larger than some others.

```
module FAIS

where
```

11.1 More on Mathematical Induction

The natural numbers form the set $\mathbb{N} = \{0, 1, 2, \ldots\}$. The Principle of Mathematical Induction allows you to prove *universal* statements about the natural numbers, that is, statements of the form $\forall n \in \mathbb{N}\, E(n)$.

Fact 11.1 (Mathematical Induction) *For every set $X \subseteq \mathbb{N}$, we have that: if $0 \in X$ and $\forall n \in \mathbb{N}(n \in X \Rightarrow n + 1 \in X)$, then $X = \mathbb{N}$.*

389

This looks not impressive at all, since it is so obviously true. For, suppose that $X \subseteq \mathbb{N}$ satisfies both $0 \in X$ and $\forall n \in \mathbb{N}(n \in X \Rightarrow n+1 \in X)$. Then we see that $\mathbb{N} \subseteq X$ (and, hence, that $X = \mathbb{N}$) as follows. First, we have as a first Given, that $0 \in X$. Then the second Given ($\forall$-elimination, $n = 0$) yields that $0 \in X \Rightarrow 1 \in X$. Therefore, $1 \in X$. Applying the second Given again ($\forall$-elimination, $n = 1$) we obtain $1 \in X \Rightarrow 2 \in X$; therefore, $2 \in X$. And so on, for all natural numbers.

Nevertheless, despite its being so overly true, it is difficult to overestimate the importance of induction. It is essentially the only means by with we acquire information about the members of the infinite set $\mathbb{N}$.

Sets versus Properties. With every property E of natural numbers there corresponds a set $\{n \in \mathbb{N} \mid E(n)\}$ of natural numbers. Thus, induction can also be formulated using properties instead of sets. It then looks as follows:

If E is a property of natural numbers such that

(i) $E(0)$,
(ii) $\forall n \in \mathbb{N}\,[E(n) \Rightarrow E(n+1)]$,

then $\forall n \in \mathbb{N}\, E(n)$.

Induction: Basis, Step, Hypothesis. According to Induction, in order that $E(n)$ is true for all $n \in \mathbb{N}$, it suffices that
(i) this holds for $n = 0$, i.e., $E(0)$ is true, and
(ii) this holds for a number $n + 1$, provided it holds for n.
As we have seen, the proof of (i) is called *basis* of a proof by induction, the proof of (ii) is called the *induction step*. By the introduction rules for $\forall$ and $\Rightarrow$, in order to carry out the induction step, you are granted the Given $E(n)$ for an arbitrary number n, whereas you have to show that $E(n+1)$. The assumption that $E(n)$ is called the *induction hypothesis*. The nice thing about induction is that this induction hypothesis comes entirely free.

Fact 11.2 (Scheme for Inductions) Viewed as a rule of proof, induction can be schematized as follows.

$$E(n)$$
$$\vdots$$
$$\frac{E(0) \quad E(n+1)}{\forall n E(n)}$$

A proof using the Principle of Induction of a statement of the form $\forall n\, E(n)$ is called a proof with induction *with respect to* n. It can be useful to mention the parameter n when other natural numbers surface in the property E.

11.1. MORE ON MATHEMATICAL INDUCTION

According to the above schema, the conclusion $\forall n\, E(n)$ can be drawn on the basis of (i) the premiss that $E(0)$, and (ii) a proof deriving $E(n+1)$ on the basis of the induction hypothesis $E(n)$.

The following strong variant of induction is able to perform tasks that are beyond the reach of the ordinary version. This time, we formulate the principle in terms of properties.

Fact 11.3 (Strong Induction) *For every property E of natural numbers: if $\forall n \in \mathbb{N}\,[\,(\forall m < n\, E(m)) \Rightarrow E(n)\,]$, then $\forall n \in \mathbb{N}\, E(n)$.*

Using strong induction, the condition $\forall m < n\, E(m)$ acts as the induction hypothesis from which you have to derive that $E(n)$.

This is (at least, for $n > 1$) stronger than in the ordinary form. Here, to establish that $E(n)$, you are given that $E(m)$ holds for *all* $m < n$. In ordinary induction, you are required to show that $E(n+1)$ on the basis of the Given that it holds for n only.

Note: using ordinary induction you have to accomplish *two* things: basis, and induction step. Using strong induction, just *one* thing suffices. Strong induction has *no* basis.

Proof. (of 11.3.) Suppose that $\forall n \in \mathbb{N}\,[\,(\forall m < n\, E(m)) \Rightarrow E(n)\,]$.   (1)

Define the set X by $X = \{n \mid \forall m < n\, E(m)\}$. Assumption (1) says that every element of X has property E. Therefore, it suffices showing that every natural number is in X. For this, we employ ordinary induction.
Basis. $0 \in X$.
I.e.: $\forall m < 0\, E(m)$; written differently: $\forall m[m < 0 \Rightarrow E(m)]$. This holds trivially since there are no natural numbers < 0. The implication $m < 0 \Rightarrow E(m)$ is trivially satisfied.
Induction step.
Assume that (*induction hypothesis*) $n \in X$. That means: $\forall m < n\, E(m)$. Assumption (1) implies, that $E(n)$ is true. Combined with the induction hypothesis, we obtain that $\forall m < n+1\, E(m)$. This means, that $n+1 \in X$. The induction step is completed.
Induction now implies, that $X = \mathbb{N}$. According to the remark above, we have that $\forall n \in \mathbb{N}\, E(n)$. ■

It is useful to compare the proof schemes for induction and strong induction. Here they are, side by side:

$$\frac{E(0) \quad \begin{array}{c} E(n) \\ \vdots \\ E(n+1) \end{array}}{\forall n\, E(n)} \qquad \frac{\begin{array}{c} \forall m < n\, E(m) \\ \vdots \\ E(n) \end{array}}{\forall n\, E(n)}$$

An alternative to strong induction that is often used is the minimality principle.

Fact 11.4 (Minimality Principle) *Every non-empty set of natural numbers has a least element.*

Proof. Assume that A is an arbitrary set of natural numbers. It clearly suffices to show that, for each $n \in \mathbb{N}$,

$$n \in A \Rightarrow A \text{ has a least element.}$$

For, if also is given that $A \neq \emptyset$, then some $n \in A$ must exist. *Thus* the implication applied to such an element yields the required conclusion.
Here follows a proof of the statement using strong induction w.r.t. n. Thus, assume that n is an arbitrary number for which:
Induction hypothesis: for every $m < n$: $m \in A \Rightarrow A$ has a least element.
To be proved: $n \in A \Rightarrow A$ has a least element.
Proof: Assume that $n \in A$. There are two cases.
(i) n is (by coincidence) the least element of A. Thus, A has a least element.
(ii) n is *not* least element of A. Then some $m \in A$ exists such that $m < n$. So we can apply the induction hypothesis and again find that A has a least element. ∎

Remark. The more usual proof follows the logic from Chapter 2. This is much more complicated, but has the advantage of showing the Minimality Principle to be *equivalent* with Strong Induction.
Strong Induction is the following schema:

$$\forall n [\forall m < n \, E(m) \Rightarrow E(n)] \Rightarrow \forall n \, E(n).$$

Since E can be any property, it can also be a negative one. Thus:

$$\forall n [\forall m < n \neg E(m) \Rightarrow \neg E(n)] \Rightarrow \forall n \neg E(n).$$

The contrapositive of this is (Theorem 2.10 p. 45):

$$\neg \forall n \neg E(n) \Rightarrow \neg \forall n [\forall m < n \neg E(m) \Rightarrow \neg E(n)].$$

Applying Theorem 2.40 (p. 63):

$$\exists n \, E(n) \Rightarrow \exists n \neg [\forall m < n \neg E(m) \Rightarrow \neg E(n)].$$

Again Theorem 2.10:

$$\exists n \, E(n) \Rightarrow \exists n [\forall m < n \neg E(m) \wedge \neg \neg E(n)].$$

11.1. MORE ON MATHEMATICAL INDUCTION

Some final transformations eventually yield:

$$\exists n\, E(n) \;\Rightarrow\; \exists n[E(n) \wedge \neg \exists m < n\, E(m)]$$

— which is Minimality formulated using properties. ∎

Exercise 11.5 Prove *Induction starting at m*, that is: for every $X \subseteq \mathbb{N}$, if $m \in X$ and $\forall n \geqslant m(n \in X \Rightarrow n+1 \in X)$, then $\forall n \in \mathbb{N}(m \leqslant n \Rightarrow n \in X)$.
Hint. Apply induction to the set $Y = \{n \mid m+n \in X\}$.

Exercise 11.6 Suppose that $X \subseteq \mathbb{N}$ is such that $1 \in X$ and $\forall n \in \mathbb{N}(n \in X \Rightarrow n+2 \in X)$. Show that every odd number is in X.

Definition 11.7 A relation $\prec$ on A is called **well-founded** if no infinite sequence $\cdots \prec a_2 \prec a_1 \prec a_0$ exists in A. Formulated differently: every sequence $a_0 \succ a_1 \succ a_2 \succ \cdots$ in A eventually terminates.

Example 11.8 According to Exercise 4.34 (p. 133), the relation $\prec$ on sets defined by: $a \prec b$ iff $\wp(a) = b$, is well-founded.
An easier example is in the next exercise.

Exercise 11.9 Show that $<$ is well-founded on $\mathbb{N}$. That is: there is no infinite sequence $n_0 > n_1 > n_2 > \cdots$.
Hint. Show, using strong induction w.r.t. n, that for all $n \in \mathbb{N}$, $E(n)$; where $E(n)$ signifies that no infinite sequence $n_0 = n > n_1 > n_2 > \cdots$ starts at n. Alternatively, use Exercise 11.10.

Exercise 11.10* **Well-foundedness as an induction principle.** Let $\prec$ be a relation on a set A.

1. Suppose that $\prec$ is well-founded. Assume that $X \subseteq A$ satisfies

$$\forall a \in A(\forall b \prec a(b \in X) \Rightarrow a \in X).$$

   Show that $X = A$.

   Hint. Show that any $a_0 \in A - X$ can be used to start an infinite sequence $a_0 \succ a_1 \succ a_2 \succ \cdots$ in A.

2. Conversely:
   Suppose that every $X \subseteq A$ satisfying $\forall a \in A(\forall b \prec a(b \in X) \Rightarrow a \in X)$ coincides with A. Show that $\prec$ is well-founded.
   Hint. Suppose that $a_0 \succ a_1 \succ a_2 \succ \cdots$, and consider the set $X = A - \{a_0, a_1, a_2, \ldots\}$.

Exercise 11.11* Let R be a relation on a set A. Recall that R^* denotes the reflexive transitive closure of R, cf. Exercise 5.46 p. 169.

1. Assume that for all $a, b_1, b_2 \in A$, if aRb_1 and aRb_2, then $c \in A$ exists such that $b_1 Rc$ and $b_2 Rc$.

   Show that R is *confluent*, that is: for all $a, b_1, b_2 \in A$, if aR^*b_1 and aR^*b_2, then $c \in A$ exists such that $b_1 R^*c$ and $b_2 R^*c$.

2. Assume that R is *weakly confluent*, that is: for all $a, b_1, b_2 \in A$, if aRb_1 and aRb_2, then $c \in A$ exists such that $b_1 R^*c$ and $b_2 R^*c$.

   A counter-example to confluence is called *bad*. That is: a is bad iff there are $b_1, b_2 \in A$ such that aR^*b_1 and aR^*b_2, and for no $c \in A$ we have that $b_1 R^*c$ and $b_2 R^*c$.

   Show: if a is bad, then a bad b exists such that aRb.

3. (In abstract term rewriting theory, the following result is known as *Newman's Lemma*.)

   Assume that R is weakly confluent.

   Furthermore, assume that R^{-1} is well-founded.

   Show that R is confluent.

   Hint. Use part 2; alternatively, use Exercise 11.10.

   Remark. That R^{-1} is well-founded is necessary here.

   For example, $R = \{(1,0), (1,2), (2,1), (2,3)\}$ is weakly confluent but not confluent.

Exercise 11.12 Suppose that $\emptyset \neq X \subseteq \mathbb{N}$, and that X is *bounded*, i.e.: that $m \in \mathbb{N}$ exists such that for all $n \in X$, $n \leqslant m$. Show that X has a *maximum*, that is: an element $m \in X$ such that for all $n \in X$, $n \leqslant m$.
Hint. Induction w.r.t. m. Thus, $E(m)$ is: every non-empty $X \subseteq \mathbb{N}$ such that $\forall n \in X \ (n \leqslant m)$, has a maximum.

Exercise 11.13 Suppose that $f : \mathbb{N} \to \mathbb{N}$ is such that $n < m \Rightarrow f(n) < f(m)$. Show by means of an induction argument that for all $n \in \mathbb{N}$: $n \leqslant f(n)$.

Exercise 11.14 Suppose that $a_0, a_1, a_2, \ldots$ is an infinite sequence of natural numbers. Prove that $i, j \in \mathbb{N}$ exist such that both $i < j$ and $a_i \leqslant a_j$.

Exercise 11.15 The function $g : \mathbb{N}^+ \times \mathbb{N}^+ \to \mathbb{N}^+$ has the following properties.

11.1. MORE ON MATHEMATICAL INDUCTION

1. If $n < m$, then $g(n,m) = g(n, m-n)$.
2. $g(n,m) = g(m,n)$.
3. $g(n,n) = n$.

Show that $g(n,m)$ is the gcd (greatest common divisor) of n and m.

Exercise 11.16* You play *Smullyan's ball game* on your own. Before you is a box containing finitely many balls. Every ball carries a natural number. Next to the box, you have a supply of as many numbered balls as you possibly need. A *move* in the game consists in replacing one of the balls in the box by arbitrarily (possibly zero, but finitely) many new balls that carry a number smaller than that on the one you replace. (Thus, 9^{9^9} balls numbered 7 can replace one ball numbered 8, but a ball numbered 0 can only be taken out of the box since there are no natural numbers smaller than 0.)

Repeat this move.

Show that, no matter how you play the game, you'll end up with an empty box eventually. (Thus, your last moves necessarily consist in throwing away balls numbered 0.)

Hint. Proof by Contradiction. Suppose that you can play ad infinitum, and that B_k is how the box looks after your k-th move. Derive a contradiction applying strong induction w.r.t. the greatest number n present on one of the balls in the box B_0 you start from. If m is the number of balls in B_0 carrying n, apply a second strong induction, now w.r.t. m.

Exercise 11.17 Implement a simplified version of Smullyan's ball game from the previous exercise, where (i) the box of balls is represented as a list of integers, (ii) it is always the *first* integer $n > 1$ that gets replaced, (iii) an integer $n > 1$ gets replaced by *two* copies of $n - 1$, (iv) the game terminates with a list consisting of just ones. The type declaration should run:

`ballgame :: [Integer] -> [[Integer]]`.

How long will it take before `ballgame [50]` terminates? Minutes? Hours? Days? Years?

Exercise 11.18 The following theorem implies that, e.g., all inhabitants of Amsterdam have the same number of hairs.
Theorem: in every set of n people, everyone has the same number of hairs.
Proof. Induction w.r.t. n.
Basis. $n = 0$ (or $n = 1$): trivial.

Induction step. Induction hypothesis: the statement holds for sets of n people.
Now assume that A is an $(n+1)$-element set of humans. Arbitrarily choose different p and q in A; we show that p and q have the same number of hairs. $A - \{p\}$ and $A - \{q\}$ have n elements, thus the induction hypothesis applies to these sets. Choose $r \in A - \{p, q\}$. Then r and q have the same number of hairs (they are both in the set $A - \{p\}$); and r and p have the same number of hairs (similar reason). Thus, p and q have the same number of hairs as well.
Explain this apparent contradiction with common sense observation.

11.2 Equipollence

In order to check whether two finite sets have the same number of elements, it is not necessary at all to count them. For, these numbers are the same iff *there is a bijection between the two sets*.

Sometimes, it is much easier to construct a bijection than to count elements. Imagine a large room full of people and chairs, and you want to know whether there are as many people as there are chairs. In order to answer this question, it suffices to ask everyone to sit down, and have a look at the resulting situation.

This observation motivates the following definition.

Definition 11.19 (Equipollence) Two sets A and B are called *equipollent* if there is a bijection from A to B.
Notation: $A \sim B$.

Example 11.20 (Trivial but Important) The set $\mathbb{N}$ is equipollent with its proper subset $\mathbb{N}^+ = \mathbb{N} - \{0\}$. For, this is witnessed by the *successor* function $n \longmapsto n + 1$. We can generate the graph of this function in Haskell by means of:

```
succs = [ (n, succ n) | n <- [0..] ].
```

Of course, that a set is equipollent with one of its proper subsets can only happen in the case of infinite sets. The example shows that the notion of equipollence can have surprising properties when applied to infinite sets.

The following theorem shows that $\sim$ is an equivalence.

Theorem 11.21 *For all sets A, B, C:*

  1. $A \sim A$ *(reflexivity)*,

11.2. EQUIPOLLENCE

  2. $A \sim B \implies B \sim A$ (symmetry),

  3. $A \sim B \wedge B \sim C \implies A \sim C$ (transitivity).

Proof.
1. 1_A is a bijection from A to itself.
2. If $f : A \to B$ is a bijection, then f^{-1} is a bijection $: B \to A$.
3. If $f : A \to B$ and $g : B \to C$ are bijections, then $g \circ f : A \to C$ is a bijection as well. (Cf. Lemma 6.36, p. 218)   ■

The standard example of an n-element set is $\{0, \ldots, n-1\}$. (Of course, $\{1, \ldots, n\}$ serves as well.) This motivates part 1 of the following definition.

Definition 11.22 (Finite, Infinite)

1. A set *has n elements* if it is equipollent with $\{0, \ldots, n-1\}$.

2. It is *finite* if $n \in \mathbb{N}$ exists such that it has n elements.

3. It is *infinite* if it is not finite.

Example 11.23
1. The set $\emptyset$ has 0 elements, hence $\emptyset$ is finite.
2. If A has n elements and x is arbitrarily chosen, then $A \cup \{x\}$ has n or $n+1$ elements (depending on whether $x \in A$). Thus, if A is finite then so is $A \cup \{x\}$.

The proof of the following unsurprising theorem illustrates the definitions and the use of induction.

Theorem 11.24 $\mathbb{N}$ *is infinite.*

Proof. We show, using induction w.r.t. n, that for all $n \in \mathbb{N}$,

$$\mathbb{N} \not\sim \{0, \ldots, n-1\}.$$

The induction step applies Exercise 11.25.
Basis.
If $n = 0$, then $\{0, \ldots, n-1\} = \emptyset$. A non-empty set like $\mathbb{N}$ cannot be equipollent with $\emptyset$.
Induction step.

Induction hypothesis: $\mathbb{N} \not\sim \{0, \ldots, n-1\}$.
To be proved: $\mathbb{N} \not\sim \{0, \ldots, n\}$.
Proof: Assume that, nevertheless, a bijection from $\mathbb{N}$ to $\{0, \ldots, n\}$ exists. According to Exercise 11.25 we may assume there is a bijection $f : \mathbb{N} \to \{0, \ldots, n\}$ such that $f(0) = n$. Its restriction $f \restriction (\mathbb{N} - \{0\})$ (Definition 6.12 p. 210) then is a bijection from $\mathbb{N} - \{0\}$ to $\{0, \ldots, n-1\}$. We have that $\mathbb{N} \sim (\mathbb{N} - \{0\})$ (This is the *Trivial but Important Example* 11.20). Conclusion (Theorem 11.21.3): $\mathbb{N} \sim \{0, \ldots, n-1\}$. But this contradicts the induction hypothesis.

∎

Exercise 11.25 Suppose that $A \sim B$, $a \in A$ and $b \in B$. Show that a bijection $f : A \to B$ exists such that $f(a) = b$.
Hint. By $A \sim B$, we have a bijection $g : A \to B$. If, by coincidence, $g(a) = b$, we let f be g. Thus, assume $g(a) = b' \neq b$. Since g is surjective, $a' \in A$ exists such that $g(a') = b$. Make a picture of the situation and look whether you can find f by suitably modifying g.

Exercise 11.26* Suppose that $A \sim B$. Show that $\wp(A) \sim \wp(B)$. Write out the bijection, using $f^* : \wp(A) \to \wp(B)$, with $f^*(X) = f[X]$, for f a bijection that witnesses $A \sim B$.

Exercise 11.27 Show that, for every set A: $\wp(A) \sim \{0, 1\}^A$.
Hint. Associate with $X \subseteq A$ its *characteristic function*, that is, the function $\chi_X : A \to \{0, 1\}$ defined by: $\chi_X(a) = 1$ iff $a \in X$. The function $X \mapsto \chi_X$ (that sends sets to functions) is the bijection you are looking for.

Exercise 11.28 Suppose that $A \sim B$. Show:

1. if A has n elements, then so has B,

2. if A is finite, then so is B,

3. if A is infinite, then so is B.

Exercise 11.29 f is a function. Show that $f \sim \mathrm{dom}(f)$.

Exercise 11.30* Suppose that R is an equivalence on A. $V = A/R$ is the corresponding quotient. Show that the set of all partitions of V is equipollent with the set of all equivalences Q on A for which $R \subseteq Q$.

11.2. EQUIPOLLENCE

Exercise 11.31* Suppose that $n, m \in \mathbb{N}$ and $n < m$. Show that $\{0, \ldots, n-1\} \not\sim \{0, \ldots, m-1\}$.
Hint. Use induction to show that $\forall n < m \; \{0, \ldots, n-1\} \not\sim \{0, \ldots, m-1\}$. Employ the method of proof of Theorem 11.24.

Exercise 11.32 Suppose that $X \subseteq \mathbb{N}$ and $m \in \mathbb{N}$ are give such that $\forall n \in X (n < m)$. Show that X is finite.

The following exercise explains how to apply induction to finite sets.

Exercise 11.33* Prove the following induction principle for finite sets.
If E is a property of sets such that

1. $E(\emptyset)$,

2. for every set A and every object $x \notin A$: if $E(A)$, then also $E(A \cup \{x\})$,

then E holds for every finite set.
Hint. Apply induction to the property E' of numbers, defined by $E'(n) \equiv \forall A[A \text{ has } n \text{ elements} \Rightarrow E(A)]$.

Exercise 11.34* Show that a subset of a finite set is finite.

Exercise 11.35* Show that the union of two finite sets is finite.

Exercise 11.36* Suppose that h is a finite injection with dom$(h) \subseteq A$ and ran$(h) \subseteq B$. Suppose that $A \sim B$. Show that a bijection $f : A \to B$ exists such that $f \supseteq h$. (What about the case where h is infinite?)
Hint. Induction on the number n of elements of h. (The case $n = 1$ is Exercise 11.25.)

Exercise 11.37* Show that a proper subset of a finite set never is equipollent with that set.

Exercise 11.38 Suppose that A and B are finite sets and $f : A \to B$ a bijection. Show:

1. $(B - A) \sim (A - B)$,

2. there exists a bijection $g : A \cup B \to A \cup B$ such that $f \subseteq g$.

Exercise 11.39* Show: a set A is finite iff the following condition is satisfied for every collection $E \subseteq \wp(A)$: if $\emptyset \in E$ and $\forall B \in E \, \forall a \in A \, (B \cup \{a\} \in E)$, then $A \in E$.

11.3 Infinite Sets

One of the great discoveries of Cantor is that for any set A, the set $\wp(A)$ is 'larger' (in a sense to made precise below) than A. This shows the existence of 'Cantor's paradise' of an abundance of sets with higher and higher grades of infinity.

At Most Equipollent To

Recall Definition 11.19: sets are *equipollent* means that there is a bijection between the two.

Definition 11.40 (At Most Equipollent To) The set A is *at most equipollent to* B if an injection exists from A into B.
  Notation: $A \preceq B$.

Example 11.41 $\mathbb{Z} \preceq \mathbb{R}^+$.

Theorem 11.42 *For every infinite set A:* $\mathbb{N} \preceq A$.

Proof. Suppose that A is infinite. An injection $h : \mathbb{N} \to A$ as required can be produced in the following way.
  Note: $A \neq \emptyset$ (since $\emptyset$ is finite). Thus, it is possible to choose an element $h(0) \in A$.
  Now, $A \neq \{h(0)\}$ (for, $\{h(0)\}$ is finite). Thus, another element $h(1) \in A - \{h(0)\}$ exist.
  Again, $A \neq \{h(0), h(1)\}$, etc.
  Going on, this argument produces different elements $h(0), h(1), h(2), \ldots$ in A; thus, the corresponding function $h : \mathbb{N} \to A$ is an injection. ∎

Thus, $\mathbb{N}$ is the "simplest" infinite set.

Exercise 11.43 Show:

  1. $A \preceq A$,

  2. $A \sim B \implies A \preceq B$,

  3. $A \preceq B \wedge B \preceq C \implies A \preceq C$,

  4. $A \subseteq B \implies A \preceq B$,

11.3. INFINITE SETS

Exercise 11.44* Show the reverse of Theorem 11.42: if $\mathbb{N} \preceq A$, then A is infinite.
Hint. Cf. the proof of Theorem 11.24, p. 397. A slight modification (that uses a modification of Exercise 11.25) shows that for all n, $\mathbb{N} \not\preceq \{0, \ldots, n-1\}$; this implies what is required.

Exercise 11.45 Suppose that $h : A \to A$ is an injection that is not surjective. Say, $b \in A - \operatorname{ran}(h)$. Define $f : \mathbb{N} \to A$ by: $f(0) = b$, $f(n+1) = h(f(n))$. E.g., $f(3) = h(f(2)) = h(h(f(1))) = h(h(h(f(0)))) = h(h(h(b)))$.
Show that $f(n)$ is different from $f(0), \ldots, f(n-1)$ ($n \in \mathbb{N}$). (Induction w.r.t. n.)
Conclusion: f is a injection, and $\mathbb{N} \preceq A$.

Exercise 11.46 Show: if $\mathbb{N} \preceq A$, then a non-surjective injection $h : A \to A$ exists.

Exercise 11.47 Show: a set is infinite iff it is equipollent with one of its proper subsets.
Hint. Use Theorem 11.42 and Exercises 11.44, 11.45 and 11.46.

Exercise 11.48 Suppose that A is finite and that $f : A \to A$. Show: f is surjective iff f is injective.
Hint. $\Leftarrow$: use Exercises 11.44 and 11.45.

Countably Infinite

The prototype of a *countably infinite* set is $\mathbb{N}$.

Definition 11.49 (Countable) A set A is *countably infinite* (or: *denumerable*) if $\mathbb{N} \sim A$.

Exercise 11.50* Show: a subset of a countably infinite set is countably infinite or finite.

Exercise 11.51 Show:

1. $\mathbb{Z}$ (the set of integers) is countably infinite,

2. a union of two countably infinite sets is countably infinite.

Theorem 11.52 $\mathbb{N}^2 = \{(n, m) \mid n, m \in \mathbb{N}\}$ *is countably infinite.*

Proof. Define $S(p) = \{(n,m) \mid n + m = p\}$.
The sets $S(p)$ are pairwise disjoint, and $\mathbb{N}^2 = S(0) \cup S(1) \cup S(2) \cup \ldots$ Moreover, $S(p)$ has exactly $p + 1$ elements: $(0,p), (1, p-1), \ldots, (p, 0)$. Verify that the function $j : \mathbb{N}^2 \to \mathbb{N}$ that is defined by $j(n,m) = \frac{1}{2}(n+m)(n+m+1) + n$ enumerates the pairs of the subsequent sets $S(p)$.
Note that the function j enumerates $\mathbb{N}^2$ as follows:
$(0,0), (0,1), (1,0), (0,2), (1,1), (2,0), (0,3), (1,2), (2,1), (3,0), \ldots$
Look at all pairs (n, m) as the corresponding points in two-dimensional space. Visualize j as a walk along these points.

Of course, there are many other "walks" along these points that prove this theorem. E.g., (visualize!)
$(0,0), (1,0), (1,1), (0,1), (0,2), (1,2), (2,2), (2,1), (2,0), (3,0), (3,1), \ldots$ ∎

Theorem 11.53 *The set of positive rationals $\mathbb{Q}^+$ is countably infinite.*

Proof. Identify a positive rational $q \in \mathbb{Q}^+$ with the pair $(n,m) \in \mathbb{N}^2$ for which $q = \frac{n}{m}$ and for which n and m are co-prime. Use Theorem 11.52 and Exercise 11.50. ∎

Exercise 11.54 Show that $\mathbb{Q}$ is countably infinite.

Exercise 11.55 Show that $\mathbb{N}^*$ is countably infinite.

Exercise 11.56 Show that a union of countably infinitely many countably infinite sets is countably infinite.

Uncountable

Definition 11.57 (Less Power Than) A set A has *power less than* B if both $A \preceq B$ and $A \not\approx B$.
Notation: $A \prec B$.

Thus:
$$A \prec B \iff A \preceq B \wedge A \not\approx B.$$

Example 11.58 $\{0, \ldots, n-1\} \prec \mathbb{N}$ (Theorem 11.24); $\mathbb{N} \prec \mathbb{R}$ (Theorem 11.60).

Warning. That $A \prec B$ implies but is *not equivalent with*: there exists a non-surjective injection from A into B.

11.3. INFINITE SETS

That $A \prec B$ means by definition: $A \preceq B$ and $A \not\sim B$. That $A \preceq B$ means that an injection $f : A \to B$ exists. If, moreover, $A \not\sim B$, then no bijection exists between A and B; in particular, the injection f cannot be surjective.

Counter-examples to the converse: (Example 11.20) the successor-function $n \mapsto n + 1$ is a non-surjective injection $: \mathbb{N} \to \mathbb{N}$, but of course it is false that $\mathbb{N} \prec \mathbb{N}$; the identity function $1_\mathbb{N}$ is a non-surjective injection from $\mathbb{N}$ into $\mathbb{Q}$, but we do not have (Exercise 11.54), that $\mathbb{N} \prec \mathbb{Q}$.

Definition 11.59 (Uncountable) A set A is *uncountable* in case $\mathbb{N} \prec A$.

The following is Cantor's discovery.

Theorem 11.60 $\mathbb{R}$ *is uncountable.*

Proof. (i) That $\mathbb{N} \preceq \mathbb{R}$ is clear (Exercise 11.43.3, $\mathbb{N} \subseteq \mathbb{R}$). (ii) It must be shown, that no bijection $h : \mathbb{N} \to \mathbb{R}$ exists. In fact, there is no surjection from $\mathbb{N}$ to $\mathbb{R}$. That is:
Claim. To every function $h : \mathbb{N} \to \mathbb{R}$ there exists a real r such hat $r \notin \mathrm{ran}(h)$.
Proof. Suppose that $h : \mathbb{N} \to \mathbb{R}$. Write down the decimal expansion for every real $h(n)$: $h(n) = p_n + 0.r_0^n r_1^n r_2^n \cdots$, where $p_n \in \mathbb{Z}$, $p_n \leqslant h(n) < p_n + 1$, and decimals $r_i^n \in \{0, 1, 2, \ldots, 9\}$. (E.g., $-\sqrt{2} = -2 + 0, 15 \cdots$)

For every n, choose a digit $r_n \in \{0, 1, 2, \ldots, 9\}$ such that $r_n \neq r_n^n$. The real $r = 0, r_0 r_1 r_2 \cdots$ then differs from $h(n)$ in its n-th decimal digit ($n = 0, 1, 2, \ldots$).

However, even if $p_n = 0$, this does not imply $r \neq h(n)$. For, a real can have *two* decimal notations. E.g., $0.5000\cdots = 0.4999\cdots$. But, a tail of zeros vs. a tail of nines is the only case for which this happens. So, this problem can be avoided if we require r_n to be different from 0 and 9. ∎

Recall, that $\wp(A) = \{X \mid X \subseteq A\}$ is the collection of all subsets of A (Definition 4.22, p. 130). The powerset operation produces in a direct way sets of greater power.

In particular, not *every* uncountable set is equipollent with $\mathbb{R}$.

Theorem 11.61 $A \prec \wp(A)$.

Proof. (i) The injection $a \mapsto \{a\}$ from A into $\wp(A)$ shows that $A \preceq \wp(A)$. (ii) To show that $A \not\sim \wp(A)$, we exhibit, as in the previous proof, for every function $h : A \to \wp(A)$, an element $D \in \wp(A) - \mathrm{ran}(h)$.

Such an element can be simply described in this context: we take $D = \{a \in A \mid a \notin h(a)\}$.

If we would have that $D \in \mathrm{ran}(h)$, then D would be the value of some $d \in A$. Now there are two cases: either $d \in D$, or $d \notin D$.

If $d \in D$, then, by definition of D, we would have that $d \notin h(d) = D$: contradiction. If $d \notin D$, then, since $D = h(d)$, we would have that $d \notin h(d)$, and hence $d \in D$; again a contradiction.

Conclusion: $D \notin \mathrm{ran}(h)$. ∎

Corollary 11.62 *1.* $\mathbb{N} \prec \wp(\mathbb{N}) \prec \wp(\wp(\mathbb{N})) \prec \cdots$,

2. for every set A there exists a set B such that $A \prec B$.

Exercise 11.63 Show:

1. $A \not\prec A$,

2. $A \preceq B \iff A \prec B \vee A \sim B$,

3. $A \prec B \wedge B \sim C \implies A \prec C$.

4. What is wrong with the following "*proof*" for $2 \Rightarrow ?$:
   Given is that $A \preceq B$. Suppose that $f : A \to B$ is an injection.
   (a) f is (by accident) surjective. Then $A \sim B$.
   (b) f is not surjective. Then $A \not\sim B$, hence $A \prec B$.

Exercise 11.64 Show: if A is finite, then $A \prec \mathbb{N}$.
(The converse of this is true as well)

Exercise 11.65 Show: the real interval $(0, \frac{2}{9}] = \{r \in \mathbb{R} \mid 0 < r \leqslant \frac{2}{9}\}$ is uncountable.

Exercise 11.66 A is a set. Define $h : A \to \wp(A)$ by $h(a) = \{a\}$. Determine $\{a \in A \mid a \notin h(a)\}$.

Exercise 11.67 Show that $\mathbb{N} \prec \{0,1\}^\mathbb{N}$.

Exercise 11.68* Show that $\mathbb{N} \prec \mathbb{N}^\mathbb{N}$. ($\mathbb{N}^\mathbb{N}$ is the set of all functions $: \mathbb{N} \to \mathbb{N}$.)
Hint. Produce, for every function $\varphi : \mathbb{N} \to \mathbb{N}^\mathbb{N}$, a function $f \in \mathbb{N}^\mathbb{N} - \mathrm{ran}(\varphi)$.

Exercise 11.69 Suppose that $h : \mathbb{N} \to \mathbb{Q}$ is surjective and that we apply the procedure from the proof of Theorem 11.60 to h to produce a real r. Is r a rational or not?

*Cantor-Bernstein Theorem

The following result is of fundamental importance in the theory of equipollence.

Theorem 11.70 (Cantor-Bernstein) $A \preceq B \land B \preceq A \implies A \sim B$.

The proof of this theorem is delegated to the exercises.

Note that we have two examples of uncountable sets: $\mathbb{R}$ (Theorem 11.60) and $\wp(\mathbb{N})$ (Theorem 11.61). It turns out that these sets are of the same magnitude.

Theorem 11.71 $\mathbb{R} \sim \wp(\mathbb{N})$.

Proof. We show that $\mathbb{R} \preceq \wp(\mathbb{Q}) \sim \wp(\mathbb{N}) \sim \{0,1\}^{\mathbb{N}} \preceq \mathbb{R}$. From this, the Cantor-Bernstein Theorem produces the required conclusion.
1. $\mathbb{R} \preceq \wp(\mathbb{Q})$. The function $r \longmapsto \{q \in \mathbb{Q} \mid q < r\}$ is an injection from $\mathbb{R}$ into $\wp(\mathbb{Q})$.
2. $\wp(\mathbb{Q}) \sim \wp(\mathbb{N})$. Choose a bijection $h : \mathbb{Q} \to \mathbb{N}$ (Exercise 11.54, p. 402). Now $X \longmapsto h[X]$ is a bijection between $\wp(\mathbb{Q})$ and $\wp(\mathbb{N})$ (Exercise 11.26, p. 398).
3. $\wp(\mathbb{N}) \sim \{0,1\}^{\mathbb{N}}$. Cf. Exercise 11.27.
4. $\{0,1\}^{\mathbb{N}} \preceq \mathbb{R}$. Associate with $h : \mathbb{N} \to \{0,1\}$ the real (in the interval $[0, \frac{1}{9}]$) that has decimal expansion $0, h(0)h(1)h(2)\cdots$. ∎

Continuum Problem and Hypothesis.
Since $\mathbb{N} \prec \mathbb{R}$, it is tempting to ask whether sets A exist such that $\mathbb{N} \prec A \prec \mathbb{R}$. (If so, such an A exists for which $A \subseteq \mathbb{R}$.) This question is Cantor's *Continuum Problem*. (The continuum is the set $\mathbb{R}$.)

Cantor's *Continuum Hypothesis* asserts that such a set does not exist.

The usual set-theoretic axioms cannot answer the question. Gödel proved in 1938 that the axioms cannot show that the Continuum Hypothesis is false. Cohen proved in 1963 that the axioms cannot show that the Continuum Hypothesis is true. Indeed, as far as the axioms are concerned, the power of $\mathbb{R}$ can be unimaginably big and there can be arbitrarily many sets $A, B, C, \ldots \subseteq \mathbb{R}$ such that $\mathbb{N} \prec A \prec B \prec C \prec \cdots \prec \mathbb{R}$.

Example 11.72 We show that $[0,1] \sim [0,1)$. Although these intervals differ in only one element, establishing a bijection between them is far from trivial. (The reader who doubts this should give it a try before reading on.) Let $f : [0,1] \to [0,1)$ be an arbitrary injection; say, $f(x) = \frac{1}{2}x$. Consider the following function $h : [0,1] \to [0,1)$: h sends 1 to $f(1) = \frac{1}{2}$, $\frac{1}{2}$ to $f(\frac{1}{2}) = \frac{1}{4}$, $\frac{1}{4}$ to $f(\frac{1}{4}) = \frac{1}{8}$ etc.; on other arguments $r \neq 2^{-n}$ in $[0,1]$, you let $h(r) = r$.

Verify that h is bijective. To check injectivity, let $r \neq s \in [0,1]$. If neither of r, s is of the form 2^{-n}, then by definition of h we have $h(r) = r \neq s = h(s)$. If both are of the form 2^{-n}, say $r = 2^{-i}$ and $s = 2^{-j}$ with $i \neq j$, then, by definition of h, $h(r) = 2^{-i-1} \neq 2^{-j-1} = h(s)$. If one of r, s is of the form 2^{-n} and the other is not, then one of $h(r), h(s)$ is of the form 2^{-n} and the other is not, so again $h(r) \neq h(s)$. This shows that h is injective. For surjectivity, let $r \in [0,1)$. If r is not of the form 2^{-n} then $h(r) = r$, so there is an $s \in [0,1]$ with $h(s) = r$. If r is of the form 2^{-n}, say $r = 2^{-i}$ with $i > 0$, then $h(2^{-i+1}) = 2^{-i} = r$. Again, there is an $s \in [0,1]$ with $h(s) = r$.

Lemma 11.73 *If $A \preceq B \subseteq A$, then $A \sim B$.*

Exercise 11.74* Prove the Lemma.
Hint. Generalize the solution for Example 11.72.

Exercise 11.75* Prove Theorem 11.70.
Hint. Apply Lemma 11.73 to the composition of the two functions that are given.

Exercise 11.76 Show the following variations on the fact that $[0,1] \sim [0,1)$:

1. $[0,1] \sim [0, \frac{2}{3})$,
2. $\{(x,y) \mid x^2 + y^2 \leqslant 1\} \sim \{(x,y) \mid x^2 + y^2 < 1\}$,
3. $\{(x,y) \mid x^2 + y^2 \leqslant 1\} \sim \{(x,y) \mid |x|, |y| < \frac{1}{2}\}$.

Exercise 11.77* Suppose that $A \subseteq \mathbb{R}$ is finite or countably infinite. Show that $\mathbb{R} - A$ is uncountable. Can you show that $(\mathbb{R} - A) \sim \mathbb{R}$?

Exercise 11.78 Show that $(\mathbb{R} - \mathbb{Q}) \sim \mathbb{R}$.

11.4 Cantor's World Implemented

The following program illustrates that $\mathbb{N}^2$ is denumerable:

```
natpairs = [(x,z-x) | z <- [0..], x <- [0..z]]
```

This gives:

11.4. CANTOR'S WORLD IMPLEMENTED

```
FAIS> natpairs
[(0,0),(0,1),(1,0),(0,2),(1,1),(2,0),(0,3),(1,2),(2,1),
(3,0),(0,4),(1,3),(2,2),(3,1),(4,0),(0,5),(1,4),(2,3),
(3,2),(4,1),(5,0){Interrupted!}
```

Exercise 11.79 Implement the function `pair :: (Int,Int) -> Int` that is the inverse of `natpairs`. It should hold for all natural numbers n that `pair (natpairs !! n) = n`.

Exercise 11.80 Implement a function `natstar :: [[Int]]` to enumerate $\mathbb{N}^*$ (cf. Exercise 11.55).

The following code illustrates that $\mathbb{Q}$ is denumerable:

```
rationals = [ (n,m) | (n,m) <- natpairs, m /= 0, gcd n m == 1 ]
```

This gives:

```
FAIS> rationals
[(0,1),(1,1),(1,2),(2,1),(1,3),(3,1),(1,4),(2,3),(3,2),(4,1),(1,5),
(5,1),(1,6),(2,5),(3,4),(4,3),(5,2),(6,1),(1,7),(3,5),(5,3),(7,1),
(1,8),(2,7),(4,5),(5,4),(7,2),(8,1),(1,9),(3,7),(7,3),(9,1),(1,10),
(2,9),(3,8),(4,7),(5,6),(6,5),(7,4),(8,3),(9,2),(10,1),(1,11),(5,7),
(7,5),(11,1),(1,12),(2,11),(3,10),(4,9),(5,8),(6,7),(7,6),(8,5),(9,4),
(10,3),(11,2),(12,1),(1,13),(3,11),(5,9),(9,5),(11,3),(13,1),(1,14),
(2,13),(4,11),(7,8),(8,7),(11,4),(13,2),(14,1),(1,15),(3,13),(5,11),
(7,9),(9,7),(11,5),(13,3),(15,1),(1,16),(2,15),(3,14),(4,13),(5,12),
(6,11),(7,10),(8,9),(9,8),(10,7),(11,6),(12,5),(13,4),(14,3),(15,2),
(16,1),(1,17),(5,13),(7,11),(11,7),(13,5),(17,1),(1,18),(2,17),(3,16),
(4,15),(5,14),(6,13),(7,12),(8,11),(9,10),(10,9),(11,8),(12,7),(13,6),
(14,5),(15,4),(16,3),(17,2),(18,1),(1,19),(3,17),(7,13),(9,11),(11,9),
(13,7),(17,3),(19,1),(1,20),(2,19),(4,17),(5,16),(8,13),(10,11),
(11,10),(13,8),(16,5){Interrupted!}
```

The following code illustrates that $\{True, False\}^{\mathbb{N}}$ is not denumerable:

```
diagonal :: (Integer -> [Bool]) -> Integer -> Bool
diagonal f n = not ((f n)!!(fromInteger n))

f :: Integer -> [Bool]
f 0 = cycle [False]
f (n+1) = True : f n
```

Now `[ f n | n <- [0..] ]` is a list of streams of booleans, all different, and `diagonal f` is a new stream of booleans, different from all members of

`[ f n | n <- [0..] ]`.

Here is an illustration for initial segments of the lists:

```
FAIS> [ take 11 (f n) | n <- [0..10] ]
[[False,False,False,False,False,False,False,False,False,False,False],
 [True,False,False,False,False,False,False,False,False,False,False],
 [True,True,False,False,False,False,False,False,False,False,False],
 [True,True,True,False,False,False,False,False,False,False,False],
 [True,True,True,True,False,False,False,False,False,False,False],
 [True,True,True,True,True,False,False,False,False,False,False],
 [True,True,True,True,True,True,False,False,False,False,False],
 [True,True,True,True,True,True,True,False,False,False,False],
 [True,True,True,True,True,True,True,True,False,False,False],
 [True,True,True,True,True,True,True,True,True,False,False],
 [True,True,True,True,True,True,True,True,True,True,False]]
FAIS> [ diagonal f n | n <- [0..10] ]
[True,True,True,True,True,True,True,True,True,True,True]
```

11.5 *Cardinal Numbers

By Theorem 11.21, equipollence is an equivalence on the collection of all sets.

Definition 11.81 (Cardinal Number) A *cardinal number* is an equivalence class w.r.t. equipollence.

$|A|$ denotes the cardinal number of the set A modulo $\sim$.

The following is immediate (cf. Lemma 5.82, p. 188):

Lemma 11.82 $|A| = |B| \iff A \sim B.$

Usually, the cardinal number $|\{0, \ldots, n-1\}|$ of the n-element sets is identified with the natural number n.

Aleph-zero. $\aleph_0 = |\mathbb{N}|.$[1]

The concept of a cardinal number can be considered as a generalisation of that of a natural number.

---

[1] א (aleph) is the first letter of the Hebrew alphabet.

11.5. CARDINAL NUMBERS

It is possible to generalize the definitions of addition, multiplication and exponentiation to cardinal numbers, and to prove natural laws that generalize those for the natural numbers.

The operations are defined as follows. $|A| + |B| = |A \cup B|$ (provided that $A \cap B = \emptyset$: only if A and B are disjoint sets of n resp. m elements does it follow that their union has $n + m$ elements), $|A| \times |B| = |A \times B|$ and $|A|^{|B|} = |A^B|$.

Thus, by Theorem 11.71, we have that $|\mathbb{R}| = 2^{\aleph_0}$.

The cardinal $\aleph_0$ is the starting point of an infinite series of cardinal numbers called *alephs*: $\aleph_0 < \aleph_1 < \aleph_2 < \cdots < \aleph_\omega < \cdots$; ($\aleph_\omega$ is the smallest cardinal bigger than every $\aleph_n$).

Using cardinals sometimes makes for amazingly compact proofs. An example is the following theorem.

Theorem 11.83 $\mathbb{R} \times \mathbb{R} \sim \mathbb{R}$, *i.e.: there are as many points in the plane as on a line.*

Proof.

$$\begin{aligned} |\mathbb{R} \times \mathbb{R}| &= |\mathbb{R}| \times |\mathbb{R}| \\ &= 2^{\aleph_0} \times 2^{\aleph_0} \\ &= 2^{\aleph_0 + \aleph_0} \\ &= 2^{\aleph_0} \\ &= |\mathbb{R}|. \end{aligned}$$

The third equality uses the rule $n^p \times n^q = n^{p+q}$, the fourth that $\aleph_0 + \aleph_0 = \aleph_0$, cf. Exercise 11.51, p. 401. ∎

*Further Exercises

Exercise 11.84 Suppose that $A_1 \sim A_2$ and $B_1 \sim B_2$. Show:

1. if $A_1 \cap B_1 = A_2 \cap B_2 = \emptyset$, then $A_1 \cup B_1 \sim A_2 \cup B_2$,

2. $A_1 \times B_1 \sim A_2 \times B_2$,

3.* $A_1^{B_1} \sim A_2^{B_2}$ (Hint: it does not do to say $|A_1^{B_1}| = |A_1|^{|B_1|} = |A_2|^{|B_2|} = |A_2^{B_2}|$, for we don't have a rule of exponentiation for cardinal numbers as yet. Instead, show how to establish a bijection between $A_1^{B_1}$ and $A_2^{B_2}$.)

Exercise 11.85 Suppose that $A_1 \preceq A_2$ and $B_1 \preceq B_2$. Show:

1. if $A_2 \cap B_2 = \emptyset$, then $A_1 \cup B_1 \preceq A_2 \cup B_2$,

2. $A_1 \times B_1 \preceq A_2 \times B_2$,

3. $\wp(A_1) \preceq \wp(A_2)$,

4.* if $A_2 \neq \emptyset$, then $A_1^{B_1} \preceq A_2^{B_2}$ (Hint: Use the fact that, since $A_2 \neq \emptyset$, you can pick $a \in A_2$ for the definition of the injection that you need.)

Exercise 11.86 Give counter-examples to the following implications:

1. $A_1 \prec A_2 \Rightarrow A_1 \cup B \prec A_2 \cup B$ ($A_1 \cap B = A_2 \cap B = \emptyset$),

2. $A_1 \prec A_2 \Rightarrow A_1 \times B \prec A_2 \times B$,

3. $A_1 \prec A_2 \Rightarrow A_1^B \prec A_2^B$,

4. $A_1 \prec A_2 \Rightarrow B^{A_1} \prec B^{A_2}$.

Exercise 11.87 Show:

1. if $B \cap C = \emptyset$, then $A^{B \cup C} \sim A^B \times A^C$,

2. $(A \times B)^C \sim A^C \times B^C$

3.* $(A^B)^C \sim A^{B \times C}$. (Hint: use the currying operation.)

Exercise 11.88 Show ($n \geqslant 1$):

1. $\{0,1\}^{\mathbb{N}} \sim \{0,\ldots,n\}^{\mathbb{N}} \sim \mathbb{N}^{\mathbb{N}} \sim \mathbb{R}^{\mathbb{N}} \sim \mathbb{R}$,

2. $\{0,1\}^{\mathbb{R}} \sim \{0,\ldots,n\}^{\mathbb{R}} \sim \mathbb{N}^{\mathbb{R}} \sim \mathbb{R}^{\mathbb{R}} \sim (\wp(\mathbb{R}))^{\mathbb{R}} \sim (\mathbb{R}^{\mathbb{R}})^{\mathbb{R}}$.

Exercise 11.89 Show, for all $n \in \mathbb{N}^+$: $\mathbb{N}^n \sim \mathbb{N}$ (n.b.: $\mathbb{N}^{\mathbb{N}} \not\sim \mathbb{N}$) and $\mathbb{R}^n \sim \mathbb{R}$ (n.b.: $\mathbb{R}^{\mathbb{N}} \sim \mathbb{R}$).

Exercise 11.90 Show: $\{(x,y) \in \mathbb{R}^2 \mid x^2 + y^2 \leqslant 1 \land y > 0\} \sim \{(x,y) \in \mathbb{R}^2 \mid 1 < y \leqslant 2\}$.

Exercise 11.91 Show: if A is infinite and B finite, then $(A-B) \cup (B-A) \sim A$.

The Greek Alphabet

Mathematicians are in constant need of symbols, and most of them are very fond of Greek letters. Since this book might be your first encounter with this new set of symbols, we list the Greek alphabet below.

name	lower case	upper case
alpha	α	
beta	β	
gamma	γ	Γ
delta	δ	Δ
epsilon	ε	
zeta	ζ	
eta	η	
theta	θ	Θ
iota	ι	
kappa	κ	
lambda	λ	Λ
mu	μ	
nu	ν	
xi	ξ	Ξ
pi	π	Π
rho	ρ	
sigma	σ	Σ
tau	τ	
upsilon	υ	Υ
phi	φ	Φ
chi	χ	
psi	ψ	Ψ
omega	ω	Ω

Bibliography

[AHV95] S Abiteboul, R. Hull, and V. Vianu. *Foundations of Databases.* Addison Wesley, 1995.

[Bab61] C. Babbage. *On the Principles and Development of the Calculator.* Dover, 1961. Edited and with an introduction by P. Morrison and E. Morrison.

[Bab94] C. Babbage. *Passages from the Life of a Philosopher.* Rutgers University Press and IEEE-Press, New Brunswick, New Jersey and Piscataway, New Jersey, 1994. Edited with a new introduction by Martin Campbell-Kelly. Originally published 1864.

[Bal91] V. K. Balakrishnan. *Introductory Discrete Mathematics.* Dover, 1991.

[Bar84] H. Barendregt. *The Lambda Calculus: Its Syntax and Semantics (2nd ed.).* North-Holland, Amsterdam, 1984.

[Bir98] R. Bird. *Introduction to Functional Programming Using Haskell.* Prentice Hall, 1998.

[BM96] J. Barwise and L. Moss. *Vicious Circles: On the Mathematics of Non-wellfounded Phenomena.* CSLI Publications, 1996.

[Bry93] V. Bryant. *Yet another introduction to analysis.* Cambridge University Press, 1993.

[Bur98] Stanley N. Burris. *Logic for Mathematics and Computer Science.* Prentice Hall, 1998.

[CG96] J.H. Conway and R.K. Guy. *The Book of Numbers.* Springer, 1996.

[CR78]   R. Courant and H. Robbins. *What is Mathematics? An Elementary Approach to Ideas and Methods.* Oxford University Press, Oxford, 1978.

[CrbIS96]   R. Courant and H. Robbins (revised by I. Stewart). *What is Mathematics? An Elementary Approach to Ideas and Methods (Second Edition).* Oxford University Press, Oxford, 1996.

[Doe96]   H.C. Doets. *Wijzer in Wiskunde.* CWI, Amsterdam, 1996. Lecture notes in Dutch.

[DP02]   B.A. Davey and H.A. Priestley. *Introduction to Lattices and Order (Second Edition).* Cambridge University Press, Cambridge, 2002. First edition: 1990.

[DvDdS78]   K. Doets, D. van Dalen, and H. de Swart. *Sets: Naive, Axiomatic, and Applied.* Pergamon Press, Oxford, 1978.

[Ecc97]   P. J. Eccles. *An Introduction to Mathematical Reasoning.* Cambridge University Press, 1997.

[EFT94]   H.-D. Ebbinghaus, J. Flum, and W. Thomas. *Mathematical Logic.* Springer-Verlag, Berlin, 1994. Second Edition.

[Euc56]   Euclid. *The Thirteen Books of the Elements, with Introduction and Commentary by Sir Thomas L. Heath.* Dover, 1956.

[Fok00]   W. Fokkink. *Introduction to Process Algebra.* Springer, 2000.

[GKP89]   R.L. Graham, D.E. Knuth, and O. Patashnik. *Concrete Mathematics.* Addison Wesley, Reading, Mass, 1989.

[Han04]   C. Hankin. *An Introduction to Lambda Calculi for Computer Scientists*, volume 2 of *Texts in Computing.* King's College Publications, London, 2004.

[Har87]   D. Harel. *Algorithmics: The Spirit of Computing.* Addison-Wesley, 1987.

[HFP96]   P. Hudak, J. Fasel, and J. Peterson. A gentle introduction to Haskell. Technical report, Yale University, 1996. Available from the Haskell homepage: http://www.haskell.org.

[Hin97]   J. Roger Hindley. *Basic Simple Type Theory.* Cambridge University Press, 1997.

BIBLIOGRAPHY 415

[HO00] C. Hall and J. O'Donnell. *Discrete Mathematics Using A Computer*. Springer, 2000.

[HR00] M. Huth and M. Ryan. *Logic in Computer Science: Modelling and Reasoning about Systems*. Cambridge University Press, 2000.

[HT] The Haskell Team. The Haskell homepage. http://www.haskell.org.

[Hud00] P. Hudak. *The Haskell School of Expression: Learning Functional Programming Through Multimedia*. Cambridge University Press, 2000.

[Jon03] S. Peyton Jones, editor. *Haskell 98 Language and Libraries; The Revised Report*. Cambridge University Press, 2003.

[JR+] Mark P. Jones, Alastair Reid, et al. The Hugs98 user manual. http://www.haskell.org/hugs/.

[Kar97] J. Karczmarczuk. Generating power of lazy semantics. *Theoretical Computer Science*, 187, 1997.

[Knu92] D.E. Knuth. *Literate Programming*. CSLI Lecture Notes, no. 27. CSLI, Stanford, 1992.

[Lar34] D. Lardner. Babbage's calculating engine. *Edinburgh Review*, 1834.

[McI99] M.D. McIlroy. Power series, power serious. *Journal of Functional Programming*, 9:323–335, 1999.

[McI00] M.D. McIlroy. The music of streams. *Elsevier Preprints*, 2000.

[Mil99] R. Milner. *Communicating and Mobile Systems: the π Calculus*. Cambridge University Press, 1999.

[NK04] R. Nederpelt and F. Kamareddine. *Logical Reasoning: A First Course*, volume 3 of *Texts in Computing*. King's College Publications, London, 2004.

[Ore88] O. Ore. *Number Theory and its History*. Dover, 1988.

[Pol57] G. Polya. *How to Solve It. A New Aspect of Mathematical Method*. Princeton University Press, Princeton, 1957.

[RL99] F. Rabhi and G. Lapalme. *Algorithms: a Functional Programming Approach*. Addison-Wesley, 1999.

[Rus67] B. Russell. Letter to Frege. In J. van Heijenoord, editor, *From Frege to Gödel*, pages 124–125. Harvard University Press, 1967.

[Rut03] J.J.M.M. Rutten. Behavioural differential equations: a coinductive calculus of streams, automata, and power series. *Theoretical Computer Science*, 308:1–53, November 2003.

[SKS01] A. Silberschatz, H.F. Korth, and S. Sudarshan. *Database System Concepts (4th edition)*. McGraw-Hill, 2001.

[Tho99] S. Thompson. *Haskell: the craft of functional programming (second edition)*. Addison Wesley, 1999.

[Vel94] D.J. Velleman. *How to Prove It. A Structured Approach.* Cambridge University Press, Cambridge, 1994.

Index

(a,b), 133
$:=$, 32
$:\equiv$, 59
A/R, 189
A^c, 128
R^{-1}, 159
Δ_A, 159
$\cap$, 125
$\cup$, 125
dom (R), 158
$\emptyset$, 124
$\wedge$, 29, 31
$\vee$, 29, 31
$\Leftrightarrow$, 29, 35
$\neg$, 29, 30
$\oplus.n$, 241
$\wp(X)$, 130
$\Rightarrow$, 29, 33
ran(R), 158
$\{a\}$, 123
$\{x \in A \mid P\}$, 116
$\{x \mid P\}$, 116
$|a|$, 188
$\sum_{k=1}^{n} a_k$, 53, 235
&&, 16, 31, 137
||, 32
(.), 68
(.*), 345
(op x), 20
(x op), 20
(x1,x2), 136
+, 6

--, 120
->, 8, 139
., 68, 217
/, 14
/=, 5, 122
::, 8
:l, 2
:r, 5
:t, 9
<, 122
<+>, 36
<=, 122
<=, 12
<=>, 36
=, 5
==, 5, 122
==>, 33
<, 8
>=, 7
@, 171
[a], 17
[n..m], 118
$\mathbb{Z}_n$, 190
\\, 147
\, 47
_, 137
$\binom{n}{k}$, 338
$\bot$, 47, 84
dom (f), 202
$\equiv_n$, 185
$\lambda x.t$, 57
$(\bmod\ n)$, 185

417

ran(f), 202
$\{{n \atop k}\}$, 192
⊤, 47
^, 6, 117
$d|n$, 55
e, 216
$n!$, 209
|, 6, 240
~, 396

'abc'-formula, 58, 318
absReal, 202
add, 260
addElem, 147
adjustWith, 331
Alexander the Great, v
algebra
    of sets, 125
all, 67
and, 41, 261
antisymmetric relation, 163
any, 67
approx, 364
approximate, 312
apprx, 312
Apt, K.R., ix
arbitrary object, 91
assignment statement, 23
asymmetric relation, 163
average, 14
axiomatics, 112

Babbage, C., 325
backsubst, 333
ballgame, 395
bell, 193
Bell numbers, 193
Benthem, J. van, viii
Bergstra, J., viii
bijection, 214
bijective, 215
bijectivePairs, 215

binary, 282
binding power of operators, 33
binomial theorem, 338
BinTree, 251
bisimulation, 369
black hole, 270
Bool, 8, 30
Boolean function, 39
brackets, 211
Brouwer, L.E.J, 29
Brunekreef, J., ix

Cantor's
    — continuum hypothesis, 405
    — theorem, 403
Cantor, G., 112
Cantor-Bernstein theorem, 405
cardinal number, 408
case, 139
cat, 258
Catalan numbers, 248, 383
chain, 165
characteristic function, 178
chr, 222
Church, A., 1, 63
class, 122
class, 222
clock, 356
closed form, 208
closures of a relation, 167
co-domain of a function, 205
co-image, 210
co-prime, 285
coprime, 285
Cohen, P.J., 405
coImage, 211
coImagePairs, 211
coinduction, 370
cols, 330
comp, 218, 377
compare, 138

INDEX

comparison property, 165
completeness, 165
complex numbers, 313
`complR`, 174
composition of functions, 68, 217
confluence, 394
congruence, 230
conjunction, 31
constructor, 11
constructor identifier, 11
`continue`, 327
continuity, 64, 309
continuum problem, 405
contradiction, 47, 48
contraposition, 35
converse, 35
conversion, 14
convolution, 345
Coquand, T., ix
corecursion, 352
corecursive definition, 352
countable, 401
curry, 179
Curry, H.B., 1, 179
curry3, 205
CWI, ix

data, 30, 136, 141, 251
database query, 141
`decExpand`, 301
`decForm`, 301, 302
`decodeFloat`, 305
deduction rule, 78
`default`, 352
`delete`, 146
`deleteSet`, 152
delta, 346
denumerable, 401
`deriv`, 348
deriving, 137
destructive assignment, 23

`diagonal`, 407
`difLists`, 326
`difs`, 324
disjunction, 31
`display`, 154
`div`, 19
divergence, 120
divides, 5, 179
domain, 158
`Double`, 17, 305

`e2o`, 385
`echelon`, 332
echelon matrix form, 330
`elem`, 120, 147
`elem'`, 146
`elemIndex`, 302
`eliminate`, 333
`else`, 224
`emptySet`, 152
`encodeFloat`, 306
`Enum`, 222
`enum_2`, 106
EQ, 138
Eq, 122, 240
eq, 180
eq1, 292
`equalSize`, 186
equation guarding, 6
equational reasoning, 23
equipollent, 396
`equiv2listpart`, 194
`equiv2part`, 194
equivalence, 35
— class, 187
— relation, 183, 226
`equivalence`, 184
`equivalence'`, 184
Eratosthenes
sieve of —, 103
error, 8, 224

Erven, T. van, ix
Euclid, 101, 284
Euclid's GCD algorithm, 284
Euler, L., 103
even, 81, 218
evens, 81, 352
evens1, 116
evens2, 117
every, 68
exception handling, 224
exclaim, 246
exp, 216
expn, 246
exponent, 306

fac, 209
fac', 209
False, 4, 30
fasterprimes, 106
fct2equiv, 226
fct2list, 203
fct2listpart, 228
Fermat, P. de, 102
fib, 248
fib', 248
Fibonacci numbers, 247, 248, 353
field, 295
filter, 21
fixity declaration, 33
flip, 180
Float, 305
floatDigits, 305
Floating, 305
floatRadix, 305
Fokkink, W., ix
foldl, 262
foldn, 245
foldr, 260
foldr1, 261
foldT, 255
forall, 11

Fraenkel, A., 112
fromEnum, 222
fromTower, 274
fst, 136
function, 201
    domain, 202
function composition, 68, 217
fundamental theorem
    of algebra, 314
    of arithmetic, 287

Gödel, K., 405
Gaussian elimination, 328
gcd, 285
genDifs, 326
genMatrix, 330
GIMPS, 106
GNU's Not Unix, 352
Goldreyer, D., 28
Goris, E., ix
greatest common divisor
    definition, 283
    Euclid's algorithm for —, 284
    properties, 394
GT, 138
gt1, 292
guard, 7
guarded equation, 6

Haan, R. de, ix
halting problem, 119
hanoi, 269
hanoi', 273
hanoiCount, 273
Haskell, 1
head, 139
Heman, S., ix
hex, 283
Hoogland, E., ix

id, 203
idR, 171

INDEX   421

Iemhoff, R., ix
if then else, 224
iff, 30
ILLC, ix
image, 210
image, 211
imagePairs, 211
implication, 33
import, 72
in, 14
indegree, 178
induction, 233, 234, 389
    strong, 391
infinity
    of primes, 102
infix, 4
infix, 33
infix notation, 20
infixl, 154
infixr, 154
init, 140
injection, 214
injective, 214
injectivePairs, 215
injs, 216
inorder tree traversal, 255
inR, 174
insertSet, 152
inSet, 151
instance
    of a type class, 122
instance, 137
Int, 11
int, 377
Integer, 8, 11
integer partition, 196
integers, 287
Integral, 11
integral rational functions, 323
intersect, 147
intransitive relation, 164

intToDigit, 282
intuitionism, 29
inverse, 180
invR, 174
irrational numbers, 304
irrationality
    of $\sqrt{2}$, 303
irreflexive relation, 163
irreflR, 175
isAlpha, 226
isEmpty, 152
iterate, 307, 327, 353

Jansen, J., ix
Jong, H. de, ix
Jongh, D. de, ix
Just, 225

Kaldewaij, A., ix

labeled transition system, 355
lambda abstraction, 47, 57, 143, 207
lambda term, 57
last, 140
law
    associativity, 45, 241, 243
    commutativity, 45, 241, 243
    contraposition, 45
    DeMorgan, 45, 129
    distribution, 45
    distributivity, 243
    dominance, 47
    double negation, 45
    excluded middle, 47
    idempotence, 45
    identity, 47
lazy list, 22, 103
lazy pattern, 361
LD, 4
ldp, 22
ldpf, 22

leaf tree, 257
LeafTree, 257
left triangular matrix form, 330
len, 258
length, 14
leq, 243
leq1, 292
lessEq, 180
let, 14
lexicographical order, 138
limit, 309
linear relation, 165
Linux, 25
list, 15, 136
list comprehension, 41, 53, 116
list2fct, 203
list2set, 152
listpart2equiv, 194
listPartition, 193
listRange, 204
lists, 136
listValues, 204
ln, 261
ln', 266
load Haskell file, 2
logBase, 274
LT, 138
Lucas numbers, 383

Mackie, I., ix
Main>, 5
mantissa, 305
map, 20, 147, 266
mapLT, 257
mapR, 258
mapT, 255
Matrix, 330
matrix, 330
maxBound, 204
Maybe, 225
maybe, 225

mechanic's rule, 307
mechanics, 307
mechanicsRule, 307
Menaechmus, v
Mersenne, M., 103
min, 12
minBound, 204
mkStdGen, 356
mlt, 260
mnmInt, 12
mod, modulo, 186
module, 2
modulo, 186
Modus Ponens, 80
mult, 246
mult1, 292
mySqrt, 312

n-tuples, 135
Napier's number, 216, 384
natpairs, 406
natstar, 407
Natural, 239
natural logarithm, 216
natural number, 239
naturals, 116, 352
necessary condition, 35
negate, 217
negation, 30
Newman's Lemma, 394
Newman, B., 28
Newton's method, 307
Newton, I., 338
newtype, 151
next, 327
nextD, 326
nondeterminism, 356
not, 30
notElem, 147
Nothing, 225
Nualláin, B. Ó, ix

nub, 141
null, 140
Num, 122

o2e, 385
odd, 218
odds, 352
odds1, 117
oddsFrom3, 106
of, 139
one-to-one, 214
ones, 123, 352
onto, 214
Oostrom, V. van, ix
open problems, 107, 109
operation, 229
operator precedence, 42
or, 261
Ord, 122, 138
ord, 222
order
    partial —, 165
    strict partial —, 165
    total —, 165
Ordering, 138
otherwise, 7
outdegree, 178
overloading, 123

p2fct, 335
pair, 407
pairs, 133
paradox
    halting —, 119
    Russell —, 118
part2error, 225
partial functions, 224
partial order, 165
partition, 187, 226
    integer —, 196
Pascal's triangle, 340
Pascal, B., 340

pattern matching, 12, 139, 244
perms, 217
Platonism, 28
plus, 240, 246
plus1, 291
po-set reflection, 190
polynomial, 314, 323
polynomials and coefficient lists, 336
Ponse, A., ix
postorder tree traversal, 255
power series, 376
powerList, 148
powerSet, 152
pre, 247
pre-order, 164
pred, 222
prefix, 4
prefix, 16
prefix notation, 20
Prelude, 2, 25
preorder tree traversal, 255
primCompAux, 138
prime, 22
prime factorization algorithm, 19
prime pair, 108
prime'', 161
prime0, 7
primePairs, 108
primes
    definition, 59
    Mersenne —, 103
primes, 104
primes', 354
primes0, 22
primes1, 22
principle
    comprehension —, 112
    minimality —, 392
Process, 356
product, 209

product of sets, 133, 228
propositional function, 39
ptd, 358

quasi-order, 164
quotient, 189
quotient, 244
quotRem, 280

random numbers, 356
random streams, 356
Random.hs, 351
randomInts, 356
randomRs, 356
ranPairs, 204
Rational, 14, 333
rationals, 293
    countability of —, 402
rationals, 407
rclass, 189
rclosR, 177
reals, 304
    uncountability of —, 403
recip, 307
recurrence, 208
recursion, 239
recursive definition, 7, 240
reduce1, 292
reflect, 257
reflexive
    closure, 167
    relation, 162
    transitive closure, 167
reflR, 174
relation
    equivalence —, 183
    antisymmetric —, 163
    asymmetric —, 163
    between, 158
    domain, 158
    from—to, 158
    intransitive —, 164

    irreflexive —, 163
    linear —, 165
    range, 158
    reflexive —, 162
    symmetric —, 163
    transitive —, 163
relatively prime, 285
reload Haskell file, 5
rem, 5
remainder, 244
removeFst, 13
reserved keywords, 11
restrict, 210
restrictPairs, 210
restrictR, 177
rev, 263
rev', 263
rev1, 265
Rodenburg, P., ix
rose tree, 258
rows, 330
royal road to mathematics, v
Russell, B., 56, 118
Rutten, J., ix

sclosR, 177
sections, 20
Set a, 151
Show, 240
showDigits, 282
showSet, 152
sieve, 104, 354
sieve of Eratosthenes, 103
sieve', 354
significand, 306
sin, 216
singleton, 123
sink, 178
small_squares1, 117
small_squares2, 118
Smullyan's ball game, 395

INDEX

Smullyan, R., 395
snd, 136
solveQ, 318
solveQdr, 58
solveSeq, 334
solving quadratic equations, 58, 318
some, 68
soundness, 165
source, 178
space leak, 270
split, 256
splitAt, 302
splitList, 140
sqrt, 58, 216
sqrtM, 307
srtInts, 14
start, 356
stirling, 193
Stirling set numbers, 193
stream, 352
stream bisimulation, 371
strict partial order, 165
String, 15
stringCompare, 226
sub domain, 52
substitution principle, 47
subtr, 244, 247
subtr1, 291
succ, 222
successor, 240
sufficient condition, 35
sum, 14, 53, 235
sumCubes, 238
sumCubes', 238
sumEvens, 236
sumEvens', 236
sumInts, 236
sumOdds, 235
sumOdds', 235
sumSquares, 237

sumSquares', 237
surjection, 214
surjective, 214
surjectivePairs, 215
Swaen, M., ix
symmetric
    closure, 167
    difference, 129
    relation, 163
symR, 175

tail, 139
take, 105
takeWhile, 307
tan, 216
tclosR, 177
Terlouw, J., ix
theFibs, 353
then, 224
theNats, 353
theNats1, 353
theOdds, 353
theOnes, 353
toBase, 283
toEnum, 222
toRational, 14
total order, 165
totalR, 174
toTower, 274
tower of Hanoi, 266
transClosure', 183
transition system, 355
transitive
    — closure, 167
    — relation, 163
transR, 175
tree
    leaf —, 257
    rose —, 258
tree traversal, 255
trivially true statements, 34

Tromp, J., ix
True, 4, 30
truncate, 274
truth function, 39
Turing, A., 63
type, 8, 17, 52, 119, 136
    conversion, 14
    declaration, 8
    judgment, 9
    variables, 17
type, 141

uncountable, 402
uncurry, 179
uncurry3, 205
undefined, 120, 363
union, 146
Unix, 25

Van Benthem, J., ix
variable identifier, 11
vending, 358
Venema, Y., ix
Vinju, J., ix
Visser, A., ix
Vries, F.J. de, ix

weak confluence, 394
Wehner, S., ix
well-founded, 239, 393
where, 2, 14
Who is Afraid of Red, Yellow
    and Blue, 28
wildcard, 11, 137
Windows, 25

z, 344
Zermelo, E., 112
zipWith, 331, 353

Printed in the United States
102718LV00003B/52-54/A